T0282592

Ontogeny, functional ecology, and evolution of bats

The study of animal development has deep historical roots in codifying the field of evolutionary biology. In the 1940s, evolutionary theory became engulfed by analyses of microevolutionary genetics and the study of development became focused on mechanisms, forsaking the evolutionary implications of ontogeny. Recently, ontogeny has resurfaced as a significant component of evolutionary change and also of population and community dynamics.

Ontogeny, Functional Ecology, and Evolution of Bats is a unique reference work by bat biologists who emphasize the importance of understanding ontogeny in analyses of evolution and ecology. In addition, the developmental underpinnings of specialized morphology, physiology, and behavior are elucidated, and the strong influence of ecology on the ontological niche of juvenile bats is illustrated. This book is an essential reference, not only for bat biologists, but for anyone working in the fields of ecology, developmental biology, evolution, behavior and systematics.

RICK A. ADAMS is an Associate Professor of Zoology at the University of Wisconsin, Whitewater. He is Founder and President of the Colorado Bat Society, a nonprofit conservation foundation. His research focuses on postcranial skeletogenesis in bats, the ontogeny of flight and niche space in juvenile bats, temporal spacing at waterholes and roost site affiliations of Colorado bat species, and bat biodiversity in Montserrat and Antigua in the Caribbean as affected by natural disasters such as volcanic eruptions and hurricanes.

SCOTT C. PEDERSEN is an Assistant Professor at South Dakota State University in Brookings, South Dakota. He is a vertebrate morphologist with interests in the evolution of the skull as an acoustical horn in bats from a developmental perspective. He has also been studying the biodiversity and biogeography of the bat fauna of the Lesser Antilles. Recently, Dr Adams has made valuable contributions to Dr Pedersen's ongoing, long-term study of the bats of Antigua and Montserrat.

Ontogeny, functional ecology, and evolution of bats

Edited by

RICK A. ADAMS and
SCOTT C. PEDERSEN

CAMBRIDGE
UNIVERSITY PRESS

CAMBRIDGE UNIVERSITY PRESS
Cambridge, New York, Melbourne, Madrid, Cape Town, Singapore, São Paulo, Delhi

Cambridge University Press
The Edinburgh Building, Cambridge CB2 8RU, UK

Published in the United States of America by Cambridge University Press, New York

www.cambridge.org
Information on this title: www.cambridge.org/9780521626323

First published 2000
This digitally printed version 2008

A catalogue record for this publication is available from the British Library

Library of Congress Cataloguing in Publication data

Ontogeny, functional ecology, and evolution of bats / edited by Rick A. Adams and
Scott C. Pedersen.
    p.   cm.
Includes indexes.
ISBN 0 521 62632 3 (hc).
1. Bats – Development.   2. Bats – Evolution.   I. Adams, Rick A. (Rick Alan)
II. Pedersen, Scott C. (Scott Campbell), 1960–
QL737.C5   O58   2000
599.4'138–dc21   99–057156

ISBN 978-0-521-62632-3 hardback
ISBN 978-0-521-08735-3 paperback

Contents

Contributors

Rick A. Adams
Department of Biological Sciences, University of Wisconsin –
Whitewater, Whitewater, WI 53190, USA

Kunwar P. Bhatnagar
Anatomical Sciences and Neurobiology, School of Medicine, University
of Louisville, Louisville, KY 40292, USA

Dianne L. Davis
School of Medicine, University of Louisville, Louisville, KY 40292, USA

John W. Hermanson
Department of Anatomy, College of Veterinary Medicine, Cornell
University, Ithaca, NY 14853, USA

Gareth Jones
School of Biological Sciences, University of Bristol, Bristol, B58 1UG, UK

Kamar B. Karim
Nagpur University Campus, Department of Zoology, Nagpur,
Maharashtra, India

Scott C. Pedersen
Department of Biology & Microbiology, South Dakota State University,
Brookings, SD 57007, USA

Carleton J. Phillips
Department of Biological Sciences, Texas Tech University, Lubbock, TX
79409, USA

Roger L. Reep
Department of Physiological Sciences, College of Veterinary Medicine,
University of Florida, Gainesville, FL 32610, USA

Nancy B. Simmons
Department of Mammalogy, American Museum of Natural History, Central Park West at 79th Street, New York, NY 10024, USA

Katherine M. Thibault
Department of Biology, University of New Mexico, Albuquerque, NM 87131, USA

Marianne Vater
Allgemeine Zoologie, Universität Potsdam, Lennestrasse 7a, 14471 Potsdam, Germany

John R. Wible
Section of Mammals, Carnegie Museum of Natural History, 5800 Baum Blvd, Pittsburgh, PA 15206, USA

RICK A. ADAMS & SCOTT C. PEDERSEN

1

Integrating ontogeny into ecological and evolutionary investigations

As far as morphological change is concerned, evolution acts by altering development.

Liem & Wake (1985)

Given that resource utilization abilities and predation risk are generally related to body size, many species will undergo extensive ontogenetic shifts in food and habitat use. Such shifts create a complex fabric of ecological interactions in natural communities.

Werner & Gilliam (1984)

The field of 'integrative biology' utilizes multidisciplinary approaches to establish a more complete and, therefore, insightful interpretation of an organism's biology. Certainly there is nothing unprecedented about integrative strategies in seeking to understand the natural world. In fact, the major structuring of Western science 2500 years ago by Aristotle and others involved those simultaneously interested in a wide range of topics from gross anatomy, physiology, chemistry, classification of organisms, astronomy, philosophy, and earth sciences. Even as late as the 18th and 19th centuries, integrative biology was alive and well and most investigators saw little need for establishing hard boundaries between subdisciplines. For example, Darwin was a master of integrative biology who, in forming his dynamic view of life, embraced studies in embryology, paleontology, functional morphology, animal husbandry, heredity, biogeography, ecology, earth sciences, and even physics. The 20th century witnessed the establishment of strong subdisciplinary boundaries that effectively compartmentalized biology into numerous factions whose investigators rarely collaborated. Despite this divisive trend, some biologists continued to seek interdisciplinary connections.

The 'Grand Synthesis' was ushered in by Fisher (1930) and others who sought to bridge the artificial schism between genetics and evolutionary theory. Evolution and genetics were united into the field of population genetics which allowed unprecedented insights into microevolutionary change. More recently, the fusion of ecology and functional morphology into the field of ecomorphology has garnered significant support in recent treatments by Ricklefs & Miles (1994), Wainwright and Reilly (1994), and others. Analogously, ecology and physiology were fused into the field of ecophysiology. The beauty of such an integrative approach is that it elucidates the evolutionary underpinnings of present ecological patterns by quantifying morphological correlates of niche space under the presumption that natural selection favors character displacement over competition, and that this displacement is measurable.

Our overall goal in this volume is to add to the growing body of integrative approaches to biological research by presenting a variety of works that incorporate ontogenetic data into studies of systematics, functional morphology, ecology, and evolution. We maintain that integrating an ontogenetic perspective into investigations of complex systems provides a more insightful and balanced interpretation because 1) it is selection on developmental variation that produces phenotypic variation among adults in populations, 2) commonality in developmental patterns may indicate common ancestry (and lack thereof may be indicative of convergence), and 3) preadult individuals directly influence the dynamics of populations and communities through time.

Discussions concerning the theoretical and empirical relationships between ontogeny and evolution have a long history that is both broad and deep (see reviews by Gould 1984; Hall 1992). Clearly, variation in developmental programs may generate novel adaptive character states upon which natural selection may act, i.e., diversity among adult forms begins with divergent developmental patterns (Müller 1990; Raff 1996; McNamara 1997; Klingenberg 1998). Genetic variation is subsequently translated through the process of ontogeny where selection may favor novel morphologies that, in turn, are inherited as evolutionary changes within populations. Thus, natural selection is manifest along the continuum from organogenesis to adulthood; morphogenetic variability is filtered throughout the developmental/adaptive program.

The pervasive force behind the evolution of morphological innovation lies in developmental plasticity and the ability of morphogenetic systems to accommodate change, while simultaneously maintaining functional relationships among subcomponents. The morphological, physiological, and behavioral results of ontogeny, as manifested in adults, result from two factors: 1) those that are intrinsic – character states that are inherited unchanged from ancestors, or are the distinctive products of emergent properties generated from tissue interactions and/or heterochrony (Klingenberg 1998), and 2) those that are extrinsic – characters generated by and strongly influenced by environmental forces (Lauder 1982). Intrinsic and extrinsic factors interact in myriad ways during ontogeny, resulting in morphological variation among juveniles. This variation is then filtered by natural selection and what remains become adults. The crucial point here is that natural selection acts upon successful integration and function during development in essentially the same way as it selects for adaptive morphologies in adults. Furthermore, it can be argued that selective pressures are greatest during development and, therefore, the key to evolutionary changes in morphology are best understood by the study of preadult organisms (Raff 1996; McNamara 1997). *Unfortunately, it is the dynamic nature of morphogenesis in relation to ecological and evolutionary theory that is so frequently overlooked.* Indeed, the adult is often considered as an immutable construct into which all of the various organ systems must somehow be packaged during development. In kind, juveniles typically are presented as prefunctional, preadult forms, rather than portrayed as evolutionarily important, adaptive stages that manifest their own niche, known as the ontogenetic niche (Werner & Gilliam 1984; Coppinger & Smith 1990).

The intuitive kinship between ontogeny and evolution has a historical partnership as represented in some of the earliest naturalistic writings (Gould 1984). Recent attempts have been made to integrate morphology, developmental, evolutionary biology, and phylogenetics into a unified theory of the evolution of biological form (Klingenberg 1998). Publications by Nitecki (1990), Raff (1996), and McNamara (1997) reignited an interest in the intrinsic power of ontogeny as an instigator of evolutionary novelty. What remains underappreciated, however, is how the interrelationship between juveniles and adults affects population/community dynamics as well as ecosystem evolution.

Because emphasis is placed mostly upon juvenile survivorship to a reproductive age, little attention is given to character differences between juveniles and adults. Furthermore, variation among juveniles, and the role that these differences play in the 'ecological theater' are frequently ignored. This is, in part, because the juvenile period is often an ephemeral component of the life cycle, and therefore hard to measure and is easily omitted from traditional theory. Most telling in this regard is the implied, if not formally defined, distinction given to adult form as *the* significant unit of a species' niche. Ostensibly, an understanding of the full mosaic of an organism's niche is crucial in interpreting the rules that govern population, assemblage, and community structure (Werner & Gilliam 1984; Liem & Wake 1985). Typically, little or no consideration is given to the juvenile niche, its adaptive nature, or its place in understanding ecological and evolutionary patterns (Werner & Gilliam 1984; Werner 1999). Indeed, resource partitioning among juveniles results in niche fragmentation often to the extent that each age-class functions as an ecological species (Tchumy 1982). We are not suggesting that adult-biased measures of natural history are in any way invalid, but we consider the inclusion of the ontogenetic perspective essential to understanding fully the many aspects of evolutionary and ecological theory.

Darwin, the master of synthesis, would probably applaud recent attempts to resynthesize the major subdisciplines of biology – the intended scope of this book. This volume is unique in that we have brought together a diverse group of bat biologists and asked them to discuss the importance of investigating the underlying developmental patterns of form, function, ecology, evolution, and systematics within the context of their own specialized fields. We have tried to present the current state of research and understanding of chiropteran development. For example, several of the authors provide insights into the ongoing mono- versus diphyletic origin of bats debate by integrating developmental data with patterns observed in morphology, genetics, and fossils. Readers of this volume will, in some cases, be provided with empirical approaches to studies in ecology and evolution from an ontogenetic perspective. In other cases, the reader will find an update on our current understanding in that field and a wealth of suggestions for further research that should help provide insights into some of the more vexing questions in bat biology. We view this compendium as a broadly based contribution to the field of bat biology, and we encourage integration of an ontogenetic framework into all studies of evolution and ecology.

STRUCTURE AND OVERVIEW

In this volume, we provide an up-to-date account of ontogenetic studies of the Chiroptera, and how these studies relate to the ecology and evolution of this diverse and highly successful group. In addition, we reassert the importance of ontogenetic data in studies of systematics, evolutionary diversity, and ecological patterns. We promote ontogeny as a fundamental link between the vastly different time scales of ecological and morphological change.

We begin with Simmons's review (Chapter 2) of chiropteran systematics to establish a taxonomic foundation for the book. She starts with a discussion concerning the range of data sets and analyses used to construct phylogenies of bats, and explains the analytical methods as well as the strengths and weaknesses of each. Simmons comments on the difficulties faced by systematists in discerning the relationships among taxa and proposes the integration of data from fossils, ontogeny, morphology, and molecular biology in systematic investigations, especially some of the more vexing questions in bat biology such as the 'monophyly versus diphyly' debate. By summarizing the literature of the current state of understanding of the relationships within families of microchiropterans, she discusses the inferences of evolutionary patterns as derived from bat phylogenies. Following Simmons's review, the text is arranged roughly in developmental order. Karim & Bhatnagar (Chapter 3) review the literature and compare patterns of implantation, placentation, and early development among extant bat species representing many families, and discuss the wide variation present. Beginning with a discussion of what is known about ovum maturation, fertilization, and preimplantation of the embryo, the authors complete this chapter with a discussion of phylogenetic considerations of fetal membrane characters. In Chapter 4, Reep & Bhatnagar tackle questions of the role that neuronal connections play in discussions of bat evolution and emphasize the importance of developmental data to this debate. New data are presented on layer VII of the cerebral cortex, as an example of an adult trait whose variation has specific developmental connotations. Reep & Bhatnagar also incorporate what is known about brain development in bats with comparative analyses of similarities and differences among other mammalian groups. Underscoring the role that developmental studies can play in elucidating the apparently convergent visual pathways in pteropodids and primates, they also assess the role of developmental studies in understanding differential enlargement of particular brain regions in taxa of differing foraging strategies, and the relationship of these patterns to general trends in mammals. Vater (Chapter 5) begins with a brief note on biosonar and after describing the basic anatomy and evolution of the bat cochlea, she discusses the ontogeny of echolocation and the developmental and functional feedback link between the larynx and hearing system in a section entitled 'Who is leading development?' Vater also relates the ontogeny of hearing and vocalization with what is known of cochlear and neurological development and function in bats. Pedersen (Chapter 6) describes the morphogenesis and anatomy of the chiropteran skull from the perspective that it has evolved first and foremost to function as an acoustical horn adapted for echolocation. He provides compelling statistical support for how the origin and evolution of nasal- versus oral-emitting taxa was driven by divergent developmental pathways early in ontogeny. Pedersen argues that the evolutionary divergence between emission types is most likely a result of selective forces acting upon the form and function of the pharynx during echolocation rather then selection on cranial shape or head posture *per se*. He casts his data into a developmental landscape model depicting divergent developmental pathways between nasal- and oral-emitters. Wible & Davis (Chapter 7) use *Megaderma lyra* as a model to provide new data on the ontogeny of the chiropteran basicranium and show how important ontogenetic studies are for understanding the details of adult basicranial anatomy, especially in identifying the components of the tympanic floor and roof. They highlight how developmental phenomena (sequence, number, pattern) are underutilized and potentially very important taxonomic characters. By using basicranial ontogeny, Wible & Davis provide information for a dynamic interpretation of static adult morphology and for assessing the interrelationships among taxa. They also discuss differences and similarities in form of the basicranium between micro- and megachiropterans. In Chapter 8, Phillips begins with the premise that mammalian dental morphology is a direct reflection of an animal's diet and evolutionary history. He utilizes dental eruption patterns to decipher evolutionary mechanisms and pathways and compares the evolution of dental morphology in micro- and megachiropterans. Phillips concludes that whereas microchiropterans have retained an evolutionarily conserved morphology, megachiropterans are relatively unconstrained evolutionarily and have lost the detailed coronal morphologies of their ancestors. He goes on to assert that biochemical studies of oral environment and the digestive tract are important in understanding dental morphology and evolution. Adams (Chapter 9) describes the ontogeny of the handwing in bats and provides analysis of previously unpublished data for both micro- and megachiropteran handwing development, including comparative growth trajectories. He also incorporates postnatal development of the wing and flight ability into an age-specific model of resource partitioning and survivorship among juveniles and adults. He demonstrates the ontogenetic niche for *Myotis lucifugus* by quantifying shifts in niche dimensions with changes in wing-loading and aspect ratio. These data are integrated and displayed in the form of an adaptive landscape model illustrating when the strongest selective pressures on juveniles may occur, challenging survivorship at the transition points among juvenile niche spaces. In Chapter 10, Adams & Thibault provide previously unpublished data on the growth and development of the hindlimb in micro- and megachiropterans. In addition, they discuss the functional morphology, evolution and phylogenetics of the calcar as described in other studies and add a previously unpublished ontogenetic perspective to deciphering its evolutionary and systematic pathways in the diphyletic versus monophyletic argument of mega- and microchiropterans origins. Hermanson reviews the literature on what is known concerning the ontogeny of flight muscles and describes the bio-chemical maturation of skeletal (flight) muscle in bats in Chapter 11. He evaluates this literature in comparison with data collected by him and others on the muscle biology of locomotion in an array of vertebrates, and uses it as a springboard to advance new considerations in discussions of the evolution of flight in mammals. Hermanson re-evaluates his 'toggle-switch' hypothesis in light of new data and develops an ontogenetic model for the evolution of flight. In Chapter 12, Jones discusses the ontog-eny, evolution, and phylogeny of social behavior in bats. Beginning with a discussion of social interactions among pups, he leads the reader through the significance of the many aspects of maternal care and allomaternal care. Jones reports on aspects of social learning, the ontogeny of emer-gence times and foraging behavior, and evidence of maternal care after weaning.

In summary, we see ontogeny as a fundamental link between the vastly different time-scales of ecological and evolutionary change and, therefore, the integration of morphogenetic analyses are essential to understanding complex biological systems. Unfortunately, as biologists we tend to become overly focused within our own subdiciplines, and readjust-ing our sights to integrating studies from other fields will require special attention. Perhaps this volume will be heuristic in stimulating more biolo-gists to integrate an ontogenetic perspective into research programs.

REFERENCES

Coppinger, R. P. & Smith, C. K. (1990) A model for understanding the evolution of mammalian behavior. In *Current Mammalogy*, ed. H. H. Genoways, pp. 335–74. New York: Plenum Press.

Darwin, C. (1859) *On the Origin of Species by Means of Natural Selection*. London: J. Murray.

Fisher, R. A. (1930) *The Genetical Theory of Natural Selection*. Oxford: Clarendon Press.

Gould, S. J. (1984) *Ontogeny and Phylogeny*. Cambridge: Belknap Press of Harvard University Press.

Hall, B. K. (1992) *Evolutionary Developmental Biology*. London: Chapman & Hall.

Hutchinson, G. E. (1965) *The Ecological Theater and the Evolutionary Play*. New Haven and London: Yale University Press.

Klingenberg, C. P. (1998) Heterochrony and allometry: the analysis of evolutionary change in ontogeny. *Biological Review*, **73**, 79–123.

Lauder, G. V. (1982) Historical biology and the problem of design. *Journal of Theoretical Biology*, **97**, 57–67.

Liem, K. F. & Wake, D. B. (1985) Morphology: current approaches and concepts. In *Functional Vertebrate Morphology*, ed. M. Hildebrand, D. M. Bramble, K. F. Leim & D. B. Wake, pp. 366–77. Cambridge: Belknap Press of Harvard University Press.

McNamara, K. J. (1997) *Shapes of Time: The Evolution of Growth and Development*. Baltimore: Johns Hopkins University Press.

Müller, G. B. (1990) Novelty: a side effects hypothesis. In *Evolutionary Innovations*, ed. M. H. Nitecki, pp. 99–132. Chicago: University of Chicago Press.

Nitecki, M. H. (ed.) (1990) *Evolutionary Innovations*. Chicago: University of Chicago Press.

Raff, R. R. (1996) *The Shape of Life*. Chicago: University of Chicago Press.

Ricklefs, R. E. & Miles, D. B. (1994) Ecological and evolutionary inferences from morphology: an ecological perspective. In *Ecological Morphology*, ed. P. C. Wainwright & S. M. Reilly, pp. 13–41. Chicago: University of Chicago Press.

Tchumy, W. O. (1982) Competition between juveniles and adults in age-structured populations. *Theoretical Population Biology*, **21**, 225–68.

Wainwright, P. C. & Reilly, S. M. (1994) *Ecological Morphology: Integrative Organismic Biology*. Chicago: University of Chicago Press.

Werner, E. E. (1999) Ecological experiments and a research program in community ecology. In *Experimental Ecology*, ed. W. J. Resetarits, Jr. & J. Bernardo, pp. 3–26. Oxford: Oxford University Press.

Werner, E. E. & Gilliam, J. F. (1984) The ontogenetic niche and species interactions in size-structured populations. *Annual Review of Ecology and Systematics*, **15**, 393–25.

2

Bat phylogeny: an evolutionary context for comparative studies

INTRODUCTION

Evolution is one of the unifying theories of modern biology. The fact that organisms (including bats) can and have evolved requires us to consider the historical origins of traits when seeking to understand the similarities and differences among taxa. Evolution may act upon any aspect of organisms, including their biochemistry, ontogeny, morphology, ecology, and behavior. Understanding modern patterns of diversity clearly requires an evolutionary perspective.

One of the principal methods available for addressing evolutionary questions is phylogenetic analysis. As generally understood, phylogenetic analysis comprises the gathering and analysis of data to generate and test hypotheses of phylogenetic relationships, usually (but not necessarily) using cladistic methods. A second step, which may be pursued regardless of the source of phylogenetic trees, consists of mapping taxon characteristics onto trees to investigate patterns of evolution of these features. Traits that can be considered in phylogenetic analyses range from DNA sequences and morphological features to behavioral characteristics; the data employed usually reflect both the nature of the questions being asked and the interests and expertise of the researcher.

Phylogenetic studies of bats are being published at an ever-increasing rate, and well-supported phylogenies for many groups are now available. These hypotheses of evolutionary relationships offer unprecedented opportunities for reconstructing historical patterns of change in different bat lineages. In keeping with the nature of this volume, the goals of this chapter are to 1) discuss briefly the types of data now being used to build bat phylogenies and the methods used to evaluate levels of perceived support for alternative phylogenetic hypotheses, 2) summarize the present state of understanding of bat relationships at several taxonomic levels, and 3) provide examples of how bat phylogenies can be used to provide evolutionary interpretations of morphological, ecological, and behavioral traits. Tree-building methods (e.g., character coding, weighting, optimality criteria, algorithms) will not be discussed here; for useful reviews of these and related topics, see Swofford & Olsen (1990), Hillis *et al.* (1993), Simmons (1993*a*), and Swofford *et al.* (1996).

DATA USED TO BUILD PHYLOGENETIC TREES OF BATS

The range of data that have been employed in analyses of bat relationships includes most of the data sources known to systematists. As would be expected given the history of systematics, many studies have been based on morphological data. However, biochemical, chromosomal, and DNA data have come to play an increasingly important role in bat phylogenetics. Many recent studies of bat relationships have been based on nucleotide sequence data from mitochondrial and nuclear genes. Other phylogenetically informative data have come from analysis of rDNA restriction sites, satellite DNA, amino acid sequences, DNA hybridization data, immunological distance data, karyotypes, and allozymes. No ecological or behavioral data have been used to help build phylogenetic trees of bats, but studies experimenting with the use of components of echolocation call structure as phylogenetic characters are now under way (Simmons & Kalko 1998).

Each type of data has its own strengths and weaknesses. Morphology is an appealing source of phylogenetic information because structure and development can be used to form hypotheses of homology, and also because many features (e.g., craniodental and pelage characters) can be easily surveyed in standard museum specimens. Highly variable characters (e.g., those that vary within species) and those that are clearly homoplastic (e.g., features that are similar in gross appearance or function but have very different internal structures) can be eliminated or otherwise dealt with prior to analysis. However, morphological characters, like all other types of data, are subject to homoplasy that frequently cannot be detected prior to phylogenetic analysis (see discussion in Simmons 1993*a*). In addition, morphological data sets are often too small to provide adequate resolution of relationships in many interesting clades, particularly speciose groups in which taxa are distinguished principally by subtle size and shape variation that cannot be easily subdivided into discrete characters.

Nucleotide sequence data provide an increasingly important source of phylogenetic information. Strengths of sequence data include the large numbers of potentially informative characters, and the ability to compare organisms that are morphologically very different (e.g., bats and whales) or very similar (e.g., species of *Myotis*). However, analysis of sequence data is becoming increasingly complex as more is learned about the structure and function of different parts of the genome (see review in Swofford *et al.* 1996). There is at present considerable disagreement concerning methods of establishing homologies (e.g., alignment), dealing with obviously variable levels of homoplasy (e.g., different saturation levels for different codon positions, different likelihood of transitions versus transversions, gap costs, etc.), and alternative methods of tree-building from sequence data (e.g., parsimony, maximum likelihood, or minimum evolution; taxonomic congruence versus character congruence). It is now clear that different genes evolve at different rates, mitochondrial genes may have different evolutionary histories than nuclear genes, and that gene trees may differ from species trees (e.g., as a result of lineage sorting). These and other issues complicate the analysis of nucleotide sequence data and interpretation of phylogenetic results.

Other sources of phylogenetic characters include amino acid sequences, rDNA restriction sites, and satellite DNA. Amino acid sequences were used for some years as an estimate of the underlying genetic code (e.g., see Pettigrew *et al.* 1989), but these data have been supplanted in recent years by nucleotide sequence data, which provide a more direct measure of DNA structure and similarity. Like amino acid sequences, restriction-site analyses provide a means of comparing segments of DNA without actually sequencing the genes that compose them. Although restriction sites may be phylogenetically informative, collecting these data is time-consuming and expensive, and restriction-site studies rarely identify enough informative characters to resolve the phylogenetic problems that they seek to address (e.g., see Baker *et al.* 1991*a*; Van Den Bussche 1992). As a result, restriction-site studies have also been largely abandoned in favor of direct sequencing. Alternatively, satellite DNA variation has been shown to provide a useful source of phylogenetic information that is relatively easy to obtain. However, usefulness of this method is somewhat limited because hypotheses of satellite DNA homology can be tested only by evaluating congruence with other data sets or by directly sequencing the satellite sequences.

DNA hybridization experiments provide distance data that describes the overall similarity of the genome of taxa included in the analysis. Likewise, immunological studies provide distance measures of similarity among taxa for particular proteins. Although these methods have proven more robust than once thought (largely as a result of improvements in analysis techniques), they suffer from three principal problems: 1) distance methods cannot distinguish between primitive and derived similarity, 2) taxa cannot be added to pre-existing hybridization or immunological studies without either conducting many additional experiments and/or conducting the work in the same laboratory under identical conditions, and 3) distance data cannot be combined with other data sets (e.g., discrete character data from morphological studies or DNA sequences) without converting the latter to distances, which reduces the resolving power of the discrete data because of problem (1) above. As a result of these difficulties, immunological and DNA hybridization studies have been largely abandoned in favor of DNA sequencing. However, cumulative results of ongoing hybridization studies continue to contribute new and interesting hypotheses about bat relationships (e.g., Kirsch *et al.* 1998).

Data from karyotypes have also played a role in recent studies of bat relationships. Standard karyotypes (i.e., diploid and fundamental numbers, chromosome morphology) are of only limited use because homologies of chromosomes and their constituent arms can be inferred only from gross structure. However, C- and G-banding of chromosomes can allow identification of homologous arms, thus improving confidence in homology assessments and increasing the number of useful characters that may be identified for use in phylogenetic analyses (e.g., Volleth & Heller 1994*a, b*). Unfortunately, major evolutionary changes in chromosomes often produce incomparable banding patterns even in closely related taxa, and (perhaps as a result) chromosome banding studies rarely produce enough characters to fully resolve relationships at any taxonomic level.

One of the first biochemical methods applied to systematics was analysis of allozymes (allelic isozymes) obtained from protein electrophoretic studies. While these data preserve some phylogenetic information, allozyme studies are hampered by small numbers of potentially informative characters and high levels of within-species polymorphism. Large numbers of individuals of each species must be surveyed to determine which alleles occur in each species, and even with large samples there is no way to tell if additional polymorphisms have been overlooked. Although methods of cladistic analysis of allozyme data have improved in recent years with the use of step matrices (e.g., Mabee & Humphries 1993; Mardulyn & Pasteels 1994), the combination of low numbers of characters and high within-species polymorphism limits the value of these data for phylogenetic studies at the species level or above.

In summary, all data sources have their limitations, and there is no single 'best' data source for reconstructing bat relationships. The fit between the taxonomic problem being addressed and the data employed is of critical importance. Some systems appear to evolve too slowly to be useful indicators of relationships at lower taxonomic levels; for example, postcranial characters may not be very useful for addressing relationships among bat species because these features are often fixed within families (Simmons 1998; Simmons & Geisler 1998). Conversely, some systems evolve too quickly to be useful at higher taxonomic levels (e.g., allozymes, cytochrome *b* gene sequences). The most useful results (relatively well-resolved, well-supported trees) are usually derived from analysis of large data sets based on systems with appropriate levels of variation, or combinations of data from alternative sources that may not, by themselves, provide adequate phylogenetic signal.

EVALUATING SUPPORT FOR DIFFERENT CLADES

In the early years of phylogenetic studies of bats, cladograms were published without explicitly describing the data and methods used to build them (e.g., Smith 1976; Van Valen 1979). This practice changed within a few years, and studies published in the 1980s included explicit data summaries and matrices (e.g., Luckett 1980; Novacek 1980, 1987; Griffiths 1982; Wible & Novacek 1988; Pettigrew et al. 1989). These contributions typically concluded with a single phylogenetic tree that summarized the results of the study, usually with the derived characters supporting each branch mapped onto the tree or listed in a table. While this method of reporting character support may have been sufficient for studies based on small data sets, it is not practical for larger character sets such as those based on DNA sequences or numerous morphological characters. Estimating character support for different groupings similarly becomes more difficult when larger sets of taxa are considered, in part because homoplasy increases with the number of taxa included in an analysis (Sanderson & Donoghue 1989). In the 1990s, systematists working on larger data sets routinely began to use three methods to evaluate the amount of support for different clades in a study: decay analysis, bootstrapping, and jackknifing. The assumptions and pitfalls associated with these methods are complex (e.g., see discussion in Swofford et al. 1996), but they nevertheless provide information useful for interpreting results of phylogenetic analyses. The following discussion is based on applications of these methods in the context of parsimony analyses; however, some methods are broadly applicable to other approaches (e.g., maximum likelihood or distance methods; see Swofford et al. 1996).

Decay analyses, first implemented by Bremer (1988), examine near-most-parsimonious trees to determine how many additional steps are required to collapse a given clade. A decay value (= Bremer support value) of 1 indicates that the clade under consideration collapses (becomes non-monophyletic) in trees that are only one step longer than the most-parsimonious trees. Higher decay values indicate increasingly higher levels of support. Decay values are typically lower than the number of derived character states that appear to support a clade because more than one optimization is generally possible for at least some characters (see 'Character mapping and other methods' below). In most cases, results of a decay analysis are presented as numbers (= decay values) listed for each node in a strict consensus of the most-parsimonious trees. Clades that appear in some but not all of the most-parsimonious trees (and thus collapse in a strict consensus tree) have a decay value of 0 and are usually not labeled.

Bootstrapping and jackknifing are resampling methods that can be used to estimate the variability associated with a statistic for which the underlying sampling distribution is unknown or difficult to derive analytically (see discussion and references cited in Felsenstein 1985a, Lanyon 1985, Sanderson 1989, Swofford & Olsen 1990, and Swofford et al. 1996). In phylogenetic studies, the 'underlying distribution' consists of all possible taxonomic characters, of which the data set is presumed to be a representative sample. Both the bootstrap and the jackknife operate by repeatedly resampling the data contained in the original data set, but differ in the way in which the resampling is performed.

In a bootstrap analysis, characters (columns in the data matrix) are randomly sampled with replacement until a new data set of the same size is obtained; the taxa remain the same in all replicates. Some characters will not be included at all in a given bootstrap replicate, while others may be included two or more times. The resampling process is usually repeated at least 100 times to produce a series of replicate data sets that are each then subjected to a phylogenetic analysis. The frequencies with which clades are recovered in the resulting trees are interpreted as indicative of the degree of support for groupings in the original data set. For example, recovery of a particular monophyletic group in 95% of 100 bootstrap replicates is typically interpreted as indicating relatively strong support, while bootstrap values under 70% reflect weak support. Some authors have suggested that bootstrap results may be used to place confidence limits on phylogenies under some conditions (e.g., Felsenstein 1985a; Sanderson 1989). However, it has recently been shown that bootstrapping often gives underestimates of phylogenetic accuracy at high bootstrap values, and overestimates of accuracy at low bootstrap values (Hillis & Bull 1993). Nevertheless, bootstrapping remains a useful tool for providing rough estimates of relative support for different clades, particularly when this method is used in conjunction with other techniques (e.g, decay analysis). In most cases, bootstrap results are presented in the form of a majority-rule consensus tree on which bootstrap values (numbers between 1 and 100, representing percentages) are given for each node.

In contrast to bootstrapping, the jackknife method resamples the original data set by dropping k data points (usually taxa) at a time and recomputing the estimate (phylogeny) from the remaining $n-k$ observations. When taxa are jackknifed, k is usually set to 1 so that each of the n taxa are dropped in turn, producing a series of 'pseudoestimates' of the phylogeny. These trees are then compared to evaluate node stability. Alternatively, the jackknife method may be applied to characters instead of taxa, with k set equal to some proportion (usually 50%) of the total characters (n). Characters to be omitted from each pseudoreplicate data set are chosen randomly, the resulting data are subjected to phylogenetic analysis, and the process is repeated 100 times or more to obtain multiple trees for comparison. Jackknife results may be reported either in the form of a strict consensus tree (which includes only those clades supported in all jackknife replicates), or the percentage of replicates supporting various clades may be mapped on a majority-rule consensus tree. Many of the same assumptions and interpretive problems associated with bootstrapping also apply to jackknifing (see discussions in Swofford & Olsen 1990, Siddal 1995, and Swofford et al. 1996).

Although it is possible to apply decay, bootstrap, and jackknife methods to the same data set, in practice most workers use either one or two of these methods to evaluate their phylogenetic results. Results of different methods generally provide roughly similar (though by no means identical) estimates of clade stability when applied to the same data set.

A remaining issue concerns how one can choose the 'best' phylogeny from among studies that are based on different data sets. I have argued that character consensus ('total evidence') offers the most productive approach, and have favored combining data sets to generate a single phylogenetic hypothesis whenever possible (Simmons et al. 1991; Simmons 1993a, 1994, 1996, 1998; Simmons & Geisler 1998). However, combination of data sets raises numerous practical and philosophical problems, and data combination may not be appropriate (or even possible) under all circumstances (Bull et al. 1993; Lanyon 1993; Chippindale & Wiens 1994; Farris et al. 1994; Huelsenbeck et al. 1994, 1996; de Queiroz *et al.* 1995; Miyamoto & Fitch 1995; Huelsenbeck & Bull 1996; Siddall 1997; Messenger & McGuire 1998). Perhaps more importantly, many 'users' of phylogenies (e.g., ecologists) may not be interested in constructing new phylogenies; these workers instead need a means of choosing among those already available.

Several arbitrary and misguided methods for choosing among phylogenetic hypotheses have been employed over the years, including uncritical acceptance of newer phylogenies (as opposed to older hypotheses) and a priori preference for molecular phylogenies (as opposed to those based on morphological studies). A more rational means of evaluating and choosing among competing phylogenies is to examine relative levels of support as reflected by decay, bootstrap, or jackknife values. Entire trees can be accepted or rejected based on overall levels of support, or relative support for particular groupings can be evaluated to decide which relationships are most likely to be correct. Lanyon (1993) argued that only strongly supported portions of trees should accepted, and that disagreement between phylogenies is a significant problem only when incongruent groups each have strong support. The latter may be true, but in many cases weakly-supported nodes or hypotheses are all we have. It is frequently more productive in the long run to use one or more phylogenies – weak though they may be – rather than to abandon a phylogenetic context in comparative studies. Even a 'fruitful, wrong hypothesis' (Pettigrew 1991*b*: 231) can lead to significant scientific contributions.

CURRENT HYPOTHESES OF BAT RELATIONSHIPS

Bat monophyly

In the 1980s and early 1990s, a controversy arose in mammalian systematics regarding bat monophyly (for a review see Simmons 1994). Most biologists had long assumed that Chiroptera was monophyletic and that all bats shared a common flying ancestor, which implies that powered flight evolved only once in mammals. However, evidence from morphology of the penis and nervous system led several authors to propose that bats are actually diphyletic (e.g., Smith & Madkour 1980; Hill & Smith 1984; Pettigrew 1986, 1991*a*, 1991*b*, 1994, 1995; Pettigrew & Jamieson 1987; Pettigrew *et al.* 1989; Pettigrew & Kirsch 1995, 1998). Although several versions of bat diphyly have been suggested (for a review see Simmons 1994), the hypothesis most commonly cited suggests that Megachiroptera is more closely related to Primates and Dermoptera than it is to Microchiroptera. If this is so, Megachiroptera and Microchiroptera must have evolved from different non-flying ancestors, and therefore represent a remarkable case of convergence.

Although the bat diphyly hypothesis has been discussed extensively in the literature and still remains popular with some workers (mostly non-systematists), a growing body of data provides very strong support for bat monophyly. Data supporting chiropteran monophyly include morphological features from many organ systems (e.g., Luckett 1980, 1993; Wible & Novacek 1988; Kovtun 1989; Baker et al. 1991b; Thewissen & Babcock 1991, 1993; Kay et al. 1992; Beard 1993; Johnson & Kirsch 1993; Simmons 1993a, 1994, 1995; Wible & Martin 1993; Novacek 1994; Simmons & Quinn 1994; Miyamoto 1996; Simmons & Geisler 1998; Stafford & Thorington 1998), DNA hybridization data (e.g., Kilpatrick & Nunez 1993; Kirsch et al. 1995; Hutcheon & Kirsch 1996; Kirsch 1996), and nucleotide sequence data from numerous mitochondrial and nuclear genes (e.g., Adkins & Honeycutt 1991, 1993, 1994; Mindell et al. 1991; Ammerman & Hillis 1992; Bailey et al. 1992; Stanhope et al. 1992, 1993, 1996; Honeycutt & Adkins 1993; Knight & Mindell 1993; Novacek 1994; Allard et al. 1996; Miyamoto 1996; Porter et al. 1996). The fact that all of these diverse data sets unambiguously support bat monophyly is remarkable, and most systematists agree that bat monophyly now represents one of the most strongly supported hypotheses in mammalian systematics (Simmons 1994; Miyamoto 1996). A revised summary of 33 morphological synapomorphies diagnosing Chiroptera was provided by Simmons & Geisler (1998).

Relationships within Megachiroptera

Megachiroptera (= Pteropodidae) includes over 45 genera and 150 species (Koopman 1993). All live in the Old World, and are characterized by a predominantly frugivorous or nectarivorous diet and absence of laryngeal echolocation. Monophyly of extant pteropodids is supported by numerous derived morphological characters including dental specializations associated with frugivory (Slaughter 1970; Koopman & MacIntyre 1980), unusual structural features of the sperm (Rouse & Robson 1986), and a retinal nutrient system that is unique among vertebrates (Kolmer 1910, 1911, 1926; Fritsch 1911; Duke-Endler 1958; Pedler & Tilley 1969; Buttery et al. 1990). In addition, DNA hybridization data and nucleotide sequence data from the mitochondrial 12S rRNA and tRNA valine gene complex strongly support pteropodid monophyly (Kirsch et al. 1995, 1998; Hollar & Springer 1997; Hutcheon et al. 1998). However, recent work has suggested that at least some fossil taxa previously referred to Megachiroptera may not be closely related to this clade. Based on an analysis of limb structure, Schutt & Simmons (1998) concluded that the Oligocene fossil *Archaeopteropus* is not an early megachiropteran as traditionally thought, but instead nests among the basal branches of the lineage leading to Microchiroptera.

Phylogenetic studies of relationships among extant pteropodids effectively began with Andersen's (1912) monograph, an exceptional precladistic treatment that nevertheless distinguished between primitive and derived characters. Although over 80 years old, this work is still cited frequently and remains an important baseline for all discussions of pteropodid relationships. Most subsequent phylogenetic studies of pteropodids have been designed to test hypotheses originally proposed by Andersen.

Six data sets have been employed in the study of pteropodid relationships in recent years: morphology (e.g., Hood 1989; Springer *et al.* 1995), karyotypes (Haiduk *et al.* 1980, 1981, 1983), DNA hybridization (e.g., Kirsch *et al.* 1995; Springer *et al.* 1995), allozymes (Juste *et al.* 1997), rDNA restriction sites (e.g., Colgan & Flannery 1995), and nucleotide sequence data (Hollar & Springer 1997). Taxonomic problems addressed in these studies have included testing the monophyly of the nectar-feeding subfamily Macroglossinae, examination of relationships among African and Southeast Asian genera, and evaluation of relationships among species of *Rousettus* sensu lato.

Some phylogenetic problems have proven relatively tractable, with analyses of different data sets supporting similar answers. For example, there is a growing consensus that *Rousettus* as traditionally recognized is not monophyletic. Data as diverse as allozymes, DNA hybridization, nucleotide sequences, karyotypes, and behavior support recognition of *Lissonycteris* as distinct from *Rousettus* (Lawrence & Novick 1963; Haiduk *et al.* 1980, 1981; Kirsch *et al.* 1995; Springer *et al.* 1995; Hollar & Springer 1997; Juste *et al.* 1997). However, there is no such agreement concerning relationships of *Lissonycteris* and *Rousettus* to other genera.

Andersen (1912) divided pteropodids into two groups: Macroglossinae, which contained taxa with specializations for nectar-feeding, and Pteropodinae, which contained all other pteropodid genera (including *Harpionycteris*, which was listed as a separate subfamily in early sections of the monograph but rejected later in the text; see Andersen 1912: 803). Several studies have tested this arrangement and found that neither Macroglossinae nor Pteropodinae are monophyletic (e.g., Hood 1989; Colgan & Flannery 1995; Kirsch *et al.* 1995; Springer *et al.* 1995; Hollar & Springer 1997). Macroglossinae appears to include at least three groups that evolved independently from within non-nectarivorous pteropodid

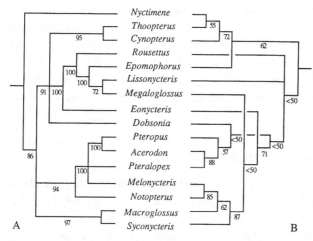

Figure 2.1. Two competing hypotheses of pteropodid relationships from Springer *et al.* (1995). The numbers below internal branches indicate the percentage of bootstrap replicates in which each clade appeared. A) Tree based on DNA hybridization data showing only those nodes that were moderately supported (bootstrap value > 70%) in four different analyses (see Kirsch *et al.* [1995] for details). B) Strict consensus tree of two equally-parsimonious trees resulting from analysis of 36 morphological characters. The trees resulting from these two analyses are completely incongruent (i.e., no clade appears in both trees). However, the level of support for clades in the DNA hybridization tree are much higher than for clades recovered from the morphological data, suggesting that the DNA data many be more informative than the morphological data in this case. Note that Macroglossinae as traditionally recognized (including *Eonycteris, Megaloglossus, Syconycteris, Macroglossus, Notopterus*, and *Melonycteris*) is monophyletic in tree B but not in tree A.

clades. Female urogenital tract morphology, DNA hybridization, and nucleotide sequences suggest that the African genus *Megaloglossus* is more closely related to a clade of endemic African taxa than it is to other macroglossines (e.g., Hood 1989; Kirsch *et al.* 1995; Springer *et al.* 1995; Hollar & Springer 1997). Analysis of DNA hybridization and restriction-site data have suggested that *Eonycteris* also belongs to this group, but is not closely related to *Megaloglossus* and thus appears to represent yet another case of convergence (Colgan & Flannery 1995; Kirsch *et al.* 1995). In contrast, an analysis of available morphological data (including Hood's urogenital tract characters) supported monophyly of Macroglossinae but not Pteropodinae, which was found to include macroglossines (Springer *et al.* 1995). Unfortunately, trees based on morphology (Fig. 2.1B) are largely incongruent with those based on DNA hybridization (Fig. 2.1A), restriction sites (Colgan & Flannery 1995) and sequence data (Hollar & Springer 1997), so no consensus has yet emerged as to just what relationships should be believed. However, bootstrap values for clades in the DNA hybridization and sequence trees are much higher than those found for clades in the morphology tree (e.g., Fig. 2.1), suggesting that the DNA data may contain stronger phylogenetic signal (and/or are more internally consistent) than the morphological data.

Eocene fossil bats

The fossil record of bats extends back at least to the early Eocene, and chiropteran fossils are known from all continents except Antarctica. Among the more interesting fossil bats are *Icaronycteris*, *Archaeonycteris*, *Hassianycteris*, and *Palaeochiropteryx*. These Eocene taxa are known from exceptionally well-preserved fossils, and they have long formed a basis for reconstructing the early evolutionary history of Chiroptera (see review in Simmons & Geisler 1998).

Unlike most other fossil bats, *Icaronycteris*, *Archaeonycteris*, *Hassianycteris*, and *Palaeochiropteryx* have not been referred to any extant family or superfamily. Smith (1977) suggested that these taxa represent an extinct clade of early microchiropterans ('Palaeochiropterygoidea'). In contrast, Van Valen (1979) argued that these fossil forms are representatives of a primitive grade, ancestral to both Megachiroptera and Microchiroptera ('Eochiroptera'). Novacek (1987) reanalyzed morphology of *Icaronycteris* and *Palaeochiropterx* and concluded that they are more closely related to extant Microchiroptera than to Megachiroptera. Most recently, Simmons & Geisler (1998) found that *Icaronycteris*, *Archaeonycteris*, *Hassianycteris*, and *Palaeochiropteryx* represent a series of consecutive sister-taxa to extant microchiropteran bats (Fig. 2.2). Based on results of a comprehensive analysis, Simmons & Geisler proposed a new higher-level classification of bats. Critical features of this classification included restriction of Microchiroptera to the smallest clade including all extant bats that use sophisticated laryngeal echolocation, and formal recognition of two more inclusive clades: Microchiropteraformes (Microchiroptera + Palaeochiropterygidae + Hassianycteridae), and Microchiropteramorpha (Microchiropteraformes + Archaeonycteridae + Icaronycteridae). Summaries of the derived characters diagnosing each of these clades can be found in Simmons & Geisler (1998).

Is Microchiroptera monophyletic?

Microchiroptera includes over 130 extant genera and 750 species (Koopman 1993). Most workers agree that Microchiroptera is monophy-

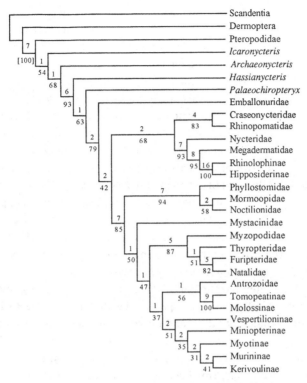

Figure 2.2. Simmons & Geisler's (1998) hypothesis of higher-level relation-ships of extant and extinct lineages of bats (redrawn from their figure 36). This is the single most-parsimonious tree that resulted from a parsimony analysis of 195 morphological characters, 12 rDNA restriction-site charac-ters, and one character based on the number of R-1 tandem repeats in the mtDNA d-loop region. The numbers below internal branches indicate the percentage of 1000 bootstrap replicates in which each clade appeared; numbers above the branches are decay values (the minimum number of additional steps required to collapse each clade). The bootstrap analysis was constrained to consider only trees in which Chiroptera was monophyletic; this was done because most of the characters that support bat monophyly were omitted from the study. Note that the bootstrap and decay values asso-ciated with nodes including fossil taxa are comparable to those associated with many extant clades.

letic, and morphological data strongly support this conclusion. In a study that did not include fossil taxa, Simmons (1998) found that microchirop-teran monophyly was supported in 100% of 1000 bootstrap replicates, and that 13 additional steps were required to render Microchiroptera non-monophyletic. These values are lower when Eocene fossil taxa are included, but Simmons & Geisler (1998) nevertheless identified 13 synapomorphies that apparently diagnose Microchiroptera: 1) presence of sophisticated laryngeal echolocation, 2) spinal cord with angle between dorsal horns of 0–25°, 3) inferior colliculus larger than superior colliculus, 4) presence of a tragus, 5) aquaeductus cochleae small or absent, 6) free premaxilla, 7) reduction to two lower premolars on each side of the jaw, 8) m. styloglossus originates from ventral surface of mid-point of stylohyal, 9) clavicle articulates with coracoid process of scapula, 10) m. spinotrapezius clearly differentiated from trapezius complex, 11) origin of m. acromiodeltoideus does not include sixth thoracic vertebra, 12) xiphisternum with prominent median keel, and 13) m. flexor digit-orum profundus does not insert on digit II of wing. It is unfortunate that many of these features cannot be scored in fossils, since some were prob-ably present in the fossil relatives of the microchiropteran crown group (e.g., *Palaeochiropteryx, Hassianycteris*; Simmons & Geisler 1998). Never-theless, all of these derived traits were apparently present in the earliest microchiropterans and absent in basal Megachiroptera.

Despite strong morphological evidence for this grouping, mono-phyly of Microchiroptera is not supported by all data sets. Recent DNA hybridization studies by Kirsch and his colleagues have suggested that Rhinolophoidea and Pteropodidae may be sister-taxa, implying micro-chiropteran paraphyly (Fig. 2.3; Pettigrew & Kirsch 1995; Kirsch 1996; Kirsch & Hutcheon 1997; Hutcheon *et al.* 1998; Kirsch *et al.* 1998). This rather startling hypothesis was also weakly supported in analyses of von Willebrand factor gene sequence data (Porter *et al.* 1996), and indirectly by discovery of a family of satellite DNA markers that apparently occur in all higher-level microchiropteran lineages except those included in Rhinolophoidea (Baker *et al.* 1997). Implications of microchiropteran par-aphyly are profound; among other things, such an arrangement implies either convergent evolution of echolocation in two lineages (in Rhinolophoidea and again in the lineage leading to the remaining micro-chiropterans), or loss of echolocation in Pteropodidae (an unlikely pos-sibility; see discussion in Simmons & Geisler 1998). In either case, much of what we surmise about bat evolution would require rethinking if Microchiroptera were paraphyletic.

The problem of reconciling the incongruence between available molecular data (which support paraphyly of Microchiroptera) and morphological data (which strongly support monophyly) was addressed by Kirsch (1996) and Kirsch & Pettigrew (1998). These authors suggested that patterns observed in the molecular data may be due to a marked AT-bias in the DNA of rhinolophoids and pteropodids. These taxa have DNA that contains more adenine (A) and thymine (T) than cytosine (C) and

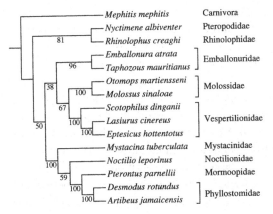

Figure 2.3. Kirsch *et al.*'s (1998) hypothesis of higher-level relationships of extant lineages of bats (redrawn from their figure 3). This tree was derived from an analysis of DNA hybridization data; numbers below internal branches indicate the percentage of 3000 bootstrap replicates in which each clade appeared. An exhaustive jackknife analysis on the same data set produced a tree that differed only in pairing emballonurids with noctilionoids plus *Mystacina*. Paraphyly of Microchiroptera and a close relationship between *Mystacina* and noctilionoids was supported in all analyses.

guanine (G), a pattern that may cause artificial grouping of these taxa in phylogenetic analyses. Kirsch (1996) noted that this bias may represent a case in which 'molecular data [are] positively misleading' about relationships. Simmons (1998) reanalyzed her morphological data to test the possibility of a sister-group relationship between Pteropodidae and Rhinolophoidea, and found that the shortest trees containing this grouping were 23 steps longer than the most parsimonious trees. All things considered, paraphyly of Microchiroptera seems unlikely; however, additional studies – including character-congruence studies combining morphological and discrete molecular data – will be necessary to adequately test this hypothesis.

Interfamilial relationships

Relationships among families of bats remained poorly understood for many decades despite advances in understanding of relationships at higher and lower taxonomic levels. While most bat species can be assigned easily to one of several monophyletic families, there has been little agreement concerning relationships among family-level taxa until recently (for a historical review see Simmons & Geisler 1998). Simmons

(1998) assembled a large data set of morphological and rDNA restriction-site characters, and used these data to construct a new higher-level phylogeny for extant bats. This data set was subsequently modified and expanded by Simmons & Geisler (1998), who additionally included fossils in their analysis (see above). Analyses of these data resulted in a relatively well-resolved, well-supported tree (Fig. 2.2). Simmons (1998) and Simmons & Geisler (1998) concluded that 1) Emballonuridae probably occupies the basal branch within Microchiroptera, and therefore Yinochiroptera sensu Koopman (1985, 1994) is not monophyletic, 2) Rhinopomatidae and Craseonycteridae are sister-taxa, 3) Rhinolophoidea, Rhinolophidae (including Hipposiderinae), Yangochiroptera, and Noctilionoidea are each monophyletic, 4) Myzopodidae, Thyropteridae, Furipteridae, and Natalidae form a clade, 5) Vespertilionidae sensu Koopman (1993, 1994) is not monophyletic, Tomopeatinae and Antrozoinae are more closely related to Molossidae than to other vespertilionids. On the basis of these results, they recommended recognition of three new superfamily-level groups: 1) Rhinopomatoidea (including Rhinopomatidae + Craseonycteridae), 2) Nataloidea (including Myzopodidae + Thyropteridae + Furipteridae + Natalidae), and 3) Molossoidea (Antrozoidae + Molossidae [including Tomopeatinae]). Vespertilionoidea was restricted to Vespertilionidae sensu stricto in anticipation that other taxa now recognized as subfamilies of the latter may be raised to family rank in future years.

Despite the strong support found for numerous clades, many groups in Simmons's (1998) and Simmons & Geisler's (1998) trees were poorly supported (e.g., decay values < 2, bootstrap values < 70), and so remain suspect. Although Mystacinidae was placed as the sister-taxon of Nataloidea + Molossoidea + Vespertilionoidea in the most-parsimonious trees, support for this placement was relatively weak (Fig. 2.2) and Mystacinidae instead grouped with Noctilionoidea in trees only one step longer. Bootstrap support for the latter arrangement was only slightly less than for the most-parsimonious placement (i.e., 38% compared with 50%). This result is particularly significant in the context of a recent DNA hybridization study by Kirsch et al. (1998), who found strong support (bootstrap value = 100) for a close relationship between Mystacinidae and Noctilionoidea (Fig. 2.3). This relationship was initially suggested by Pierson (1986) and Pierson et al. (1986) on the basis of immunological distance data. Congruence of results of studies based on such disparate data sets suggests that Mystacinidae probably represents the most basal branch within Noctilionoidea (Kirsch et al. 1998). Noctilionoidea is traditionally viewed as containing three families: Noctilionidae, Mormoopidae, and Phyllostomidae. Relationships among these groups remains somewhat problematic. Simmons (1998) and Simmons & Geisler (1998) found weak support for a sister-group relationship between Mormoopidae and Noctilionidae (Fig. 2.2), but Kirsch *et al.*'s (1998) analysis (Fig. 2.3) strongly supported a sister-group relationship between Mormoopidae and Phyllostomidae.

Relationships within Emballonuridae

Emballonuridae is a pantropical/subtropical group that includes 13 extant genera and over 45 species (Koopman 1993). Monophyly of the family is supported by morphology (Griffiths & Smith 1991; Simmons 1998), immunological distance data (Robbins & Sarich 1988), and results of DNA hybridization experiments (Kirsch *et al.* 1998). Phylogenetic analyses of Emballonuridae have included studies based on allozymes (Robbins & Sarich 1988), immunological data (Robbins & Sarich 1988), DNA hybridization (Kirsch *et al.* 1998), and morphology (Barghoorn 1977; Griffiths *et al.* 1991; Griffiths & Smith 1991). At higher taxonomic levels, there is general agreement that Neotropical emballonurids form a clade, that *Emballonura*, *Mosia*, and *Coleura* are more closely related to this clade than to other Old World taxa, and that *Taphozous* and *Saccolaimus* occupy the most basal branch in the family. However, there is little agreement concerning interrelationships of Neotropical genera or species relationship within the larger genera (e.g., *Taphozous*, *Saccolaimus*, *Emballonura*). Griffiths *et al.* (1991) argued that *Emballonura* is paraphyletic with respect to *Coleura*, but incongruence between different data sets (e.g., immunological distance data and morphology) led them to retain the latter as a distinct genus pending further study.

Relationships within Rhinopomatoidea

Rhinopomatoidea is an Old World superfamily that includes only two families (Rhinopomatidae and Craseonycteridae), each of which is monogeneric (Simmons 1998; Simmons & Geisler 1998). Monophyly of Rhinopomatoidea is strongly supported by morphological data, as is monophyly of Rhinopomatidae (Fig. 2.2; Simmons 1998; Simmons & Geisler 1998). Craseonycteridae is monotypic (Hill 1974). Relationships among the four extant species of Rhinopomatidae has not been formally investigated in a phylogenetic analysis, but a key published in a recent revision of the family by Van Cakenberghe & de Vree (1994) suggests that *Rhinopoma hardwickei* and *R. macinnesi* are sister-taxa, and that *R. muscatellum* is the sister-taxon of that clade.

Relationships within Rhinolophoidea

Many authors recognize Rhinolophidae and Hipposideridae as separate families, but there is overwhelming evidence that these groups are sister-taxa (e.g., Pierson 1986; Simmons 1998; Simmons & Geisler 1998; Kirsch et al. 1998). Simmons (1998) and Simmons & Geisler (1998) followed Koopman (1993, 1994) in recognizing Hipposiderinae as a subfamily of Rhinolophidae, a nomenclatural arrangement that provides recognition of both the similarities and differences between these clades. In addition to Rhinolophidae, two other families are included in the superfamily Rhinolophoidea: Nycteridae and Megadermatidae (Simmons 1998; Simmons & Geisler 1998). All rhinolophoids are insectivorous or carnivorous, and all are restricted to the Old World.

Rhinolophinae currently includes one genus and over 60 species (Koopman 1993). Monophyly of Rhinolophinae is supported by morphology (Bogdanowicz & Owen 1992; Simmons 1998) and immunological data (Pierson 1986). Phylogenetic relationships among rhinolophines have been investigated using allozymes (Qumsiyeh et al. 1988; Maree & Grant 1997) and morphometric data (Bogdanowicz & Owen 1992). Bogdanowicz & Owen (1992) found that analyses of their morphometric data under different sets of assumptions resulted in incongruent trees, none of which was congruent with the allozyme tree. Some clades identified in one morphometric tree are supported by limited karyotype data, but few relationships could be identified with any confidence (Bogdanowicz & Owen 1992). Results of a cladistic analysis of allozyme data from ten South African species (Maree & Grant 1997) were consistent with some of the relationships depicted in Bogdanowicz & Owen's (1992) common-part-removed tree, but small sample sizes in the allozyme study limit the usefulness of the data. Interrelationships of most species of *Rhinolophus* therefore remain largely uncertain.

Hipposiderinae is a somewhat more diverse group including nine genera and over 70 extant species (Koopman 1993). Hipposiderine monophyly is strongly supported by morphological data (Simmons 1998; Bogdanowicz & Owen 1998; Hand & Kirsch 1998). Relationships within this clade were recently investigated in two studies based on different but overlapping morphological data sets (i.e., Bogdanowicz & Owen 1998; Hand & Kirsch 1998). Both concluded that *Hipposideros* is paraphyletic, but differed in most other details of tree topology. Results of neither study were congruent with Hill's (1963) divisions of species into subgroups. Better resolution of hipposiderine relationships will probably require development of larger data sets combining more subsets of morphological data and/or use of molecular methods.

Megadermatidae currently includes four extant genera and five species (Koopman 1993), and monophyly of this family is strongly supported by morphological data (Simmons 1998). Phylogenies of Megadermatidae have been developed based on craniodental characters (Hand 1985) and morphology of the hyoid region (Griffiths et al. 1992). Unfortunately, the trees that resulted from these studies are completely incongruent, and no study has yet combined these data in a single analysis. Relationships within Megadermatidae thus remain unresolved.

Nycteridae is a small family that includes only one genus and fewer than 20 species (Koopman 1993). Monophyly of Nycteridae is strongly supported by morphological data (Griffiths 1994, 1997; Simmons 1998). Griffiths (1994) proposed a phylogeny of Nycteris based on morphology of the hyoid region and other morphological characters. While this phylogeny is not entirely congruent with results of recent revisionary studies (e.g., Van Cakenberghe & de Vree 1985, 1993; Thomas et al. 1994), it does support monophyly of several species groups including the hispida and thebaica groups.

Relationships within Noctilionoidea

As noted previously, Noctilionoidea minimally includes three Neotropical families: Noctilionidae, Mormoopidae, and Phyllotomidae. Two of these families are small; Noctilionidae comprises a single pair of sister-taxa, and Mormoopidae includes eight species. Relationships within Mormoopidae have been addressed in only two studies. Smith (1972) proposed a phylogeny of the group based on a precladistic revisionary treatment of the family. Congruent results were obtained in a subsequent study including immunological, allozyme, and morphological characters (Arnold et al. 1982). The latter study found support for monophyly of Mormoops, Pteronotus, and a clade comprising most Pteronotus species but excluding P. parnellii.

No clade of bats has received as much systematic attention as Phyllostomidae, a large family that includes over 140 extant species (Koopman 1993). Fascinated by the dietary and morphological diversity of the group, workers have employed virtually every type of data and phylogenetic method available in studies of various subsets of taxa (for a historical review see Wetterer et al. in press). Several topics have been the focus of attention, including 1) relationships of vampires (Desmodontinae) to

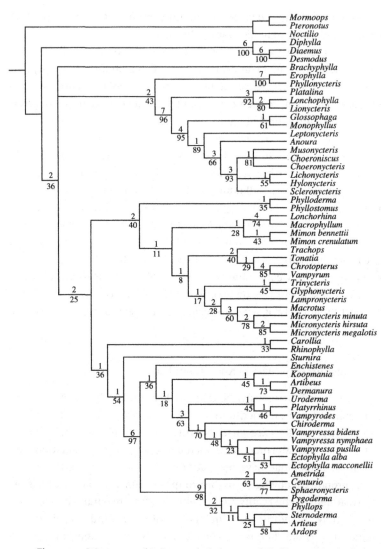

Figure 2.4. Wetterer *et al.*'s (in press) phylogeny of Phyllostomidae (redrawn from their figure 45). The tree shown is a strict consensus of 18 equally-parsimonious trees that resulted from a parsimony analysis of 137 morphological characters, 12 rDNA restriction-site characters, and one character based on presence/absence of a 900-base-pair EcoRI defined nuclear satellite DNA repeat. The numbers below internal branches indicate the percentage of 1000 bootstrap replicates in which each clade appeared; numbers above the branches are decay values (the minimum number of additional steps required to collapse each clade). Generic names follow Wetterer *et al.* (in press), who recommended the following: 1) splitting *Micronycteris* sensu lato into *Trinycteris* (=*T. nicefori*), *Glyphonycteris* (=*G. sylvestris* and *G. daviesi*),

other phyllostomids, 2) monophyly of various subfamilies and relationships among them, 3) interrelationships of stenodermatine genera, and 4) species relationships within selected genera (e.g., *Artibeus, Chiroderma, Micronycteris, Phyllostomus*).

Most problems in higher-level phyllostomid systematics boil down to a single question: how are the approximately 50 genera of phyllostomid bats related to one another? No single data subset (e.g., karyotypes, immunology, allozymes, craniodental characters, features of the female urogenital tract) has provided the resolution necessary to adequately address this problem. Baker *et al.* (1989) adopted a taxonomic-congruence approach to build a consensus tree based upon morphology of the reproductive tract, immunology, and karyology. Although several clades were identified in their study, most relationships remained unresolved due to incongruence among the data sets employed. Increased resolution was subsequently obtained by Van Den Bussche (1992), who optimized restriction-site characters onto the Baker *et al.* tree. However, Van Den Bussche's study did not include all genera, and relationships within many large groups (e.g., Stenodermatinae) remained poorly resolved and/or weakly supported.

Most recently, Wetterer *et al.* (in press) obtained much greater resolution and stronger support for many clades in a phylogenetic analysis of 150 morphological, restriction-site, karyotype, and satellite DNA characters surveyed in all phyllostomid genera. This study built upon the work of many authors who had focused on smaller subsets of taxa and characters (e.g., Benedict 1957; Strickler 1978; Straney 1980; Griffiths 1982; Haiduk & Baker 1982; Hood & Smith 1982; Owen 1987, 1991; Van Den Bussche 1991, 1992; Lim 1993; Van Den Bussche *et al.* 1993; Gimenez *et al.* 1996; Simmons 1996; Wible & Bhatnagar 1997), but it additionally included character data from several previously unexplored anatomical systems. Results of parsimony analyses (Fig. 2.4) provided support for monophyly of all traditionally recognized subfamilies (Desmodontinae, Brachyphyllinae, Phyllonycterinae, Glossophaginae [including Lonchophyllini], Phyllostominae, Carolliinae, and Stenodermatinae). Relationships within most clades were well resolved and, at least in some cases,

Figure 2.4 (*cont.*)

Lampronycteris (= *L. brachyotis*), *Neonycteris* (= *N. pusilla*), and *Micronycteris* (= *M. megalotis, M. microtis, M. hirsuta, M. minuta, M. schmidtorum, M. sanborni*), 2) retention of *Dermanura* and *Koopmania* as subgenera of *Artibeus*, and 3) inclusion of *Mesophylla* in *Ectophylla*. Revised diagnoses of some of these taxa (*Micronycteris, Trinycteris, Glyphonycteris,* and *Ectophylla*) were recently provided by Simmons & Voss (1998).

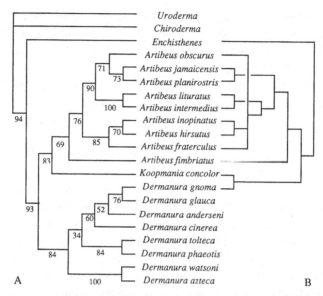

Figure 2.5. Two hypotheses of relationships among species of the *Artibeus* complex; taxonomic names follow Van Den Bussche *et al.* (1998). A) Van Den Bussche *et al.*'s (1998) hypothesis of relationships among species of *Artibeus*, *Dermanura*, and *Koopmania* (redrawn from their figure 1). This is the single most-parsimonious tree found in an analysis of 305 variable sites in the mitochondrial cytochrome *b* gene. The numbers below internal branches indicate the percentage of 1000 bootstrap replicates in which each clade appeared. Separate analysis of EcoRI sattelite DNA fragments (results not shown) provided additional support for monophyly of *Dermanura* and monophyly of the *Artibeus* + *Dermanura* + *Koopmania* clade (excluding *Enchisthenes*). B) Marques-Aguilar's (1994) hypothesis of relationships among *Artibeus* sensu stricto. This is a strict consensus of two equally-parsimonious trees derived from an analysis of 31 morphological characters (redrawn from her figures 10 and 11). No bootstrap, decay, or jackknife values were reported.

were well supported in decay and bootstrap analyses. The Wetterer *et al.* tree (Fig. 2.4) represents the most comprehensive hypothesis of phyllostomid relationships currently available.

Relationships within a number of genera have also been investigated in recent years. There is a growing consensus that *Artibeus* (including *Dermanura* and *Koopmania*, but not *Enchisthenes*) is monophyletic (Lim 1993; Van Den Bussche *et al.* 1993, 1998; Marques-Aguilar 1994; Wetterer *et al.* in press), despite some claims to the contrary (e.g., Owen 1987, 1991). Within this group, *Dermanura* and *Artibeus* each appear to be monophyletic (Fig. 2.5; Van Den Bussche *et al.* 1993, 1998; Marques-Aguilar 1994).

Although there is incongruence between species-level trees derived from cytochrome *b* sequence data (Van Den Bussche *et al.* 1998) and morphological data (Marques-Aguilar 1994), many groupings were supported in both analyses (Fig. 2.5). Cytochrome *b* data have also proven informative with respect to species-level relationships in *Phyllostomus* (Van Den Bussche & Baker 1993) and *Chiroderma* (Baker *et al.* 1994), both of which appear to be monophyletic. In contrast, some recent phylogenies have called into question the affinities of species traditionally considered congeneric. Analyses of combined morphological and restriction-site data indicate that *Micronycteris*, as traditionally recognized, is not monophyletic (Wetterer *et al.* in press). However, the subgenera of *Micronycteris* recognized by Simmons (1996) are apparently monophyletic, prompting Wetterer *et al.* to raise these taxa (*Glyphonycteris*, *Lampronycteris*, *Micronycteris*, *Neonycteris*, and *Trinycteris*) to genus rank. *Vampyressa* is also not monophyletic, but more work on species relationships must be done before the genus can be split into demonstrably monophyletic units (Wetterer *et al.* in press).

Relationships within Nataloidea

As defined by Simmons (1998), Nataloidea comprises four small families: Myzopodidae (which is monotypic), Thyropteridae (one genus and three species), Furipteridae (two genera and two species), and Natalidae (one genus and six species; Koopman 1993; Pine 1993). All of these are clearly monophyletic, and the three Neotropical taxa (Thyropteridae, Furipteridae, and Natalidae) together form a clade (Fig. 2.2; Simmons 1998; Simmons & Geisler 1998). Within this group there is strong support for a sister-group relationship between Furipteridae and Natalidae (Simmons 1998; Simmons & Geisler 1998). Relationships within the only two speciose groups, Thyropteridae and Natalidae, have never been formally analyzed.

Relationships within Molossoidea

Simmons (1998) defined Molossoidea as including two families, Antrozoidae and Molossidae. Antrozoidae comprises two species of gleaning insectivores, *Antrozous pallidus* and *Bauerus dubaiquercus*. In contrast, Molossidae (including Tomopeatinae) is a diverse group of fast-flying aerial insectivores that currently includes over 13 genera and 80 species (Koopman 1993; Sudman *et al.* 1994). Monophyly of Molossidae (including Tomopeatinae) is strongly supported by morphological data (Fig. 2.2;

Simmons 1998; Simmons & Geisler 1998). Phylogenetic relationships within Molossidae have been investigated using morphometrics (Freeman 1981), discrete morphological characters (Legendre 1984, 1985; Hand 1990), allozymes (Sudman et al. 1994), and cytochrome b gene sequences (Sudman et al. 1994). Trees and classifications derived from these studies are largely incongruent, but some groupings have been suggested by more than one data set. These include a close relationship between *Eumops*, *Molossus*, and *Promops* (supported by morphometric data and gene sequences), a close relationship between *Mormopterus* and *Nyctinomops* (gene sequences, allozymes), and a close relationship between *Chaerephon* and *Mops* (morphometric data, discrete morphological data). Relationships of these groupings to other genera (and to each other) remains unclear. Monophyly of many molossid genera remains questionable (e.g., *Tadarida*, *Mormopterus*), and relationships among Old World and New World lineages are unknown.

Relationships within Vespertilionidae

Vespertilionidae (excluding Antrozoidae and Tomopeatinae) contains almost one-third of living bat species (Koopman 1993, 1994). Five subfamilies of Vespertilionidae are currently recognized: Vespertilioninae, Myotinae, Miniopterinae, Murininae, and Kerivoulinae (Volleth & Heller 1994b; Simmons 1998; Simmons & Geisler 1998). Of these, the latter three are each clearly monophyletic (for a list of synapomorphies see Simmons 1998). Separation of Myotinae from Vespertilioninae is suggested by karyotype data (Volleth & Heller 1994b) and results of Simmons's (1998) and Simmons & Geisler's (1998) phylogenetic analyses (Fig. 2.2), but DNA hybridization data has suggested that *Myotis* nests within a clade of vespertilionines (Kirsch et al. 1998). Monophyly of Myotinae (*Myotis* + *Lasionycteris*) remains uncertain because no unambiguous morphological synapomorphies diagnose this group, and no karyotype data are available for *Lasionycteris*. Monophyly of *Myotis* is supported by derived karyotypic features (Volleth & Heller 1994b), but it is not known if any or all of these traits also occur in *Lasionycteris*. Monophyly of Vespertilioninae (excluding Myotinae) is also supported only by karyotype data (Volleth & Heller 1994b). Unfortunately, comparable karyotype data are not available for Lasiurini or Antrozoidae, leaving open the possibility that Vespertilioninae may be paraphyletic.

The most comprehensive study of vespertilionid relationships is that of Volleth & Heller (1994b), who examined banded chromosomes of Old World representatives of over 20 genera (Fig. 2.6). They found support

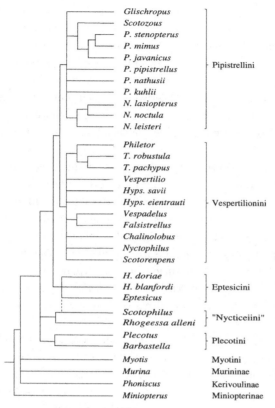

Figure 2.6. Volleth & Heller's (1994b) cladogram of Vespertilionidae based on karyological features (redrawn from their figure 7). Each clade depicted here is supported by one to five karyotypic synapomorphies; see text for discussion. Abbreviations for generic names are as follows: *H.* – *Hesperopterus*; *Hyps.* – *Hypsugo*; *N.* – *Nyctalus*; *P.* – *Pipistrellus*; *T.* – *Tylonycteris*.

for monophyly of Vespertilioninae (excluding *Myotis*), and recognized four clades within this group: Vespertilionini, Pipistrellini, Eptescini, and Plecotini. A fifth group of vespertilionines ('Nycticeiini' including *Scotophilus* and *Rhogeessa*) was tentatively recognized, but Volleth & Heller noted that this group might be paraphyletic with respect to Eptescini. Relationships between tribes was largely unclear, although support was found for a sister-group relationship between Pipistrellini and Vespertilionini (Fig. 2.6). Lasiurines were not included in Volleth & Heller's study, but limited DNA hybridization data have suggested that *Lasiurus* may be more closely related to *Eptesicus* than to *Scotophilus* (Fig. 2.3; Kirsch *et al.* 1998).

Using *Natalus* (Natalidae) and *Molossus* (Molossidae) as outgroups,

Volleth & Heller (1994b) found support for vespertilionid monophyly with Miniopterinae occupying the most basal branch in the family tree (Fig. 2.6). This contrasts somewhat with Simmons's (1998) and Simmons and Geisler's (1998) tree topology (Fig. 2.2), which placed Vespertilioninae as the basal branch. Volleth & Heller could not resolve the relative relationships of Myotinae, Murininae, and Kerivoulinae, but their results do not contradict Simmons's (1998) and Simmons & Geisler's (1998) findings that Murininae and Kerivoulinae form a clade with Myotinae as their sister-taxon.

Volleth & Heller (1994b) included 50 species in their study, and their results also have some implications for lower-level systematics of some taxa. Monophyly of *Myotis*, *Nyctalus*, and *Tylonycteris* was found to be supported by karyotypic synapomorphies, but *Pipistrellus* is paraphyletic as it minimally includes *Scotozous* and might also include *Nyctalus* and/or *Glischropus*. *Pipistrellus stenopterus* clearly groups with *Pipistrellus* and *Scotozous* and thus does not appear to belong to *Hypsugo* as proposed by Hill & Harrison (1987). Monophyly of *Eptesicus*, *Hesperopterus*, and the remaining species of *Hypsugo* was neither refuted, nor supported, by the karyotype data.

Relationships among genera and species in most vespertilionid subfamilies and tribes remain poorly resolved. Most of the widely-used classifications (e.g., Corbet & Hill 1980, 1992; Hill & Harrison 1987; Koopman 1993, 1994) have grouped taxa based on phenetic similarity rather than shared derived characters, in part because phylogenetic analyses have not been completed for many taxa. One exception is Plecotini, which has been the subject of several recent phylogenetic studies based on morphology (Frost & Timm 1992; Tumlinson & Douglas 1992; Bogdanowicz *et al.* 1998) and banded karyotypes (Frost & Timm 1992; Qumsiyeh & Bickham 1993; Volleth & Heller 1994a; Bogdanowicz *et al.* 1998). Problems with interpretation and reporting of chromosome-arm homologies have confused matters (see discussion in Volleth & Heller 1994a), but all of these studies agree that *Euderma* and *Idionycteris* are sister-taxa and that *Barbastella* belongs to Plecotini. Analysis of karyotypes (Qumsiyeh & Bickham 1993) and a recent combined-data parsimony analysis that included all relevant morphological and karyotypic characters (Bogdanowicz *et al.* 1998) additionally support placement of *Otonycteris* (usually considered a vespertilionine of unknown affinities) in Plecotini.

Species relationships of few vespertilionid genera have been investigated recently. Morales & Bickham (1995) used restriction-site data to analyze relationships within *Lasiurus*, and found support for monophyly of the genus and most currently recognized species, including separation of *L. borealis* and *L. blossevillii*. Within *Pipistrellus* sensu lato, Volleth & Heller (1994*b*) found karyotypic support for a sister-group relationship between *P. stenopterus* and *P. mimus*, with *P. javanicus* as the sister to that clade. Derived karyotypic features also provided some resolution of relationships among species of *Nyctalus*, suggesting that *N. lasiopterus* and *N. noctula* form a clade with *N. leisleri* as its sister-taxon (Volleth & Heller 1994*b*).

INFERRING PATTERNS OF EVOLUTION FROM BAT PHYLOGENIES

Character mapping and other methods

Evolutionary transformations in morphology, developmental patterns, ecologies, and behaviors can be analyzed by mapping the distribution of characters or traits onto a phylogeny. Excellent summaries of methods have been provided by Brooks & McLennan (1991), Maddison & Maddison (1992), and Swofford & Maddison (1992). The mapping process is generally guided by parsimony criteria – character transformations are optimized (placed on the tree in optimal positions) so as to minimize the number of state transformations required by the tree topology. Several optimization methods are available, including two often-used methods known as ACTRAN (accelerated transformation optimization) and DELTRAN (delayed transformation optimization).

As discussed by Simmons (1993*a*), two kinds of character transformations may be recognized during the optimization process: unequivocal transformations, which have only one parsimonious placement on the optimal tree(s), and equivocal transformations, which can be parsimoniously arranged in two or more ways. Both ACTRAN and DELTRAN place unequivocal transformations on a given tree in the same manner, but they treat equivocal transformations differently. ACTRAN forces transformations to the lowest possible points on the tree, and thus favors hypotheses of reversal over hypotheses of convergence. Conversely, DELTRAN forces transformations to the highest possible points on a tree, thus favoring hypotheses of convergence over reversal. DELTRAN is often favored in studies involving fossils because it places transformations at the minimal level at which they can be observed (i.e., in the face of missing data it does not conclude that derived states exist below the level at which they can be demonstrated).

Multiple optimizations (minimally including both ACTRAN and DELTRAN) are necessary to identify all possible parsimonious placements of transformations on a given tree. Because different optimizations may not agree in all respects, consensus optimizations often include branches for which state assignments remain uncertain. Reconstruction of transformation patterns is also complicated by cases of taxonomic polymorphism, unresolved nodes in a tree, and incomplete taxonomic sampling. For a discussion of these and other pitfalls of character-state reconstructions, see Swofford & Maddison (1992).

Other methods besides character mapping may also may be used to investigate patterns of evolution once a phylogeny is available. Covariation among characters can be investigated using computer programs like MacClade (Maddison & Maddison 1992), and comparative-method techniques such as independent contrasts can be used to test hypotheses of concerted evolution among traits (Felsenstein 1985b; Martins & Garland 1991; Garland et al. 1992). Kirsch & Lapointe's (1997) study of pteropodid evolution contains a helpful discussion of these approaches and their limitations.

Morphological and functional evolution

Evolutionary biologists have long used phylogenetic trees to infer patterns of evolutionary change in morphological systems. With respect to bats, many basic evolutionary questions may be framed in a phylogenetic context. For example, the origin of flight may be explored by tracing the evolution of wings on higher-level phylogenies of mammals. Use of virtually any recent phylogeny of mammals (see 'Bat monophyly' above) leads to the conclusion that wings only evolved once in mammals, and that these structures were present in the most recent common ancestor of Megachiroptera and Microchiroptera (Simmons 1994). If dermopterans are the sister-group of bats, then the common ancestor of both bats and dermopterans probably had small interdigital patagia that only later evolved into the 'handwing' of bats (Szalay & Lucas 1993; Simmons 1995). This scenario suggests that true powered flight evolved from gliding in the lineage leading to bats (Simmons 1995).

Phylogenies at lower taxonomic levels can be used to investigate patterns of morphological change in systems that vary among different bat groups. Character mapping on family-level phylogenies of bats have been used to explore the evolution of cochlear structure (Fig. 2.7; Novacek 1991; Simmons & Geisler 1998), pubic nipples (Simmons 1993b), the digital tendon locking mechanism (Simmons & Quinn 1994), the noseleaf (Pedersen 1995), features of the vomeronasal organ system (Wible & Bhatnagar 1997), the calcar (Simmons & Geisler 1998; Schutt & Simmons

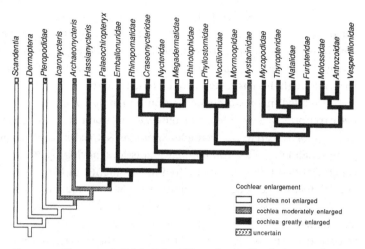

Figure 2.7. Simmons & Geisler's (1998) hypothesis of evolutionary changes in relative size of the cochlea in bats (redrawn from their figure 40). In the lineage leading to extant microchiropterans, a moderately enlarged cochlea apparently evolved prior to the divergence of *Icaronycteris*; a greatly enlarged cochlea evolved prior to the divergence of *Hassianycteris*. Reversals to the former condition apparently occurred independently in Mystacinidae and within Megadermatidae and Phyllostomidae (note that the latter taxa are marked as 'uncertain' as a result of taxonomic polymorphism). Among extant microchiropterans, moderate enlargement of the cochlea is seen in taxa that regularly use passive means (e.g.,vision, olfaction, listening for prey-generated sounds) to detect prey or other food items; obligate aerial insectivores are characterized by great enlargement of the cochlea, a feature that is not seen in any non-echolocating taxon. This pattern (interpreted in the context of other data on morphology and stomach contents; see Simmons & Geisler 1998) suggests that aerial insectivory evolved very early in the microchiropteran lineage, prior to the divergence of *Hassianycteris*. Although a moderately enlarged cochlea apparently evolved independently in non-echolocating Pteropodidae, evidence from other characters (e.g., morphology of the stylohyal) suggests that *Icaronycteris* and *Archaeonycteris* were indeed echolocating bats as suggested by Novacek (1985, 1987) and others.

1998), premaxilla articulations (Simmons & Geisler 1998), and patterns of phalangeal ossification (Simmons & Geisler 1998; Schutt & Simmons 1998). Character mapping has also been used to investigate patterns of change among genera within families. Hood (1989) and Springer *et al.* (1995) used phylogenies of Pteropodidae to investigate the origin of specializations for nectar-feeding (e.g., tongue length) in the family, and Springer *et al.* (1995) also investigated changes in tail length in pteropodids

by character mapping. Similarly, Wetterer *et al.* (in press) recently investigated patterns of change in noseleaf structure and vibrissal patterns in
phyllostomids by mapping these traits on a genus-level phylogeny of the
family. Character mapping could also be used profitably at the species
level, although this has yet to be done explicitly for bats.

Detailed analyses of character evolution in the context of a phylogenetic tree often lead to interesting observations that in turn suggest new
avenues for research. For example, Simmons & Geisler (1998) mapped
presence/absence of a calcar on their higher-level phylogeny of bats, and
concluded that this structure evolved independently in Megachiroptera
and Microchiroptera. Schutt & Simmons (1998) subsequently examined
the calcar and associated muscles in representatives of all bat families,
and found major morphological differences between the microchiropteran calcar and the equivalent structure (which they termed the 'uropatagial spur') in pteropodids. Because the uropatagial spur is always
cartilaginous and never articulates with the calcaneum, it might have
been present but not preserved in the earliest fossil microchiroptera-
morph bats. If so, we would conclude that the uropatagial spur was transformed in to a calcar, and that these structures are indeed homologous.
However, differences in musculature associated with the calcar and uropatagial spur suggest that such an evolutionary transformation would
have required numerous changes, and might therefore be considered
unlikely (Schutt & Simmons 1998). Developmental studies of the calcar
and uropatagial spur are now under way in an effort to more fully explore
this system (Adams & Thibault, this volume). Regardless of the outcome,
this example illustrates the interesting and sometimes unexpected
results of character mapping on phylogenetic trees.

Opportunities for these types of character analyses are virtually
endless. Any discrete morphological trait that varies among taxa can be
mapped onto a phylogeny, and continuous characters may also be evaluated although the methods are somewhat more complex (Maddison 1991;
Maddison & Maddison 1992). As noted earlier, covariation among characters can be investigated using computer programs like MacClade
(Maddison & Maddison 1992), although this approach has yet to be
applied in studies of bats. Independent contrasts may also be used to try
to discriminate between correlations due to phylogeny and those due to
concerted functional changes (e.g., Kirsch & Lapointe 1997).

One potential problem with using phylogenies to infer patterns of
character evolution is circular or biased reasoning. If the character(s) of
interest were used to help build a phylogeny in the first place, doesn't this
bias any subsequent attempts to use that phylogeny to investigate charac-

ter evolution? Some authors have advocated excluding characters of interest from the phylogeny-building process (e.g., Brooks & McLennan 1991), but many others consider this to be counterproductive (e.g., Maddison & Maddison 1992; Swofford & Maddison 1992; Simmons 1993a). Circular reasoning obviously can be avoided entirely if the phylogeny is derived from an independent data set (as was done when Springer *et al.* [1995] mapped tongue length on a phylogeny of pteropodids derived from DNA hybridization data), but bias can also be avoided if the data set is large enough to be little influenced by the character or characters under consideration (Simmons 1993a). In many cases, inclusion or exclusion of a few characters has little or no effect on tree topology. If so, the characters and the tree are effectively independent. If inclusion of characters of interest in an evolutionary study significantly changes the phylogeny, this raises serious doubts about the phylogeny reconstructed without them, suggesting that it is probably not strongly enough supported to provide an appropriate background for the character analysis (Maddison & Maddison 1992). It is important to use the best-supported phylogeny as a basis for character reconstruction, regardless of its source. It may be better to risk some circularity of reasoning than to miss opportunities for new insights into evolutionary patterns.

Griffiths and his colleagues have investigated evolution in hyoid morphology in many bat lineages by building phylogenies using hyoid characters and then mapping those characters back on the resulting trees (Griffiths 1982, 1994; Griffiths & Smith 1991; Griffiths *et al.* 1992). While this was indeed a somewhat circular process, Griffiths and others (e.g., Gimenez *et al.* 1996; Simmons 1998; Simmons & Geisler 1998; Wetterer *et al.* in press) have shown that many of the phylogenetic hypotheses derived from hyoid data are also supported by other data. This suggests that many of these phylogenies (e.g., those of glossophagines, lonchophyllines, nycterids, *Emballonura* sensu lato) are robust enough to provide an appropriate framework for investigating patterns of character evolution.

Evolution of development and physiology

Evolutionary changes in development and physiology can also be inferred from mapping traits on a phylogeny. Pedersen (1993, 1995) examined differences in skull ontogeny among bat lineages and found that a complex of three derived developmental traits (retention of the basicranium ventral to the cervical axis, retention of the rostrum below the basicranial axis, and caudal rotation of the lateral semicircular canals) evolved at least twice in bats, once in rhinolophoids and once in phyllostomids.

These changes are correlated with nasal emission of echolocation calls and presence of a noseleaf, suggesting that they form a linked functional complex. Recognition that the 'nasal emission' complex evolved more than once in microchiropteran bats (something that could not be inferred in the absence of a phylogeny) suggests that evolution of this complex was facilitated by developmental canalization (Pedersen 1995). This example illustrates the benefits of an evolutionary perspective for interpreting developmental patterns as well as adult morphologies.

Physiological traits can be interpreted in a phylogenetic context in the same way as other characteristics of organisms, although the nature of the data (e.g., often with continuous variation) may make character mapping somewhat difficult. McNab (1982) investigated the evolution of energetics in bats by considering patterns of variation in basal metabolic rate, temperature regulation, and dietary habits in the context of Van Valen's (1979) phylogeny of bats. Based on this analysis, McNab (1982: 186) concluded that 1) prebats were tropical, scansorial omnivores with low basal rates and without any marked propensity to enter torpor, 2) the earliest insectivorous bats were tropical in distribution with low basal rates and molossid-type thermoregulation (they may have facultatively entered torpor, especially if they had small mass), 3) with the shift of foods from omnivory or insects to meat, fruit, and nectar, either early or late in the phylogeny of bats, basal rates increased, which in turn allowed thermoregulation to become precise and reproduction to be more independent of environmental conditions, 4) the use of hibernation is a derived specialization in response to seasonal variations in the availability of food in temperate environments.

These conclusions have far-reaching implications for the interpretation of many physiological and behavioral features of bats, but they require continued testing with additional data and improved phylogenies. McNab & Bonaccorso (1995) further investigated the relationship between basal metabolic rate, body temperature, body size, and nectarivory in pteropodids by mapping variation in these traits on the Kirsch *et al.* (1995) DNA hybridization phylogeny (Fig. 2.8). Their analysis demonstrated that historical linkages among these features are probably much more complex than previously thought.

Ecological and behavioral evolution

One of the most exciting areas in phylogenetic research is the use of phylogenies for reconstructing ecological and behavioral evolution. Whole books have been devoted to this topic (e.g., Brooks & McLennan

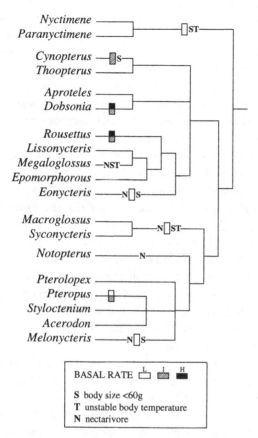

Figure 2.8. Evolution of basal metabolic rate (low, intermediate, or high), body size, body temperature, and nectarivorous habits in Pteropodidae (redrawn from McNab & Bonaccorso 1995). The phylogeny is that of Kirsch *et al.* (1995) with additional taxa included in positions consistent with presumed evolutionary relationships (e.g., following Andersen 1912).

1991), and recent issues of *Systematic Biology, Evolution,* the *Journal of Mammalian Evolution,* and other periodicals are full of examples and discussions of methodological issues. In addition to the examples discussed above, two topics concerning bat ecology and behavior have been explicitly discussed in a phylogenetic context: the evolution of feeding habits (e.g., Ferrarezzi & Gimenez 1996; Kirsch & Lapointe 1997; Wetterer *et al.* in press) and the evolution of echolocation and foraging strategies (e.g., Simmons & Geisler 1998).

In one of the most extensive phylogenetic treatments of ecological evolution in bats, Ferrarezzi & Gimenez (1996) used a series of phylogenetic trees to reconstruct the evolution of feeding patterns in Chiroptera. Beginning at the highest taxonomic levels, they accepted monophyly of Archonta, Volitantia, and Chiroptera, and used a tree from Simmons (1993a) as the basis of their reconstruction. Assuming that insectivory is primitive for Microchiroptera and herbivory (broadly defined to include frugivory) for Megachiroptera, they concluded that the feeding habit of the most recent common ancestor of bats was either herbivory (as seen in Dermoptera, the sister-taxon of bats in the tree they used) or omnivory (a possible logical intermediate between frugivory and insectivory). This conclusion contrasts sharply with previous assumptions that the earliest bats were insectivorous (e.g., Romer 1966; Gillette 1975; Colbert 1980; Findley 1993; Altringham 1996). Ferrarezzi & Gimenez noted that the assumption that insectivory is primitive for all bats is in part based on the insectivorous-type dentition found in the earliest bats (e.g., *Icaronycteris, Palaeochiropteryx*), which closely resembles that seen in many Cretaceous and Paleocene mammals. Our current understanding that these Eocene bats are really basal members of the microchiropteran lineage (Novacek 1985, 1987; Simmons & Geisler 1998) means that they are only indirectly informative about the habits of the first bats (i.e., those ancestral to both Megachiroptera and Microchiroptera).

In a second analysis that recognized more subdivisions of feeding habits (i.e., herbivory, frugivory, nectarivory, insectivory, piscivory, carnivory, and omnivory), Ferrarezzi & Gimenez (1996) used an emended version of Simmons (1998) family-level phylogeny of bats to reconstruct patterns of dietary evolution within Chiroptera (Fig. 2.9). This analysis suggested that insectivory is the ancestral habit for all major branches within Microchiroptera, including those that today include omnivorous (i.e., Mystacinidae, Phyllostomidae), carnivorous (i.e., Nycteridae, Megadermatidae, Phyllostomidae), or piscivorous members (i.e., Noctilionidae, Myotinae).

Ferrarezzi & Gimenez (1996) went on to reconstruct the evolution of feeding habits in Phyllostomidae, the family which shows the greatest diversity in dietary habits. In that analysis, they mapped seven feeding habits (strict insectivory, predominant insectivory, predominant carnivory, strict frugivory, predominant frugivory, predominant nectarivory, and sanguivory) on a summary cladogram based on the results of phylogenetic analyses by Baker *et al.* (1989), Honeycutt & Sarich (1987), Owen (1987), Lim (1993), and Gimenez *et al.* (1996). Results of their character-mapping suggested that predominant insectivory is primitive for phyllostomids, and that most of the other feeding habits evolved only once within the group. Wetterer *et al.* (in press) reached broadly similar conclu-

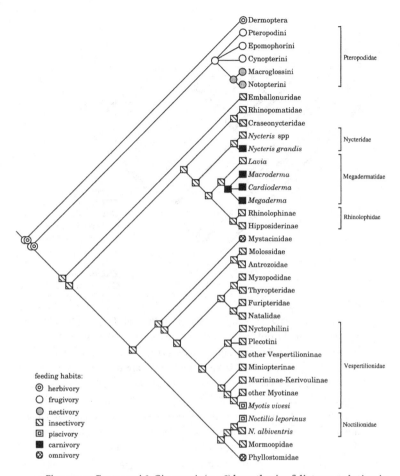

Figure 2.9. Ferrarezzi & Gimenez's (1996) hypothesis of dietary evolution in bats (redrawn from their figure 3). Tree topology was based on Simmons (1998); see text for discussion.

sions based on their revised phylogeny of Phyllostomidae (Figs. 2.4, 2.10), although differences in tree topology within subfamilies lead them to conclude that patterns of dietary evolution within a few clades may be more complex than suggested by Ferrarezzi & Gimenez (1996). For example, carnivory many have evolved twice (or there may have been a reversal) in Vampyrini, and degrees of reliance on fruit (as opposed to a mixed diet including insects and/or nectar and pollen) may have shifted several times within various branches of Stenodermatinae (Fig. 2.10).

Ferrarezzi & Gimenez (1996) took their analysis to even lower taxonomic levels in a study of desmodontines (Fig. 2.11). Using a phylogeny of

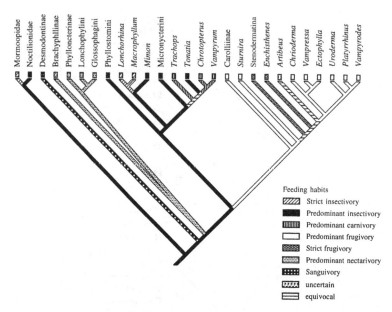

Figure 2.10. Hypothesis of dietary evolution in Phyllostomidae after Wetterer *et al.* (in press). Dietary habits have been mapped on a reduced version of the tree shown in Fig. 2.4. Categories used to describe dietary variation follow Ferrarezzi & Gimenez (1996); data are from Wetterer *et al.* (in press). Taxa described as predominant insectivores are known to supplement their diet with fruit, nectar, and pollen; predominant carnivores often eat insects; predominant frugivores sometimes feed on nectar, pollen, and insects; and predominant nectarivores often feed on pollen and sometime eat fruit and insects. Higher-level taxonomic names follow Wetterer *et al.* (in press): Desmodontinae includes *Diphylla*, *Diaemus*, and *Desmodus*; Brachyphyllinae includes only *Brachyphylla*; Phyllonycterinae includes *Phyllonycteris* and *Erophylla*; Lonchophyllini includes *Platalina*, *Lonchophylla*, and *Lionycteris*; Glossophagini includes *Glossophaga*, *Monophyllus*, *Leptonycteris*, *Anoura*, *Scleronycteris*, *Lichonycteris*, *Hylonycteris*, *Choeronycteris*, *Musonyteris*, and *Choeroniscus*; Phyllostomini includes *Phyllostomus* and *Phylloderma*; Micronycterini includes *Micronycteris*, *Lampronycteris*, *Neonycteris*, *Trinycteris*, *Glyphonycteris*, and *Macrotus*; Stenodermatina includes *Ametrida*, *Centurio*, *Sphaeronycteris*, *Pygoderma*, *Phyllops*, *Stenoderma*, *Ariteus*, and *Ardops*; *Artibeus* includes *Koopmania* and *Dermanura* as subgenera.

the three extant genera following Honeycutt *et al.* (1981) and Koopman (1988), they mapped favored prey (avian or mammalian), foraging site (arboreal or terrestrial), and specificity of plasminogen activation of saliva (activates both avian and mammalian plasminogens or just mammalian plasminogens). Their results suggest that feeding on avian prey approached in an arboreal milieu is probably primitive for vampires, and

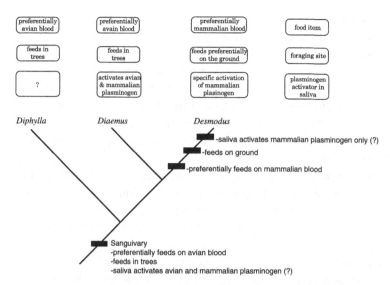

Figure 2.11. Ferrarezzi & Gimenez's (1996) hypothesis of behavioral and physiological evolution in vampire bats (Desmodontinae). Character states are listed in boxes above the taxonomic names; one optimization is given on the tree below. Note that changes in salivary plasminogen activators are somewhat uncertain because the condition in *Diphylla* is unknown.

the habits of *Desmodus* (which preys principally or entirely on mammals that are approached on the ground) are probably derived.

Dietary habits of pteropodids have also been investigated in an evolutionary context. As noted earlier, Hood (1989) and Springer *et al.* (1995) used genus-level phylogenies to investigate the origin of specializations for nectar-feeding in 'macroglossine' pteropodids. Kirsch & Lapointe (1997) took a different approach, summarizing available information on dietary habits and mapping these data on competing phylogenetic trees. They recognized four dietary classes based on different percentages of nectar in the diet (0%, 20%, 50%, 100%), and mapped distribution of these classes on alternative phylogenetic trees (e.g., Fig. 2.1) derived from the work of Andersen (1912), Springer *et al.* (1995), and Kirsch *et al.* (1995). They found high degrees of homoplasy no matter what tree or assumption set (ordered or unordered transformations) was used, and concluded that varying degrees of nectarivory had evolved multiple times in pteropodids. Their results indicated that 'mixed feeding' – including perhaps 50% nectar in the diet – may be the primitive condition for most or all pteropodid lineages (Kirsch & Lapointe 1997: 326). Additional analyses of correlations among molar measurements, mandibular measurements, and diet using independent contrasts were largely inconclusive, but did seem to suggest that 'nectarivory must have evolved several times among ptero-podids, and not always by following the same morphological trajectory' (Kirsch & Lapointe 1997: 328).

Dietary habits are not the only behavioral traits that have been inves-tigated in a phylogenetic context. Ideas about the evolution of major echo-location strategies have long been informed by phylogenetic hypotheses. For example, recognition that *Pteronotus parnellii* (a mormoopid) is probably only distantly related to rhinolophids has lead many workers to conclude that high-duty-cycle CF (constant frequency) echolocation evolved twice in bats, once in the lineage leading to Rhinolophidae and once within the genus *Pteronotus* (e.g., Pye 1980; Simmons 1980; Simmons & Stein 1980; Fenton *et al.* 1995; Simmons & Geisler 1998). Simmons (1980) presented a branching diagram of bat echolocation calls, but proposed this as a hypoth-esis of possible evolutionary patterns (i.e., directionality of change) rather than as a phylogeny. Noting that a higher-level phylogeny of bats would be need to confirm his ideas, Simmons posed several questions about the evo-lution of echolocation (e.g., are the low-intensity, multiharmonic calls of phyllostomids relatively primitive or derived?) that were not addressed again for almost 20 years. Simmons & Geisler (1998) reconstructed the early stages in the evolution of echolocation using their phylogenetic tree (Figs. 2.2, 2.7) and published reconstructions of the ecologies of Eocene bats (e.g., those of Novacek 1985, 1987, Habersetzer & Storch 1987, 1989, Norberg 1989, and Habersetzer *et al.* 1994). Simmons & Geisler concluded that the earliest echolocating bats were probably perch-hunting gleaners that used echolocation for orientation but actually located their prey by passive means (vision or listening for prey-generated sounds). Use of echolocation to detect, track, and assess moving prey evolved later; however, aerial insectivory was probably the primitive foraging habit for extant Microchiroptera. In answer to Simmons's (1980) question about phyllosto-mid echolocation, this reconstruction implies that gleaning, passive prey detection, and perch hunting among extant microchiropterans (e.g., in megadermatids, nycterids, and phyllostomids) are secondarily derived spe-cializations rather than retentions of primitive habits (Simmons & Geisler 1998). Similarities in the echolocation calls and foraging strategies of these taxa are presumably due to convergence, a conclusion that could only be reached in the context of a phylogeny.

CONCLUSIONS

Our understanding of the phylogenetic relationships of bats is improving rapidly, with phylogenies now available for groups at many taxonomic levels. New data sets and methods for estimating the degree of support for various clades are providing both revised trees and the means for evaluating competing hypotheses of relationships among bats. These advancements in turn provide phylogeny 'users' (be they systematists, ecologists, ethologists, functional morphologists, physiologists, developmental biologists, or paleontologists) with a means of choosing the most appropriate phylogeny for investigating evolutionary patterns in their data. Phylogenetic hypotheses offer a critical evolutionary framework for reconstructing historical patterns of change, and character-mapping techniques can be modified to fit almost any data set. Explicit applications of character-mapping and other methods to the study of bats have thus far been limited (virtually all are described above), but the explanatory power of these techniques is clearly enormous. Understanding the biology of living organisms requires an evolutionary perspective that can only be gained by reference to phylogeny. In the case of Chiroptera, the available phylogenetic framework is rapidly growing in complexity, detail, and strength. Creative use of these phylogenies will doubtless continue to lead to new and important insights about the biology and evolution of bats.

ACKNOWLEDGMENTS

First and foremost, I am indebted to all of the systematists working on bats – without their phylogenies, this chapter could not have been written. R. Adams, S. Pedersen, and two anonomous reviewers commented on earlier versions of this chapter, and I thank them for their helpful suggestions. Special thanks also to T. Conway, who helped with the bibliography and figures, and to P. Brunauer, who always managed to find what I needed in the library. This study was supported by NSF Research Grants DEB-9106868 and DEB-9873663.

This contribution is dedicated to the late Karl Koopman, my predecessor as curator of the American Museum of Natural History bat collections and a valued colleague. Before there were phylogenies, there was Karl – his classifications provided the groundwork for many of the phylogenetic studies described above. He is sorely missed.

REFERENCES

Adkins, R. M. & Honeycutt, R. L. (1991) Molecular phylogeny of the superorder Archonta. *Proceedings of the National Academy of Science, U.S.A.*, **88**, 10317–21.
Adkins, R. M. & Honeycutt, R. L. (1993) A molecular examination of archontan and

chiropteran monophyly. In *Primates and their Relatives in Phylogenetic Perspective*, ed. R. D. E. MacPhee, *Advances in Primatology Series*, pp. 227–49. New York: Plenum Press.

Adkins, R. M. & Honeycutt, R. L. (1994) Evolution of the primate cytochrome oxidase subunit II gene. *Journal of Molecular Evolution*, **38**, 215–31.

Allard, M. W., McNiff, B. E. & Miyamoto, M. M. (1996) Support for interordinal eutherian relationships, with an emphasis on Primates and their archontan relatives. *Molecular Phylogeny and Evolution*, **5**, 78–88.

Altringham, J. D. (1996) *Bats: Biology and Behaviour*. Oxford: Oxford University Press.

Ammerman, L. K. & Hillis, D. M. (1992) A molecular test of bat relationships: monophyly or diphyly? *Systematic Biology*, **41**, 222–32.

Andersen, K. (1912) *Calalogue of the Chiroptera in the Collection of the British Museum*, vol. 1: *Megachiroptera*, 2nd edn. London: British Museum. [Reprinted by Johnson Reprint Corporation, New York]

Arnold, M. L., Honeycutt, R. L., Baker, R. J., Sarich, V. M. & Jones, J. K. (1982) Resolving a phylogeny with multiple data sets: a systematic study of phyllostomoid bats. *Occasional Papers Museum of Texas Tech University*, **77**, 1–15.

Bailey, W. J., Slightom, J. L. & Goodman, M. (1992) Rejection of the 'flying primate' hypothesis by phylogenetic evidence from the ε-globin gene. *Science*, **256**, 86–9.

Baker, R. J., Honeycutt, R. L. & Van Den Bussche, R. A. (1991a) Examination of monophyly of bats: restriction map of the ribosomal DNA cistron. *Bulletin of the American Museum of Natural History*, **206**, 42–53.

Baker, R. J., Hood, C. S. & Honeycutt, R. L. (1989) Phylogenetic relationships and classification of the higher categories of the New World bat family Phyllostomidae. *Systematic Zoology*, **3**, 228–38.

Baker, R. J., Longmire, J. L., Maltbie, M., Hamilton, M. J. & Van Den Bussche, R. A. (1997) DNA synapomorphies for a variety of taxonomic levels from a cosmid library from the New World bat *Macrotus waterhousii*. *Systematic Biology*, **46**, 579–89.

Baker, R. J., Novacek, M. J. & Simmons, N. B. (1991b) On the monophyly of bats. *Systematic Zoology*, **40**, 216–31.

Baker, R. J., Taddei, V. A., Hudgeons, J. L. & Van Den Bussche, R. A. (1994) Systematic relationships within *Chiroderma* (Chiroptera: Phyllostomidae) based on cytochrome *b* sequence variation. *Journal of Mammalogy*, **75**, 321–7.

Barghoorn, S. F. (1977) New material of *Vespertiliavus* Schlosser (Mammalia, Chiroptera) and suggested relationships of emballonurid bats based on cranial morphology. *American Museum Novitates*, **2618**, 1–29.

Beard, K. C. (1993) Phylogenetic systematics of Primatomorpha, with special reference to Dermoptera. In *Mammal Phylogeny: Placentals*, ed. F. S. Szalay, M. J. Novacek & M. C. McKenna, pp. 129–50. New York: Springer-Verlag.

Benedict, F. A. (1957) Hair structure as a generic character in bats. *University of California Publications, Zoology*, **59**, 285–548.

Bogdanowicz, W., Kasper, S. & Owen, R. D. (1998) Phylogeny of plecotine bats: reevaluation of morphological and chromosomal data. *Journal of Mammalogy*, **79**, 78–90.

Bogdanowicz, W. & Owen, R. D. (1992) Phylogenetic analyses of the bat family Rhinolophidae. *Zeitschrift für zoologische Systematik und Evolutionsforschung*, **30**, 142–60.

Bogdanowicz, W. & Owen, R. D. (1998) In the Minotaur's labyrinth: the phylogeny of the bat family Hipposideridae. In *Bat Biology and Conservation*, ed. T. H. Kunz & P. A. Racey, pp. 27–42. Washington: Smithsonian Institution Press.

Bremer, K. (1988) The limits of amino acid sequence data in angiosperm phylogenetic reconstructions. *Evolution*, **42**, 795–03.

Brooks, D. R. & McLennan, D. A. (1991) *Phylogeny, Ecology, and Behavior.* Chicago: University of Chicago Press.

Bull, J. J., Huelsenbeck, J. P., Cunningham, C. W., Swofford, D. L. & Waddell, P. J. (1993) Partitioning and combining data in phylogenetic analysis. *Systematic Biology,* **42**, 384–97.

Buttery, R. G., Haight, J. R. & Bell, K. (1990) Vascular and avascular retinae in mammals: a funduscopic and fluorescein angiographic study. *Brain Behavior Evolution,* **35**, 156–75.

Chippindale, P. T. & Weins, J. J. (1994) Weighting, partitioning, and combining characters in phylogenetic analysis. *Systematic Biology,* **43**, 278–87.

Colbert, E. H. (1980) *Evolution of the Vertebrates: A History of the Backboned Animals Through Time,* 3rd edn. New York: John Wiley.

Colgan, D. J. & Flannery, T. F. (1995) A phylogeny of Indo-west Pacific Megachiroptera based on ribosomal data. *Systematic Biology,* **44**, 209–20.

Corbet, G. B. & Hill, J. E. (1980) *A World List of Mammalian Species.* London: British Museum (Natural History).

Corbet, G. B. & Hill, J. E. (1992) *The Mammals of the Indomalayan Region: A Systematic Review.* London: Natural History Museum Publications.

de Queiroz, A., Donoghue, M. J. & Kim, J. (1995) Separate versus combined analysis of phylogenetic evidence. *Annual Review of Ecology and Systematics,* **26**, 657–81.

Duke-Endler, W. S. (1958) *System of Opthalmology,* vol. 1: *The Eye in Evolution.* London: Henry Kimpton.

Farris, J. S., Kallersjo, M., Kluge, A. G. & Bult, C. (1994) Testing significance of incongruence. *Cladistics,* **10**, 315–19.

Felsenstein, J. (1985*a*) Confidence levels on phylogenies: an approach using the bootstrap. *Evolution,* **39**, 783–91.

Felsenstein, J. (1985*b*) Phylogenies and the comparative method. *American Naturalist,* **125**, 1–15.

Fenton, M. B., Audet, D., Obrist, M. K. & Rydell, J. (1995) Signal strength, timing, and self-deafening: the evolution of echolocation in bats. *Paleobiology,* **21**, 229–42.

Ferrarezzi, H. and Gimenez, E. A. (1996) Systematic patterns and the evolution of feeding habits in Chiroptera (Archonta: Mammalia) *Journal of Comparative Biology,* **1**, 75–94.

Findley, J. S. (1993) *Bats: A Community Perspective.* Cambridge: Cambridge University Press.

Freeman, P. W. (1981) A multivariate study of the family Molossidae (Mammalia: Chiroptera): morphology, ecology, evolution. *Fieldiana,* **7**, 1–173.

Fritsch, G. (1911) Contributions to the histology of the *Pteropus* eye. *Zeitschrift für wissenschaftliche Zoologie,* **98**, 288.

Frost, D. R. & Timm, R. M. (1992) Phylogeny of plecotine bats (Chiroptera: 'Vespertilionidae'): summary of the evidence and proposal of a logically consistent taxonomy. *American Museum Novitates,* **3034**, 1–16.

Garland, T., Harvey, P. H. & Ives, A. R. (1992) Procedures for the analysis of comparative data using phylogenetically independent contrasts. *Systematic Biology,* **41**, 18–32.

Gillette, D. D. (1975) Evolution of feeding strategies in bats. *Tebiwa,* **18**, 39–48.

Giminez, E. A., Ferrarezzi, H. & Taddei, V. A. (1996) Lingual morphology and cladistic analysis of the New World nectar-feeding bats (Chiroptera: Phyllostomidae) *Journal of Comparative Biology,* **1**, 41–64.

Griffiths, T. A. (1982) Systematics of the New World nectar-feeding bats (Mammalia, Phyllostomidae), based on the morphology of the hyoid and lingual regions. *American Museum Novitates,* **2742**, 1–45.

Griffiths, T. A. (1994) Phylogenetic systematics of Slit-faced bats (Chiroptera,

Nycteridae), based on hyoid and other morphology. *American Museum Novitates*, **3090**, 1–17.

Griffiths, T. A. (1997) Phylogenetic position of the bat *Nycteris jamanica* (Chiroptera: Nycteridae) *Journal of Mammalogy*, **78**, 106–16.

Griffiths, T. A. & Smith, A. L. (1991) Systematics of emballonuroid bats (Chiroptera: Emballonuridae and Rhinopomatidae), based on hyoid morphology. *Bulletin of the American Museum of Natural History*, **206**, 62–83.

Griffiths, T. A., Koopman, K. F. & Starrett, A. (1991) The systematic relationship of *Emballonura nigrescens* to other species of *Emballonura* and to *Coleura* (Chiroptera: Emballonuridae) *American Museum Novitates*, **2996**, 1–16.

Griffiths, T. A., Truckenbrod, A. & Sponholtz, P. J. (1992) Systematics of megadermatid bats (Chiroptera, Megadermatidae), based on hyoid morphology. *American Museum Novitates*, **3041**, 1–21.

Habersetzer, J., Richter, G. & Storch, G. (1994) Paleoecology of Early Middle Eocene bats from Messel, FRG. Apsects of flight, feeding and echolocation. *Historical Biology*, **8**, 235–60.

Habersetzer, J. & Storch, G. (1987) Klassifikation und funktionelle Flügelmorphologie paläogener Fledermäuse (Mammalia, Chiroptera) *Courier ForschungsInstitut Senkenberger*, **91**, 11–150.

Habersetzer, J. & Storch, G. (1989) Ecology and echolocation of the Eocene Messel bats. In *European Bat Research 1987*, ed. V. Hanák, T. Horácek & J. Gaisler, pp. 213–33. Prague: Charles University Press.

Haiduk, M. W. & Baker, R. J. (1982) Cladistical analysis of G-banded chromosomes of nectar feeding bats (Glossophaginae: Phyllostomidae) *Systematic Zoology*, **3**, 252–65.

Haiduk, M. W., Baker, R. J., Robbins, L. W. & Schlitter, D. A. (1981) Chromosomal evolution in African Megachiroptera: G- and C-band assessment of the magnitude of change in similar standard karyotypes. *Cytogenetics and Cell Genetics*, **29**, 221–32.

Haiduk, M. W., Robbins, L. W. & Schlitter, D. A. (1980) Karyotypic studies of seven species of African megachiropterans (Mammalia: Pteropodidae) *Annals of the Carnegie Museum*, **49**, 181–91.

Haiduk, M. W., Robbins, L. W. & Schlitter, D. A. (1983) Chromosomal banding studies and their systematic implications in African fruit bats (Chiroptera: Pteropodidae) *Annalen van Koninklijk Museum voor Midden Afrika, Zoologische Wetenschappen*, **237**, 1–8.

Hand, S. J. (1985) New Miocene megadermatids (Chiroptera: Megadermatidae) from Australia with comments on megadermatid phylogenetics. *Australian Mammalogy*, **8**, 5–43.

Hand, S. J. (1990) First tertiary molossid (Microchiroptera: Molossidae) from Australia: its phylogenetic and biogeographic implications. *Memoirs of the Queensland Museum*, **28**, 175–92.

Hand, S. J. & Kirsch, J. A. W. (1998) A southern origin for the Hipposideridae (Microchiroptera)? Evidence from the Australian fossil record. In *Bat Biology and Conservation*, ed. T. H. Kunz & P. A. Racey, pp. 72–90. Washington: Smithsonian Institution Press.

Hill, J. E. (1963) A revision of the genus *Hipposideros*. *Bulletin of the British Museum of Natural History, Zoology*, **11**, 1–129.

Hill, J. E. (1974) A new family, genus, and species of bat (Mammalia: Chiroptera) from Thailand. *Bulletin of the British Museum of Natural History, Zoology*, **27**, 301–36.

Hill, J. E. & Harrison, D. L. (1987) The baculum in the Vespertilioninae (Chiroptera: Vespertilionidae) with a systematic review, a synopsis of *Pipistrellus* and

Pye, D. (1980) Adaptiveness of echolocation signals in bats: flexibility in behavior and in evolution. *Trends in Neuroscience*, **3**, 232–5.

Qumsiyeh, M. B. & Bickham, J. W. (1993) Chromosomes and relationships of long-eared bats of the genera *Plecotus* and *Otonycteris. Journal of Mammalogy*, **74**, 376–82.

Qumsiyeh, M. B., Owen, R. D. & Chesser, R. K. (1988) Differential rates of genic and chromosomal evolution in bats of the family Rhinolophidae. *Genome*, **30**, 326–35.

Robbins, L. W. & Sarich, V. M. (1988) Evolutionary relationships in the family Emballonuridae (Chiroptera) *Journal of Mammalogy*, **69**, 1–13.

Romer, A. S. (1966) *Vertebrate Paleontology*, 3rd edn. Chicago: University of Chicago Press.

Rouse, G. W. & Robson, S. K. (1986) An ultrastructural study of megachiropteran (Mammalia: Chiroptera) spermatozoa: implications for chiropteran phylogeny. *Journal of Submicroscopic Cytology*, **18**, 137–52.

Sanderson, M. J. (1989) Confidence limits on phylogenies: the bootstrap revisited. *Cladistics*, **5**, 113–29.

Sanderson, M. J. & Donoghue, M J. (1989) Patterns of variation in levels of homoplasy. *Evolution*, **43**, 1781–95.

Schutt, W. A. & Simmons, N. B. (1998) Morphology and homology of the chiropteran calcar, with comments on the relationships of *Archaeopteropus. Journal of Mammalian Evolution*, **5**, 1–32.

Siddall, M. E. (1995) Another monophyly index: revisiting the jackknife. *Cladistics*, **11**, 33–56.

Siddall, M. E. (1997) Prior agreement: arbitration or arbitrary? *Systematic Biology*, **46**, 765–9.

Simmons, J. A. (1980) Phylogenetic adaptations and the evolution of echolocation in bats. In *Proceedings of the Fifth International Bat Research Conference*, ed. D. E. Wilson & A. L. Gardner, pp. 267–78. Lubbock: Texas Tech Press.

Simmons, N. B. (1993a) The importance of methods: archontan phylogeny and cladistic analysis of morphological data. In *Primates and their Relatives in Phylogenetic Perspective*, ed. R. D. E. MacPhee, *Advances in Primatology Series*, pp. 1–61. New York: Plenum Press.

Simmons, N. B. (1993b) Morphology, function, and phylogenetic significance of pubic nipples in bats (Mammalia: Chiroptera) *American Museum Novitates*, **3077**, 1–37.

Simmons, N. B. (1994) The case for chiropteran monophyly. *American Museum Novitates*, **3103**, 1–54.

Simmons, N. B. (1995) Bat relationships and the origin of flight. In *Ecology, Evolution and Behavior of Bats*, ed. P. A. Racey & S. M. Swift, *Symposium of the Zoological Society of London*, **67**, 27–43.

Simmons, N. B. (1996) A new species of *Micronycteris* (Chiroptera: Phyllostomidae) from northeastern Brazil, with comments on phylogenetic relationships. *American Museum Novitates*, **3158**, 1–34.

Simmons, N. B. (1998) A reappraisal of interfamilial relationships of bats. In *Bat Biology and Conservation*, ed. T. H. Kunz & P. A. Racey, pp. 3–26. Washington: Smithsonian Institution Press.

Simmons, N. B. & Geisler, J. H. (1998) Phylogenetic relationships of *Icaronycteris, Archaeonycteris, Hassianycteris*, and *Palaeochiropteryx* to extant bat lineages, with comments on the evolution of echolocation and foraging strategies in Microchiroptera. *Bulletin of the American Museum of Natural History*, **235**, 1–182.

Simmons, N. B. & Kalko, E. K. V. (1998) Is there phylogenetic signal in bat echolocation calls? *Bat Research News*, **39**, 102.

Simmons, N. B. & Quinn, T. H. (1994) Evolution of the digital tendon locking mechanism in bats and dermopterans: a phylogenetic perspective. *Journal of Mammalian Evolution*, **2**, 231–54.

Simmons, J. A. & Stein, R. A. (1980) Acoustic imaging in bat sonar: echolocation signals and the evolution of echolocation. *Journal of Comparative Physiology A*, **135**, 61–84.

Simmons, N. B. & Voss, R. S. (1998) The mammals of Paracou, French Guiana: A Neotropical lowland rainforest fauna. Part 1. Bats. *Bulletin of the American Museum of Natural History*, **237**, 1–219.

Simmons, N. B., Novacek, M. J. & Baker, R. J. (1991) Approaches, methods, and the future of the chiropteran monophyly controversy: a reply to J. D. Pettigrew. *Systematic Zoology*, **40**, 239–43.

Slaughter, B. H. (1970) Evolutionary trends of chiropteran dentitions. In *About Bats*, ed. B. H. Slaughter & W. D. Walton, pp. 51–83. Dallas: Southern Methodist University Press.

Smith, J. D. (1972) Systematics of the chiropteran family Mormoopidae. *Miscellaneous Publications of the Museum of Natural History, University of Kansas*, **56**, 1–132.

Smith, J. D. (1976) Chiropteran evolution. In *Biology of Bats of the New World Family Phyllostomatidae*, Part I, ed. R. J. Baker, J. K. Jones & D. C. Carter, *Special Publications Museum*, **10**, 46–9. Lubbock: Texas Tech University.

Smith, J. D. (1977) Comments on flight and the evolution of bats. In *Major Patterns in Vertebrate Evolution*, ed. M. K. Hecht, P. C. Goody & B. M. Hecht, *Proceedings NATO Advanced Study Institute*, pp. 427–38. New York: Plenum Press.

Smith, J. D. & Madkour, G. (1980) Penial morphology and the question of chiropteran phylogeny. In *Proceedings of the Fifth International Bat Research Conference*, ed. D. E. Wilson & A. L. Gardner, pp. 347–65. Lubbock: Texas Tech Press.

Springer, M. S., Hollar, L. J. & Kirsch, J. A. W. (1995) Phylogeny, molecules versus morphology, and rates of character evolution among fruitbats (Chiroptera: Megachiroptera) *Australian Journal of Zoology*, **43**, 557–82.

Stafford, B. J. & Thorington, R. W. (1998) Carpal development and morphology in archontan mammals. *Journal of Morphology*, **235**, 135–55.

Stanhope, M. J., Bailey, W. J., Czelusniak, J., Goodman, M., Si, J.-S., Nickerson, J., Sgouros, J. G., Singer, G. A. M. & Kleinschimdt, T. K. (1993) A molecular view of primate supraordinal relationships from the analysis of both nucleotide and amino acid sequences. In *Primates and their Relatives in Phylogenetic Perspective*, ed. R. D. E. MacPhee, *Advances in Primatology Series*, pp. 251–92. New York: Plenum Press.

Stanhope, M. J., Czelusniak, J., Si, J-S., Nickerson, J. & Goodman, M. (1992) A molecular perspective on mammalian evolution from the gene encoding interphotoreceptor retinoid binding protein, with convincing evidence for bat monophyly. *Molecular Phylogeny and Evolution*, **1**, 148–60.

Stanhope, M. J., Smith, M. R., Waddell, V. G., Porter, C. A., Shivji, M. S. & Goodman, M. (1996) Mammalian evolution and the interphotoreceptor retinoid binding protein (IRBP) gene: convincing evidence for several superordinal clades. *Journal of Molecular Evolution*, **43**, 83–92.

Straney, D. O. (1980) Relationships of phyllostomatine bats: evaluation of phylogenetic hypotheses. PhD dissertation, University of California, Berkeley.

Strickler, T. L. (1978) Functional osteology and the myology of the shoulder in the Chiroptera. In *Contributions to Vertebrate Evolution*, vol. 4, ed. M. K. Hect & F. S. Szalay, pp. 1–198. New York: S. Karger.

Sudman, P. D., Barkley, L. J. & Hafner, M. S. (1994) Familial affinity of *Tomopeas ravus* (Chiroptera) based on protein electrophoretic and cytochrome *b* sequence data. *Journal of Mammalogy*, **75**, 365–77.

Swofford, D. L. & Maddison, W. P. (1992) Parsimony, character-state reconstructions, and evolutionary inferences. In *Systematics, Historical Ecology, and North American Freshwater Fishes*, ed. R. L. Mayden, pp. 186–223. Stanford: Stanford University Press.

Swofford, D. L. & Olsen, G. J. (1990) Phylogenetic reconstruction. In *Molecular Systematics*, ed. D. M. Hillis & C. Moritz, pp. 411–501. Sunderland: Sinaeur Associates.

Swofford, D. L., Olsen G. J., Waddel, P. J. & Hillis, D. M. (1996) Phylogenetic inference. In *Molecular Systematics*, 2nd edn, ed. D. M. Hillis, C. Moritz & B. K. Mable, pp. 407–514. Sunderland: Sinaeur Associates.

Szalay, F. S. & Lucas, S. G. (1993) Cranioskeletal morphology of archontans, and diagnoses of Chiroptera, Volitantia, and Archonta. In *Primates and their Relatives in Phylogenetic Perspective*, ed. R. D. E. MacPhee, *Advances in Primatology Series*, pp. 187–226. New York: Plenum Press.

Thewissen, J. G. M. & Babcock, S. K. (1991) Distinctive cranial and cervical innervation of wing muscles: new evidence for bat monophyly. *Science*, **251**, 934–6.

Thewissen, J. G. M. & Babcock, S. K. (1993) The implications of the propatagial muscles of flying and gliding mammals for archontan systematics. In *Primates and their Relatives in Phylogenetic Perspective*, ed. R. D. E. MacPhee, *Advances in Primatology Series*, pp. 91–109. New York: Plenum Press.

Thomas, N. M., Harrison, D. L. & Bates, P. J. J. (1994) A study of the baculum in the genus *Nycteris* (Mammalia, Chiroptera, Nycteridae) with consideration of its taxonomic importance. *Bonner zoologische Beiträge*, **45**, 17–31.

Tumlinson, R. & Douglas, M. E. (1992) Parsimony analysis and the phylogeny of the plecotine bats (Chiroptera: Vespertilionidae) *Journal of Mammalogy*, **73**, 276–85.

Van Cakenberghe, V. & de Vree, F. (1985) Systematics of African *Nycteris* (Mammalia: Chiroptera) In *Proceedings of the International Symposium on African Vertebrates – Systematics, Phylogeny and Evolutionary Ecology*, ed. K.-L. Schuchmann, pp. 53–90. Bonn: Zoologisches Forschungsinstitut und Museum Alexander Koenig.

Van Cakenberghe, V. & de Vree, F. (1993) The systematic status of southeast Asian *Nycteris* (Chiroptera: Nycteridae) *Mammalia*, **57**, 227–44.

Van Cakenberghe, V. & de Vree, F. (1994) A revision of the Rhinopomatidae Dobson 1872, with the description of a new subspecies. *Senkenbergiana Biologica*, **73**(1–2), 1–24.

Van Den Bussche, R. A. (1991) Phylogenetic analysis of restriction site variation in the ribosomal DNA complex of New World leaf-nosed bat genera. *Systematic Zoology*, **40**, 420–32.

Van Den Bussche, R. A. (1992) Restriction site variation and molecular systematics of New World leaf-nosed bats. *Journal of Mammalogy*, **73**, 29–42.

Van Den Bussche, R. A. & Baker, R. J. (1993) Molecular phylogenetics of the New World bat genus *Phyllostomus* based on cytochrome *b* DNA sequence variation. *Journal of Mammalogy*, **74**, 793–802.

Van Den Bussche, R. A., Baker, R. J., Wichman, H. A. & Hamilton, M. J. (1993) Molecular phylogenetics of Stenodermatini bat genera: congruence of data from nuclear and mitochondrial DNA. *Molecular Biology and Evolution*, **10**, 944–59.

Van Den Bussche, R. A., Hudgeons, J. L. & Baker, R. J. (1998) Phylogenetic accuracy, stability, and congruence: relationships within and among the New World bat genera *Artibeus*, *Dermanura*, and *Koopmania*. In *Bat Biology and Conservation*, ed. T. H. Kunz & P. A. Racey, pp. 59–71. Washington: Smithsonian Institution Press.

Van Valen, L. (1979) The evolution of bats. *Evolutionary Theory*, **4**, 104–21.

Volleth, M. & Heller, K.-G. (1994a) Karyosystematics of plecotine bats: a reevaluation of chromosomal data. *Journal of Mammalogy*, **75**, 416–19.

Volleth, M. & Heller, K.-G. (1994b) Phylogenetic relationships of vespertilionid genera (Mammalia: Chiroptera) as revealed by karyological analysis. *Zeitschrift für zoologische Systematik und Evolutionsforschung*, **32**, 11–34.

Wetterer, A. L., Rockman, M. V. & Simmons, N. B. (In press) Phylogeny of phyllostomid bats: data from diverse morphological systems, sex chromosomes, and restriction sites. *Bulletin of the American Museum of Natural History*.

Wible, J. R. & Bhatnagar, K. P. (1997) Chiropteran vomeronasal complex and the interfamilial relationships of bats. *Journal of Mammalian Evolution*, **3**, 285–14 (dated 1996 but issued in 1997).

Wible, J. R. & Martin, J. R. (1993) Ontogeny of the tympanic floor and roof in archontans. In *Primates and their Relatives in Phylogenetic Perspective*, ed. R. D. E. MacPhee, *Advances in Primatology Series*, pp. 111–48. New York: Plenum Press.

Wible, J. R. & Novacek, M. J. (1988) Cranial evidence for the monophyletic origin of bats. *American Museum Novitates*, **2911**, 1–19.

3

Early embryology, fetal membranes, and placentation

INTRODUCTION

Bats comprise a very large and diverse group of mammals, second only to rodents in numbers of species. Over 950 species of bats are so far described and often a misunderstanding is perpetrated that generalities exist within the taxonomic units, be they between suborders, families, subfamilies, or within genera. It is not uncommon to find appreciable differences relating to morphology, behavior, ecology, reproduction, and placentation among bat species. Mossman (1987) emphasized the value of using fetal membrane characters as phylogenetic indicators among the major groups of eutherian mammals, and attributed their value to conservatism as compared to the developmental and morphological characters of the individual. However, as Luckett (1980) pointed out, fetal membrane data are not clear enough in assessing chiropteran phylogeny. An attempt is made in this overview to update information and present diversities portrayed by bats with regard to preimplantation development, implantation, fetal membranes, and placentation.

OVUM MATURATION, FERTILIZATION, AND PREIMPLANTATION DEVELOPMENT

Development of the embryo

The chiropteran ovum (Fig. 3.1A, B) resembles that of other mammals (Blandau 1961). It is either round or oblong in shape, microlecithal, measures 65–75 μm in diameter (Gopalakrishna et al. 1974) with an eccentric nucleus, and is surrounded by an eosinophilic and PAS-positive zona pellucida. After ovulation the ovum shrinks. It maintains its size until hatching. Data on ovum maturation and fertilization are available for relatively few species of bats (Rasweiler 1979). The first meiotic division has been observed in mature preovulatory follicles of

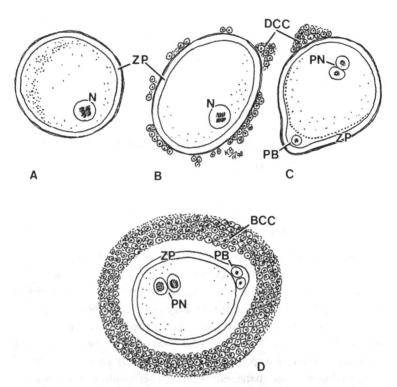

Figure 3.1. The pronuclear ovum from the oviduct of *Miniopterus schreibersii* (A) and oviductal ampulla of *Desmodus rotundus* (B). The ovum is surrounded by a zona pellucida (ZP) with the nucleus (N) at the metaphase of second meiotic division located towards the periphery. Degenerating cumulus oophorus cells (DCC) surround the ovum in *D. rotundus*. The fertilized egg from the oviduct of *Myotis lucifugus* (C) and from the periovarian space of *Pipistrellus mimus mimus* (D). The egg surrounded by the zona pellucida (ZP) shows two pronuclei (PN ♂ and ♀) and the polar bodies (PB). While few degenerating cumulus oophorus cells (DCC) surround the egg in *M. lucifugus*, a ball of cumulus oophorus cells (BCC) surrounds the egg in *P. mimus*.

Carollia perspicillata and *Desmodus rotundus* (de Bonilla & Rasweiler 1974, Quintero & Rasweiler 1974), *Noctilio albiventris* (Rasweiler 1977), and few other species. Elimination of the first polar body and the formation of the second meiotic spindle have been observed in preovulatory follicles, e.g., in *Vesperugo noctula* (van der Stricht 1909). As in other mammals, the second maturation division commences but is arrested at the metaphase stage. Ovulation occurs at this stage. At ovulation the ovum is surrounded by cumulus cells, e.g., in *Myotis lucifugus* (Wimsatt 1944a). Ovulation is spontaneous in *Pteropus* spp. (Martin *et al.* 1987), *Glossophaga soricina*, and *Molossus ater*, while it is induced in *Carollia* and few other species.

Fertilization

This subject is reviewed by Krutzsch & Crichton (1991). The second maturation/meiotic division is completed after the ovum is fertilized. The second polar body is released. Fertilized eggs with two pronuclei have been reported for *Myotis lucifugus* (Fig. 3.1C; Wimsatt 1944*b*), *Pipistrellus mimus* (Fig. 3.1D; Karim 1975) and some other vespertilionids. In bats, fertilization usually occurs in the infundibulum of the oviduct. Another site reported for fertilization is the oviductal ampulla. However, in *Pipistrellus* spp. (Uchida 1953, Karim 1975) sperm entry and fertilization occur in the periovarian space. Two different devices may protect bat eggs from polyspermy, i.e, the zona block as in *P. mimus* (Karim 1975) and a vitelline block found in *Miniopterus schreibersii* (Mori & Uchida 1981).

Cleavage

After fertilization the egg undergoes division (Fig. 3.2A, B, C, D; Fig. 3.3A, B), cleavage being holoblastic and equal. Development up to the two-cell stage in the periovarian space has been observed in *Pipistrellus* (Fig. 3.2B). While undergoing cleavage the fertilized egg is transported through the oviduct. In bats bred in captivity, the sojourn of the embryo in the oviduct is relatively short. In the little brown bat *Myotis lucifugus* (Rasweiler 1979) oviduct traveling time for the embryo is 4–5 days. The tubal journey is prolonged in *Glossophaga* and *Desmodus*, requiring 12–17 days (Quintero & Rasweiler 1974). Tubal blastocysts are observed as late as day 16 postcoitum in *Carollia* and day 22 in *Desmodus* (Rasweiler 1979). In some bats, e.g., *Glossophaga soricina*, *Noctilio albiventris* and *Peropteryx kappleri*, degenerating ova are often retained by the oviduct rather than being transported into the uterus.

Preimplantation development of the embryo

Considerable diversity exists among bats concerning preimplantation development of the embryo (Karim 1976; Rasweiler 1979; Gopalakrishna & Karim 1980). Information on preimplantation is available for nearly 50 species of bats representing nine families. The embryo develops into a unilaminar blastocyst while still in the oviduct in *Rousettus leschenaulti* (Karim 1976; Fig. 3.4). The unilaminar blastocyst of

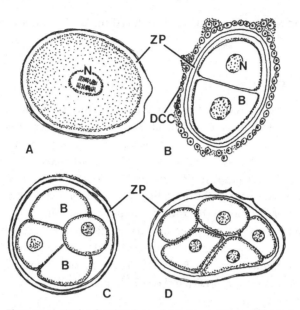

Figure 3.2. A) The fertilized egg from the oviduct of *Myotis lucifugus* and *Pipistrellus mimus* with the nucleus (N) in the first cleavage spindle. The egg is surrounded by a zona pellucida (ZP). B) The two-cell stage from the ampulla of *Peropteryx kappleri* and the periovarian space of *Pipistrellus mimus*. The egg comprises two blastomeres (B) each with a nucleus (N) and is surrounded by the zona pellucida (ZP) beyond which degenerating cumulus cells (DCC) are present. C) Four-cell stage, a mammalian tetrahedron (arrangement of four blastomeres [B] in the form of a cross) from the intramural portion of the oviduct of *M. lucifugus* and the uterine lumen of *Scotophilus heathi*. The egg is surrounded by a zona pellucida (ZP). D) Ten-celled embryo (all cells are not in view) surrounded by the zona pellucida (ZP) from the oviductal ampulla of *Desmodus rotundus*.

bats, with the exception of *N. albiventris* (Rasweiler & Badwaik 1996) resembles a typical mammalian blastocyst. The preimplantation blastocysts of *N. albiventris* lack a typical inner cell mass. Instead, the trophoblastic cells are distributed as either a partial or complete monolayer with the characteristic embryonic cell mass forming only after the attachment of the blastocyst to the uterine wall (Rasweiler & Badwaik 1996) Wimsatt (1954, 1975) noted that the blastocyst of *D. rotundus* is unusual in exhibiting the differentiation of endoderm prior to its passage into the uterus.

In Rhinolophidae, Vespertilionidae, and Molossidae, the embryos pass into the uteri during early cleavage or at the morula stage (Rasweiler 1993). A bilaminar blastocyst lying free in the uterus has been observed in some species, e.g., *R. leschenaulti* (Fig. 3.5), and *D. rotundus*. The free passage

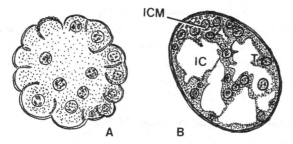

Figure 3.3. A) 65-cell embryo from the uterine lumen of *Myotis lucifugus lucifugus*. The embryo has hatched out of the zona pellucida. B) A late uterine morula of *M. lucifugus* showing intercellular clefts (IC), forerunners of the blastocyst cavity. Early differentiation of inner cell mass (ICM) and the trophoblast (T) is visible.

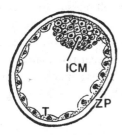

Figure 3.4. A unilaminar blastocyst from the oviduct of *Rousettus leschenaulti*. The blastocyst comprises of a single layer of trophoblast (T) with the inner cell mass (ICM) attached to one pole of the blastocyst. The zona pellucida (ZP) surrounds the trophoblast.

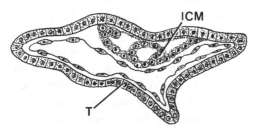

Figure 3.5. A free uterine bilaminar blastocyst of *Rousettus leschenaulti*. The inner cell mass (ICM) shows intercellular spaces – the forerunners of the primitive amniotic cavity. (T) trophoblast.

of the blastocyst in the uterine lumen prior to implantation is limited in duration; extension of this period constitutes the phenomenon of delayed implantation, as in *Miniopterus* (Bernard 1980). Free bilaminar blastocysts in the uterine lumen collected from bats in the wild cannot be assessed accurately for duration of the delay.

Formation of endoderm – bilaminar blastocyst

Generally in bats, as in most mammals except marsupials, the primitive endoderm (hypoblast) in unilaminar blastocysts is formed by the delamination and migration of cells from the ventral surface of the inner cell mass. The endoderm spreads peripherally to line the interior of the blastocyst cavity creating the yolk sac. Thus, a unilaminar blastocyst develops into one that is bilaminar.

In *Glossophaga soricina, Haplonycteris fischeri* (Heideman 1989), and *Carollia perspicillata* (Badwaik *et al.* 1997), the primitive endoderm also differentiates on the outer surface of the inner cell mass as in *D. rotundus* (Wimsatt 1954) where it was designated as a precociously formed extraembryonic mesoderm. The endoderm in the blastocyst of *G. soricina* and *C. perspicillata* (Badwaik *et al.* 1997) has an unusual, reticulated appearance (endodermal meshwork) which is rare among mammals, having been previously reported only in the human, chimpanzee, and *Galago* (see Badwaik *et al.* 1997).

Fate of the zona pellucida

Using appropriate fixatives, investigators have observed the zona pellucida surrounding early embryos both in the oviduct and uterine lumen (Wimsatt 1944a; Rasweiler 1979, 1993; Gopalakrishna & Karim 1980). The process and stage of development of zona loss varies in vespertilionids. In *M. lucifugus*, the zona distends, progressively thins and finally disappears as the blastocyst expands (Wimsatt 1944a). In *Pipistrellus pipistrellus* (Potts & Racey 1971), the zona pellucida begins to fragment as early as the uterine morula stage, but remnants survive until the primitive streak stage. In *N. albiventris, G. soricina, C. perspicillata,* and *D. rotundus*, zona loss occurs before the passage of the blastocyst into the uterus (Rasweiler 1979). Zona-encased blastocysts have been observed in *N. albiventris, G. soricina* and *C. perspicillata*, yet in *D. rotundus*, the zona is apparently eliminated at the morula or the early blastocyst stage. Discarded zonae pellucidae have been noted in the uteri of *Miniopterus australis* which contained a unilaminar blastocyst (Richardson 1977). Mechanical

Figure 3.6. The location of implantation sites along the longitudinal axis of the bicornuate uterus. A) Implantation site at the cranial end of the uterus (U), B) Implantation site squarely at the middle region of the uterus, C) Implantation site towards the vaginal end (V) of the uterus.

factors such as stretching may play a role in loss of zonae from early embryos (Rasweiler 1990). The other mechanism could be through lysis as a result of blastocyst expansion prior to hatching.

Implantation

A literature survey on implantation in bats (Wimsatt 1975; Rasweiler 1979; Mossman 1987; Karim 1986; Heideman *et al.* 1993; Heideman & Powell 1998) reveals that there are considerable species variations as described below:

Implantation sites vary between species with a bicornuate or simplex uterus

The blastocyst implants close to the utero-tubal junction or extreme cranial end of the uterine horn (Fig. 3.6A) usually ipsilateral to the new corpus luteum as in *H. fischeri* (Heideman 1989), *Pteropus* spp. (Pow & Martin 1994), *Taphozous melanopogon*, and *Tadarida aegyptiaca* (Sandhu 1986). The blastocyst implants at about the middle of the uterine horn (Fig. 3.6B) in *M. lucifugus, D. rotundus* (Wimsatt 1954), and *Molossus ater* (Rasweiler 1990, 1993), and even close to the vaginal end of the uterine horn (Fig. 3.6C), e.g., in *Scotophilus wroughtoni* (Gopalakrishna 1949). *Glossophaga* and *Carollia* have a simplex uterus with relatively narrow tubular segments interposed between the distal end of each oviduct and the uterine cavity. These tubular segments are lined with endometrium and contain uterine glands. These tubular segments appear to be homologous with the cranial ends of the uterine horns in species having bicornuate uteri and have therefore been referred to as 'intramural uterine cornua.' Implantation in both *Glossophaga* (Fig. 3.7A) and *Carollia* occurs in one of

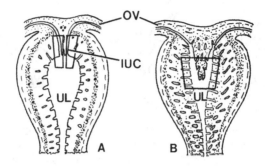

Figure 3.7. Schematic drawings of the simplex uterus of *Glossophaga soricina* (A) and *Carollia* spp. (B) showing implantation sites (boxed). IUC, intramural uterine cornu; OV, oviduct; UL, uterine lumen. (Redrawn from Rasweiler 1979.)

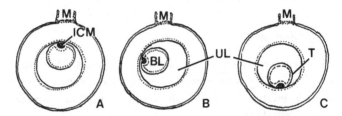

Figure 3.8. Drawings to show the site of implantation of the blastocyst in relation to the mesometrium. A) mesometrial implantation; B) lateral implantation; C) antimesometrial implantation. BL – blastocyst, ICM – inner cell mass, M – mesometrium, T – trophoblast, UL – uterine lumen.

these cornua, but in *Carollia*, implantation sometimes takes place in the main uterine cavity (Fig. 3.7B) in immediately adjacent areas of the endometrium. In *G. soricina* the blastocyst does not pass into the main uterine cavity.

The site of implantation in the uterus relative to its mesenteries

The site of implantation varies, being either mesometrial, antimesometrial, or lateral (Fig. 3.8 A, B, C).

The point of the embryo that effects attachment/implantation

Whereas in most bats the blastocyst expands to come into contact with the uterine wall circumferentially (Fig. 3.9A), in vespertilionid bats,

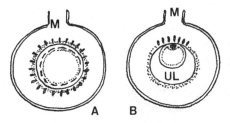

Figure 3.9. Drawings to show the point of the embryo that effects initial trophoblastic attachment. (A) circumferential, (B) embryonic.

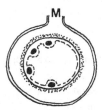

Figure 3.10. Drawing to show the variable orientation of the epi-blast/embryonic shield during implantation. For mesometrial, antimeso-metrial and lateral orientation of epiblast, see Figure 3.8A, B, C.

the blastocyst attaches to the uterine wall by its embryonic pole (Fig. 3.9B). The blastocyst of *M. ater* attaches to the uterine wall by its embry-onic pole and lateral surface (Rasweiler 1990). In *N. albiventris* the initial attachment of the blastocyst is bipolar with the presumptive embryonic pole becoming attached to the cranial end of a prominent endometrial ridge that runs from the lateral to the antimesometrial side of each uterine horn; the presumptive abembryonic pole becomes attached to the endometrium at the cranial end of the gravid uterine horn (Rasweiler 1979).

The orientation of the epiblast/embryonic shield

Apart from mesometrial (Fig. 3.8A), antimesometrial (Fig. 3.8B), and lateral orientation (Fig. 3.8C) of the epiblast, variable orientation of the epiblast/embryonic shield (Fig. 3.10) has been observed in *Megaderma* (Gopalakrishna & Khaparde 1972), *Rhinopoma hardwickei* (Karim 1986), *H. fulvus* (Gopalakrishna & Karim 1973), *H. lankadiva* (Khan 1996), and *Molossus ater* (Rasweiler 1990).

Kirby (1971) noted that the embryonic mass in the blastocyst of mice migrates actively inside the blastocyst covering. While the same

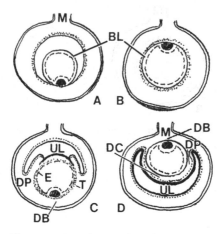

Figure 3.11. Drawings to show the depth of implantation. A) and B) superficial; C) partially interstitial; D) interstitial, BL – blastocyst, DB – decidua basalis, DC – decidua capsularis, DP – decidua parietalis, E – endoderm, M – mesometrium, T – trophoblast, UL – uterine lumen.

hypothesis was suggested for the variable orientation of the embryonic mass in *Megaderma lyra* (Gopalakrishna & Khaparade 1972) and other species, Ramakrishna (1977) ruled out the possibility of rotation of the embryonic mass in *Rhinolophus rouxi* since the blastocyst was in advanced stages of implantation, and the embryonic disc had already become flattened. Likewise, Sapkal (1981) found evidence against the rotation of the embryonic mass in implanted blastocysts of *Taphozous melanopogon*. Rasweiler & Badwaik (1996) suggested that the orientation of the embryonic mass may be accomplished by preferential proliferation of the inner cells at the presumptive embryonic pole from the 'improperly' positioned early clusters of cells. Although the inner cell migration hypothesis was rejected for mice by marker experiments (Gardner 1975, 1990), Rasweiler & Badwaik (1996) do not negate this as the possible mechanism of inner cell mass orientation for *N. albiventris*.

The depth of implantation

The degree of intimacy of the blastocyst with the uterine wall (Fig. 3.11 C, D) varies among species. Superficial implantation (Fig. 3.11A, B) is most common. Partially interstitial implantation (Fig. 3.11C) occurs in *Pteropus giganteus* and *N. albiventris*. Complete interstitial implantation (Fig. 3.11D) occurs in *G. soricina*, *C. perspicillata*, *D. rotundus* and *Thyroptera tricolor* (Fig. 3.10).

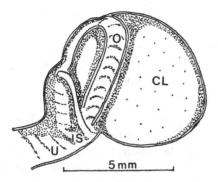

Figure 3.12. Camera lucida drawing of the female reproductive tract of *Rousettus leschenaulti* in early pregnancy (free bilaminar blastocyst in uterus). The implantation site (IS) anatomically appears as a swollen bulb at the cranial end of the uterus. CL – corpus luteum, O – oviduct, U – uterus.

Physiological relationships during implantation

Most bats in which implantation takes place near the cranial end of the uterus (anatomically this is seen as a swollen bulb in early pregnancy; Fig. 3.12), also exhibit preferential stimulation of the oviduct (e.g., *N. albiventris* and *Peropteryx* sp., Pow & Martin 1994). In *Rhinopoma hardwickei* (Karim & Fazil 1987), ovulation takes place from either ovary with equal frequency with the progestational reaction being localized to a small cranial segment of the uterus. Estrous changes are augmented in the uterine cornu on the side of ovulation, whereas the contralateral cornu reverts to an anestrous condition after implantation. In *Molossus ater*, implantation is initiated approximately half-way between the cranial end of the right uterine horn and the point at which it joins the left horn distally (Rasweiler 1990).

Why attachment occurs at a mid-horn position is not clearly understood. Movement of the embryo to this location may be due to myometrial contractions, the initiation of luminal closure, or luminal fluid movements. *M. ater* is unusual in exhibiting the spontaneous development of decidual reaction during the luteal phase of the non-pregnant cycle (Rasweiler 1991a). In *Rousettus leschenaulti* (Karim 1971) the mesometrial part and in *Hipposideros lankadiva* (Khan 1996) the antimesometrial part of the uterine lumen forms an implantation chamber, whereas the rest of the uterine lumen remains slit-like. The endometrial reaction is usually localized, or most pronounced, at the prospective site of implantation which may be due to the local delivery of ovarian steroids to the site via the uterine artery following a countercurrent exchange between it and the ovarian venous, or lymphatic, drainage (Rasweiler 1978).

Recent studies on *Pteropus* sp. (Pow 1992) have established that 1) the ovarian artery vascularizes the cranial end of the ipsilateral uterine horn in these species, 2) the anatomical relationships between ovarian venous and arterial vessels are usually intimate in the presumed region of steroid transfer, and 3) ^{3}H-estradiol microinjected into the ovary of *P. poliocephalus* is preferentially transported to the ipsilateral uterine horn. It is not known if uteri of bats with mid- or distal horn implantation sites are also stimulated by similar mechanisms. According to Rasweiler (1993), this would raise the possibility that implantation sites in these species are vascularized primarily by the uterine artery. The degree to which the uterus is supplied by the ovarian and/or uterine arteries on each side is known to vary among species (Pow 1992). In the absence of some intimate anatomical relationships between the ovarian and uterine vessels (e.g., arterial anastomoses), blood delivered to either side of the uterus via the uterine arteries should contain equal concentrations of ovarian steroids (Rasweiler 1993).

Delays in embryonic development

Two types of delays in embryonic development occur in bats:

Developmental diapause or delayed implantation

The free uterine blastocyst enters a period of developmental quiescence and remains unimplanted for a variable length of time with prolongation of the gestation period. For example, a maximum period of 4 months of developmental diapause/delayed implantation has been reported for *Miniopterus* from Africa (Bernard 1980). The gestation period of *M. schreibersii* lasts for 8 months.

Delayed development

The blastocyst implants on schedule, after which embryonic development slows, or is retarded for a variable length of time. A maximum length of 8 months of retarded embryonic development is reported in the megachiropteran *Haplonycteris fischeri* (Heideman 1989). This species has a gestation period of 11.5 months, the longest recorded for any bat. Amongst the Microchiroptera, *Hipposideros lankadiva* (Khan 1996) exhibits a 5-month period of retarded embryonic development after implantation, whereas, *Macrotus californicus* (Bradshaw 1962) exhibits a delay of 4.5 months. Both microchiropteran species have a long gestation period of 9 months.

Embryogenesis in bats, as in other mammals (Perry 1981), involves the development of the amnion, yolk sac, allantois, and chorion. A non-cellular, homogeneous, eosinophilc and PAS-positive accessory membrane (Reichert's membrane) has been described in a few species of bats.

General pattern of development of fetal membranes

A common pattern of development of the fetal membranes is exhibited in all bats so far studied (Gopalakrishna & Karim 1979, 1980) except for specific variations mentioned otherwise. The morphogenesis of the fetal membranes has been traditionally examined in the five distinct stages of development: primitive streak/somite stage, allantoic diverticulum stage, early limb-bud stage, late limb-bud stage, and term pregnancy.

In some species, changes are introduced in the fetal membranes during late pregnancy. Arrangement of membranes in *Rousettus leschenaulti* (Fig. 3.13A–F), *Rhinopoma hardwickei* (Fig. 3.14A, B) and *Pipistrellus mimus* (Fig. 3.15A–D) are depicted. Modifications of these arrangements occur in other species.

The fetal membranes

Amnion

The amnion is a thin, nonvascular membrane lining the amniotic cavity and the outer aspect is surrounded by mesoderm. In some bats the amnion partly becomes secondarily vascularized when it comes into contact with the vascularized wall of the allantois to form an amnio-allantoic membrane. Four methods of amniogenesis have been reported in bats (Gopalakrishna & Karim 1979, 1980).

Amniogenesis by cavitation

This process involves both the apoptosis of cells within the inner cell mass and the coalescence of spaces that develop between the epiblast or by the radial rearrangement and polarization of epiblast in the inner cell

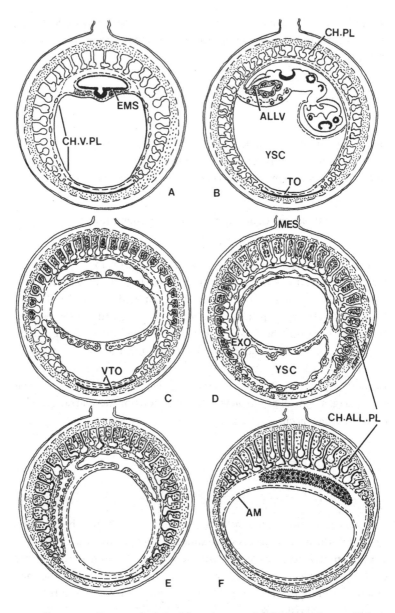

Figure 3.13. *Rousettus leschenaulti*, arrangement of fetal membranes. A) late neural groove/primitive streak stage; B) allantoic diverticulum stage; C) early limb-bud stage; D) late limb-bud stage; E) mid-pregnancy stage; F) term pregnancy. Abbreviations for Figs. 3.13–3.19: ALLM – allantoic mesenchyme, ALLV – allantoic vesicle, ALLP – allantoic placenta, AM – amnion, AMF – amniotic folds, BME – basement membrane of maternal endothelium, CHALLPL – chorioallantoic placenta, CHPL – chorionic placenta, CHVPL –

mass giving rise to the amniotic cavity. The epiblastic roof of the cavity so formed becomes thin with the expansion of the epiblast into a disc. The persistent epiblastic roof gets an investment of mesoderm to form the definitive amnion. This process was observed in *H. fischeri* (Heideman 1989), *T. tricolor* (Wimsatt & Enders 1980), *N. albiventris* (Rasweiler & Badwaik 1996) and *C. perspicillata* (Badwaik *et al.* 1997).

Primordial amniotic cavity formed by cavitation, the definitive amnion by folding

The epiblastic roof of the primordial amniotic cavity is lost, creating a trophoepiblastic cavity. The definitive amnion is subsequently formed by folds from the margins of the epiblast over which the mesoderm extends. This mode of amniogenesis is described for *Taphozous melanopogon* and *Chaerephon plicata* (Sandhu 1986) and *Molossus ater* (Rasweiler 1990).

Amniogenesis by folding

The inner cell mass transforms directly into an epiblastic plate. Between the epiblast and the overlying cytotrophoblast, the trophoepiblastic cavity is formed. Subsequent folds develop from the margins of the epiblastic plate which are reinforced by the extraembryonic mesoderm, thereby establishing the definitive amnion as in *R. hardwickei* (Karim & Fazil 1985) and *Hipposideras lankadiva* (Khan 1996).

Amniogenesis by coalescence

An unusual method of amniogenesis has been described for *Cynopterus sphinx gangeticus* (Moghe 1956) and for *C. marginatus*. In these two species the primordial amniotic cavity develops as a result of apoptosis of cells within the inner cell mass, which is pushed deeper into the blastocyst

Figure 3.13 (*cont.*)
choriovitelline placenta, E– endoderm, EMS – embryonic shield, EAD – endodermal allantoic duct, ENAC – endodermal acini, EXO – exocoelom, FB – fetal blood, FC – fetal capillary, HEN – hypertrophied endoderm, IM – interstitial membrane, MB – maternal blood, ME – maternal endothelium, MES – mesometrium, RM – Reichert's membrane, ST – syncytiotrophoblast, T – trophoblast, TEC – trophoectodermal cavity, TO – trilaminar omphalopleure, VTO – vascular trilaminar omphalopleure, YSC – yolk-sac cavity, YSG – yolk-sac gland, YSS – yolk-sac splanchnopleure.

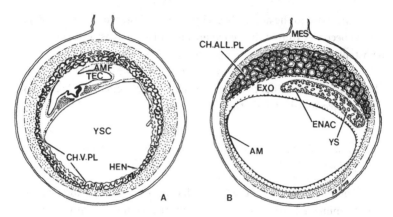

Figure 3.14. *Rhinopoma hardwickei*, arrangement of fetal membranes. A) Neural groove stage; B) at term pregnancy.

cavity by a proliferation of precociously-formed extraembryonic mesoderm. A mass of spongy tissue intervenes between the hollow embryonic mass and the cytotrophoblast layer. Folds develop from the margins of the roof of the primordial amniotic cavity forming a second cavity. For a short time, therefore, these two cavities stack over one another. Soon the two cavities become confluent with the breakdown of the roof. The roof of the secondary amniotic cavity constitutes the ectodermal component of the definitive amnion after it is enveloped by the extraembryonic mesoderm.

Yolk sac

The development of the yolk sac in bats displays three main trends. (1) The yolk sac splanchnopleure invaginates over the abembryonic bilaminar or trilaminar omphalopleure which persists throughout gestation. The omphalopleures may be in contact with the uterine wall or remain freely hanging in the uterine lumen. With the progressive invagination of the yolk sac splanchnopleure, the yolk sac cavity becomes reduced to streak-like spaces. The endodermal cells of the yolk sac hypertrophy as the vitelline vessels lengthen and the splanchnopleure separates. Toward the end of gestation the endoderm remains as a single layer of hypertrophied cuboidal or columnar cells lining the remnants of the yolk sac cavity. The mesodermal cells undergo enormous hypertrophy remaining as large vacuolated cells projecting into the exocoelom. This mode of development of the yolk sac is exhibited by *T. tricolor* (Wimsatt & Enders 1980), *Mormoops megalophylla*, and *Pteronotus davyi* (Gopalakrishna *et al.* 1992). (2) The yolk sac splanchnopleure separates from the chorion, progressively undergoes collapse and is withdrawn from its original opposition to finally lie

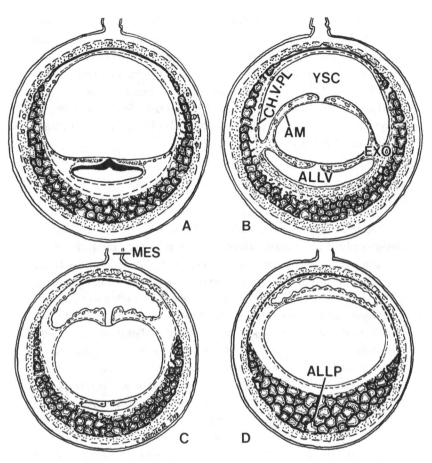

Figure 3.15. *Pipistrellus mimus mimus*, arrangement of fetal membranes. A) Late neural groove stage; B) early limb-bud stage; C) late limb-bud stage; D) term pregnancy.

adjacent to the chorioallantoic-placental disc. The yolk sac lumen, however, persists as narrow streak-like spaces. Both endodermal and mesodermal cells of the separated yolk sac splanchnopleure undergo hypertrophy thereby giving the collapsed yolk sac a glandular appearance. This developmental trend of the yolk sac has been described for *R. hardwickei* (Karim & Fazil 1985), *Molossus ater* (Rasweiler 1990, 1991a), and other species. In *H. lankadiva* (Khan 1996), although such a trend is exhibited, the yolk sac splanchnopleure does not undergo a shift in its location because the collapsed yolk sac splanchnopleure occurs on the mesometrial and lateral sides during morphogenesis. (3) The yolk sac splanchnopleure completely separates from the chorion, progressively draws from the abembryonic (antimesometrial side) to the embryonic pole (mesometrial side) of the chorionic sac and abuts the chorioallantoic placenta. Approaching the end of gestation, it is converted into a solid, richly vascular, glandular and endocrine structure with both the endodermal and mesodermal cells undergoing hypertrophy. The endodermal cells become grouped into clusters of acinar-like structures surrounded by columnar mesodermal cells. The yolk sac lumen in most of the acini is obliterated. Such a morphogenesis and histogenesis of the yolk sac is exhibited by *Pteropus poliocephalus* and *P. scapulatus* (Hughes 1989).

Allantois

The allantois in bats arises as a hindgut diverticulum during the formation of the primitive streak and early limb-bud stages. This stage is designated as the 'allantoic diverticulum stage'. The allantois lies free in the exocoelom temporarily and is lined by the endoderm, itself surrounded by allantoic mesenchyme, in which lie the allantoic blood capillaries with nucleated fetal corpuscles. The vesicle reaches its maximum size when the fetus has entered the limb-bud stage and when the placenta is either diffuse or cup-shaped. With the consolidation of the placenta into a disc, or double disc, the allantoic vesicle becomes progressively reduced until it either completely disappears (as in most bats), or persists as a small streak-like space within the mesenchyme beneath the placental discs as in *H. ater* (Inamdar 1986) and *H. lankadiva* (Khan 1996).

The chiropteran allantoic vesicle as compared to that of Carnivora, Perissodactyla, and Artiodactyla (Mossman 1987) is of moderate size. Exceptions are seen in *Natalus stramineus* (Mossman 1987), *T. tricolor* (Wimsatt & Enders 1980) and *M. ater* (Rasweiler 1990) in which the allantoic vesicle is a small sac. The allantoic vesicle in *D. rotundus murinus* (Wimsatt 1954) is small, nonvesicular, and transient. In *G. soricina* (Hamlett 1935) the allantois is not distinct.

Although the allantoic vesicle becomes obliterated with advancement of pregnancy, in some bats such as in *P. giganteus*, a narrow endodermal allantoic duct, or solid endodermal cord, persists in the umbilical cord. The endodermal allantoic duct is absent in the umbilical cord of *Taphozous melanopogon* (Sandhu 1986), *Chaerophon plicata* and *Pipistrellus* sp. (Gopalakrishna & Karim 1979).

Umbilical cord

The umbilical cord is inserted centrally on the fetal side of the placental disc or in between the two discs in instances where the chorioallantoic placenta is double discoid as in the hipposiderid bats (Khan 1996).

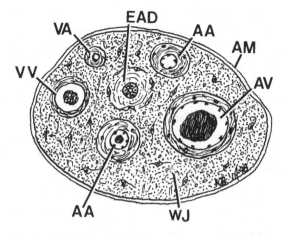

Figure 3.16. Schematic drawing of cross-section of the umbilical cord of *Rhinopoma hardwickei*. Note the five vessels: VA – vitelline artery, AV – allantoic vein, AA – two allantoic arteries, VV – single vitelline vein, EAD – endodermal allantoic duct, WJ – Wharton's jelly.

The length of the umbilical cord varies among species, but consistently shows five blood vessels: two allantoic arteries, one allantoic vein, one vitelline artery, and one vitelline vein, lying suspended within the mucoid connective tissue (Fig. 3.16). The endodermal allantoic duct may be present or absent.

PLACENTATION

To date, the placental ultrastructure of some 20 species of bats representing nine families has been reported. In bats, during gestation several different types of placentae are formed in a chronological sequence with more than one type present at a particular stage of development, but occurring in different regions of the uterus. For example, the early primitive streak stage is characterized by the presence of trophoblastic placenta and non-vascular yolk sac placenta; the late primitive streak/somite stage and allantoic diverticulum stages are characterized by the presence of chorionic and choriovitelline placentae whereas, the early limb-bud stage is characterized by the presence of chorioallantoic and choriovitelline placentae. The different types of placenta formed are:

Trophoblastic placenta or 'preplacenta'

It is formed soon after the implantation of the blastocyst. The syncytial trophoblast invades the uterine endometrium. The proximal segment of the uterine glands breaks down or disintegrates, the syncytial trophoblastic cells merge with the decidual cells and thus with the invading tissues a symplasmic syncytiotrophoblastic shell – the trophoblastic placenta – is formed in which maternal capillaries lie lined by endothelium. In bats with circumferential attachment of the blastocyst the trophoblastic placenta is diffuse, forming on all sides of the implantation chamber, e.g., in R. hardwickei (Karim & Fazil 1985), Peropteryx kappleri (Rasweiler 1982), Nycteris thebiaca (Gopalakrishna et al. 1995), H. ater (Inamdar 1986) H. lankadiva (Khan 1996) and Miniopterus schreibersii natalensis (van der Merwe 1982).

Yolk sac placenta

The bilaminar blastocyst becomes trilaminar with the differentiation of mesoderm below the epiblast and its migration to the lateral sides and to the abembryonic pole of the blastocyst. The vespertilionid bats are an exception since the mesoderm extends to about two-thirds the distance of the yolk sac wall, the abembryonic segment being a bilaminar omphalopleure which hangs freely in the uterine lumen. Thus at this stage, a major portion of the trophoblastic placenta becomes converted into the non-vascular yolk sac placenta. The placenta on the dorsal aspect of the epiblast continues to be the trophoblastic placenta.

Choriovitelline placenta

Formation of the vitelline blood capillaries with nucleated fetal corpuscles in the mesoderm of the yolk sac wall converts the non-vascular yolk sac placenta into the vascular or choriovitelline placenta. The choriovitelline placenta is well developed and extensive during the early limb-bud stage of development. The formation and extension of the exocoelom in the mesoderm of the yolk sac wall separates the vascular splanchnopleure from the chorion, thereby progressively abolishing the choriovitelline placenta from the embryonic to the abembryonic region.

Chorionic placenta

At the late primitive streak/somite stage and the allantoic diverticulum stage, formation of the exocoelom above the epiblast separates the amnion from the chorion. Hence the trophoblastic placenta becomes converted into the chorionic placenta.

Chorioallantoic placenta

The allantois bridges across the exocoelom and establishes contact with the fetal border of the chorionic placenta at the early limb-bud stage converting it into the chorioallantoic placenta. Thus the early limb-bud stage is characterized by the presence of both choriovitelline and chorioallantoic placentae formed at different regions of the uterus. The extension of the exocoelom abolishes the choriovitelline placenta, and, concurrently, there is expansion of the allantoic vesicle, and hence the entire placenta is converted into the chorioallantoic placenta which is vascularized by the allantoic blood vessels. The chorioallantoic placenta, during its early formation, is either diffuse or cup-shaped, but with the advancement of pregnancy, becomes consolidated into a disc (Rasweiler & Badwaik 1999) or a double disc type. The conversion of the cup-shaped placenta into a disc is brought on by the progressive destruction and disappearance of placental tubules near the margins of the placental cup. This has been observed in H. *fulvus* (Karim 1972, 1976) and R. *hardwickei* (Karim & Fazil 1985). Finally consolidation of the placenta towards its definitive position occurs.

Miniopterus schreibersii is unique among bats to develop three types of chorioallantoic placentae – primary, secondary, and tertiary (Peyre & Malassine 1969; Gopalakrishna & Chari 1985; Bhiwgade *et al.* 1992). The final location of the chorioallantoic placenta within the uterus relative to its mesenteries varies among species of bats. In *Taphozous melanopogon* (Sandhu 1986) the mesometrial moiety becomes converted into an hematoma.

Histogenesis of the chorioallantoic placenta

With blastocyst implanted at the maternal border of the syncytiotrophoblastic shell, necrotic patches of the trophoblast indicate destructive and invasive activity. Progressively the trophoblast destroys almost three-fourths of the thickness of the endometrial stroma leaving only a strip of uninvaded endometrium below the uterine myometrium, termed the junctional zone. The presence of lightly staining large mono-, bi-, and multinucleate endometrial cells have been reported in *Pteropus giganteus* (Karim & Bhatnagar 1996) and H. *lankadiva* (Khan 1996).

The basal cytotrophoblastic layer becomes activated, its cells divide and form hillocks which penetrate into the syncytiotrophoblastic shell. The cytotrophoblastic hillocks hollow out (forming the chorionic villi) at their fetal border, and thus the maternal capillaries, invested by the syncytiotropoblast and cytotrophoblast, appear as the placental tubules (*H. bicolor*, Gopalakrishna & Moghe 1960). During early stages of their formation, the placental tubules are small, but as pregnancy advances they become distinct and hang from the uterine wall. Finally the placental tubules undergo ramification and anastomosis to establish the definitive placental labyrinth. The maternal vascularization of the placenta is invested by maternal efferent vessels which penetrate the uterine wall. The allantoic mesenchyme along with fetal/allantoic capillaries with nucleated corpuscles invade the chorionic villi and the meshes of the placental labyrinth, establishing the fetal vascularization of the placenta (Gopalakrishna & Badwaik 1990).

Interhemal membrane

The structure of the interhemal membrane, or the placental membrane, of the chorioallantoic placenta varies among families of Chiroptera, being endotheliochorial in some and hemochorial in others (Wimsatt 1958; Gopalakrishna & Karim 1979, 1980; Luckett 1980, 1993; Mossman 1987; Bhiwgade 1990; Rasweiler 1991*b*; Karim & Bhatnagar 1996). The layering of the trophoblast in the endotheliochorial and hemochorial placenta also varies among families (Enders 1965; Karim & Bhatnagar 1996), being dichorial when both syncytio- and cytotrophoblast persist or monochorial when either syncytio- or cytotrophoblast persists.

Endotheliochorial or vasochorial placenta

This is one in which the maternal endothelium persists and the maternal capillary wall is in contact with the trophoblast (Fig. 3.17; Grosser 1909, 1927; Enders 1965). When both syncytio- and cytotrophoblast are present, the placenta is endotheliodichorial, and when either syncytio- or cytotrophoblast persists the placenta is endotheliomonochorial.

Hemochorial placenta

This is one in which the trophoblast is bathed with maternal bood (Fig. 3.18; Grosser 1909, 1927; Enders 1965). With both syncytio- and cytotrophoblast, the placenta is hemodichorial and with either syncytio- or cytotrophoblast, the placenta is hemomonochorial. A hemodichorial placenta is seen in *Rousettus leschenaulti*, *Cynopterus sphinx*, and *P. giganteus*

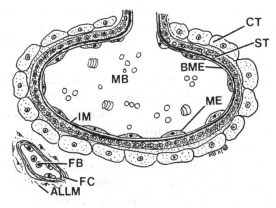

Figure 3.17. Placental tubules towards fetal border of placenta, drawn from
Pteropus giganteus: vasodichorial/endotheliodichorial placenta.

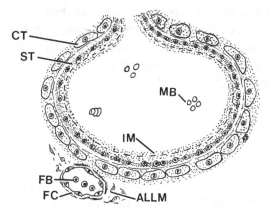

Figure 3.18. Placental tubules towards fetal border of placenta, drawn from
Pteropus giganteus: hemodichorial placenta.

(Karim & Bhatnagar 1996), and in *Pteronotus parnellii* (Badwaik & Rasweiler
1998). Microstructure of the chorioallantoic placenta of *Mormoops megalophylla* and *P. davyi* (Gopalakrishna *et al.* 1992) is described as hemomonochorial with only syncytiotrophoblast present.

Interstitial membrane

A significant and constant component of the chorioallantoic placenta is a reticulate PAS- positive, azan-positive, eosin-negative and electron-dense interstitial membrane (Karim & Bhatnagar 1996). Two types of

interstitial membranes have been described. In the endotheliochorial placenta the interstitial membrane lies in close proximity to the endothelial cells, whereas in the hemochorial placenta, the interstitial membrane lies embedded in a homogeneous eosinophilic or aniline-blue-positive layer. In *Pteropus giganteus*, it has been observed that the interstitial membrane is thicker in the placental tubules towards the fetal border of the placenta (Karim & Bhatnagar 1996). In *Myotis lucifugus*, Enders & Wimsatt (1968) stated that the interstitial membrane, which is present earlier as the basement membrane of the endothelium receives a covering of cytoplasm from the syncytiotrophoblast so that it is converted into an intrasyncytial lamina perforated only at irregular intervals by syncytial tongues which spread out over its upper surface forming a nearly continuous layer. The discontinuous 'homogeneous substance' described in the chorioallantoic placenta of *Tadarida brasiliensis* (Stephens 1969) has the same topography as the interstitial membrane (Gopalakrishna & Karim 1979, 1980). A discontinuous interstitial membrane has been described in *R. leschenaulti* (Karim *et al.* 1978; Bhiwgade 1990), *H. speoris* (Kothari & Bhiwgade 1992) and in the secondary placenta of *Miniopterus schreibersii* (Bhiwgade *et al.* 1992). A continuous interstitial membrane has been described for *Rhinolophus rouxi* (Bhiwgade 1990).

Origin of interstitial membrane

Wimsatt (1958), who first recognized the presence of the interstitial membrane in the placental tubules of many species of bats, deduced that the material of this membrane may be wholly or partly derived from the syncytiotrophoblast. Gopalakrishna & Karim (1980) observed that the thickness of the interstitial membrane appears to be approximately proportional to the caliber of the maternal vascular channels, which undergo dilation as pregnancy advances. These facts also indicate the possibility that the interstitial membrane may not be entirely the remnant of the basement membrane of the endothelium of the maternal blood capillaries in the placenta. Further, since the only other structure which is topographically adjacent to the interstitial membrane is the syncytiotrophoblast, it is highly suggestive that the syncytiotrophoblast is in some way responsible for the increase in thickness of the interstitial membrane. However, these facts do not preclude the possibility that the basement membrane of the original endothelial lining may also become incorporated in the definitive interstitial membrane of the definitive/mature placenta of all bats. In *Myotis lucifugus* (Enders & Wimsatt 1968) and *D. rotundus* (Bjorkman & Wimsatt 1968) it was observed that the

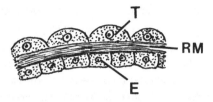

Figure 3.19. Reichert's membrane: drawing showing the noncellular, homogeneous membrane formed in between the endoderm and the trophoblast.

interstitial membrane may have a dual origin, only part of it being contributed by the trophoblast.

Reichert's membrane

A noncellular, homogeneous, eosinophilic and PAS-positive accessory membrane named Reichert's membrane (Fig. 3.19) has been
described in bats (Reichert 1861). The Reichert's membrane usually develops between the parietal endoderm and the trophoblast on the abembryonic pole of the implanting bilaminar blastocyst. The time of the first
appearance of the membrane and the period of its persistence varies.
Reichert's membrane has also been observed dorsal and lateral to the
inner cell mass in the phyllostomid bats. Both the endoderm and cytotrophoblast probably contribute to the formation of Reichert's membrane in *Carollia perspicillata* (Badwaik *et al.* 1997). Ultrastructurally,
Reichert's membrane is most intimately associated with the cytotrophoblast of the early blastocyst. In the parietal wall of the yolk sac it consists
of two layers. One layer appears to be the basal lamina of the trophoblast,
whereas the other layer may be of endodermal (maternal) origin because
similar material is commonly observed in the endodermal cells and in
the cisternae of their rough endoplasmic reticulum. In *Noctilio* (Rasweiler
& Badwaik 1996), the epiblast, endoderm, and cytotrophoblast may all
participate in producing parts of Reichert's membrane. Reichert's membrane probably plays an important mechanical role in tethering the inner
cell mass and the embryonic shield to the preplacenta and in directing
the invasion of the preplacenta by the mesoderm.

Accessory placental structures

Accessory placental structures have been described in bats. The
occurrence of a villous syndesmochorial placenta on the maternal border
of the main placenta has been reported in *M. lucifugus* (Wimsatt 1945) and

Megaderma lyra (Gopalakrishna & Khaparde 1978). In the emballonurid bats (Rohatgi *et al.* 1992), the mesometrial moiety of the chorioallantoic placenta becomes converted into a sac-like hematoma during the second half of gestation. Maternal erythrocytes are actively ingested and phagocytosed by the cells of the chorion which is directly bathed in the extravasated maternal blood in the hematoma. A triangular accessory hemochorial placenta protrudes from the maternal border in between the two placental discs of the chorioallantoic placenta in *Hipposideros* (Karim 1972; Khan 1996).

Giant cells

Any enlarged cell of fetal or maternal origin is referred to as a giant cell (Mossman 1987). Two types of giant cells, which may be mono-, bi-, or multinucleate, are found in bats. Decidual (endometrial-maternal origin) giant cells have been observed in the junctional zone of the placenta after implantation of the blastocyst. Trophoblastic giant cells (fetal origin) are associated with the placenta or placental tubules, e.g., cytotrophoblastic giant cells of *H. fulvus* (Karim 1973) and *H. lankadiva* (Khan 1996). Multinucleate giant cells (Wimsatt 1954) arise by mitotic division. Glycogen and glycoprotein granules have been observed in the giant cells of *Desmodus* (Wimsatt 1954). Functions attributed to the giant cells include transport of glycogen and mucopolysaccharides across the placental barrier and involvement in the progressive destruction of endometrial stroma, thus assisting in the firm implantation of the embryo.

PHYLOGENETIC CONSIDERATIONS

Reproduction in bats is a subject which can shed light on the ontogeny and phylogeny interfamilially, as well as in relation to other groups such as dermopterans, insectivores, and primates. The comparative study of the very primitive fetal membranes is almost a requirement when genera such as *Taphozous, Brachyphylla, Pteronotus, Noctilio,* and *Miniopterus* and their interfamilial relationships are considered (Bhatnagar *et al.* 1996; Wible & Bhatnagar 1996).

For many years there was a considerable debate concerning the systematic relationships of the family Noctilionidae with some considering this family to be in the superfamily Emballonuroidea and others suggesting it to be more closely allied with what was then referred to as the superfamily Phyllostomatoidea (Forman *et al.* 1989). Partly in an effort to resolve this issue, comparative studies were carried out on aspects of the

female reproductive biology of *Noctilio* and representatives of Emballonuridae and Phyllostomidae (Rasweiler 1993). This work established that
Noctilio and the phyllostomid bats exhibit many reproductive similarities
and, during early pregnancy, share some uniquely derived, homologous
reproductive characteristics (synapomorphies). Because these reproductive characteristics are not possessed by the emballonurid bat *Peropteryx
kappleri*, the noctilionids would appear to be most closely related to the
phyllostomids. The latter two families are now classified as belonging to
the same superfamily, the Noctilionoidea (Koopman 1994).

Based on the results of comparative and cytochemical studies of the
chorioallantoic placental membrane in 23 species of bats representing
nine families, Wimsatt (1958) deduced an interesting sequence of
morphological changes between more ancestral and more specialized
families (Fig. 32 in Wimsatt [1958]; 17 chiropteran families arranged
systematically according to Miller [1907]; Simpson [1945]). Likewise, based
on embryological data, Gopalakrishna and Karim (1980: 38) presented a
schematic representation of the relationships of the chiropteran families
which corresponds very nearly to the taxonomic hierarchy given by
Simpson (1945), except for the position of Noctilionidae, Megadermatidae, and Molossidae, and the addition of a new family, the Miniopter-
idae. According to Gopalakrishna & Chari (1983), *Miniopterus schreibersii*
exhibits embryological features which not only are unlike those of any
vespertilionid bat, but taken together, are unlike those of any other
mammal.

Wimsatt (1958) observed that the maternal endothelium of the
definitive placental stages is most prominent in the more ancestral families, in which it is characteristically hypertrophied and syncytial. In the
derived families the endothelium becomes attenuated, and the nuclei are
flattened and sparsely distributed. In these derived, extant families, the
endothelium either persists as a thin enucleated lamina (Vespertilionidae), or disappears altogether (Molossidae) in the later stages.
Ultrastructural examination of the interhemal membrane of the chorioallantoic placenta of 20 species of bats does not justify the relationship
alluded to by Wimsatt (1958). In the three pteropodids (*R. leschenaulti*,
Cynopterus sphinx, and *Pteropus giganteus*), the interhemal membrane is
hemodichorial. In some of the microchiropteran families (Emballonuridae, Rhinopomatidae, Megadermatidae, and Rhinolophidae), the interhemal membrane is endotheliochorial and successively in more derived
extant families (Phyllostomidae, Molossidae, Vespertilionidae, and Thyropteridae) the interhemal membrane is hemochorial. In *H. lankadiva* and
H. speoris the interhemal membrane is endotheliochorial, and in *H. fulvus*

it is hemochorial. No explanation can be given for such differences in the various hipposiderid species. *M. schreibersii* is unique among bats to develop different types of placentae; the interhemal membrane of the primary and tertiary placentae is hemodichorial, whereas the secondary placenta is endotheliomonochorial with only the cytotrophoblast persisting (Bhiwgade *et al.* 1992).

According to Luckett (1980) the morphotypic condition for both chiropteran suborders is a labyrinthine endotheliochorial placenta, with a prominent 'interstitial membrane' separating the maternal endothelium from the syncytiotrophoblast. Luckett (1993) has shown that in both Mega- and Microchiroptera, there is ontogenetic and ultrastructural evidence for the occurrence of labyrinthine endotheliochorial and labyrinthine hemochorial placentae (for review see Enders & Wimsatt 1968; Gopalakrishna & Karim 1980). He further emphasized that ultrastructural study of changing tissue relationships during ontogeny in the vespertilionid *Myotis lucifugus* (Enders & Wimsatt 1968) provides valuable evidence for the probable phylogenetic transformation of the endotheliochorial to hemochorial condition during chiropteran evolution.

CONCLUSIONS

With obvious interspecies differences in their functional morphology and physiology, bats can not be classified yet into taxonomic groups based upon reproductive strategies and characters which are so highly variable, e.g., body weights not corresponding with gestation periods because both short and long gestations occur regardless of body size; developmental delays occurring in some species; the process of ovum development and implantation differing greatly amongst families, even between species, as does the underlying endocrinology. The influence of geographical location also imposes considerable variability and presents further difficulties in attempting to establish taxonomic relationships based upon reproduction.

One major reason for the difficulty in assigning taxonomic relationships in bats, based on reproductive traits, seems to be that too few species have been thoroughly and systematically investigated. Another reason seems to be that apparently we do not know which reproductive paradigms we should be looking into. The advent of molecular systematics should prove to be directly helpful in this respect. Hayssen *et al.* (1993) list about 500 species of bats for which data on breeding biology are recorded, yet these data in most cases are extremely meager addressing minimal and partial observations. There is an urgent need for systematic studies of reproductive biology of bats in the absence of which our ignorance will simply continue.

ACKNOWLEDGMENTS

The authors are grateful to Susan Hodge for preparing and formatting numerous drafts of this manuscript and to Benjamin Tucker for help in putting the References together. Comments from two anonymous reviewers greatly improved an earlier draft of this manuscript.

REFERENCES

Badwaik, N. K. & Rasweiler, J. J. IV (1998) The interhaemal barrier in the chorioallantoic placenta of the Greater Mustache Bat, *Pteronotus parnellii*, with observations on amplification of its intrasyncytial lamina. *Placenta*, **19**, 391–401.

Badwaik, N., Rasweiler, J. J. IV & Oliveira, S. F. (1997) Formation of reticulated endoderm, Reichert's membrane, and amniogenesis in blastocysts of captive-bred short-tailed fruit bats, *Carollia perspicillata*. *Anatomical Record*, **247**, 85–101.

Bernard, R. T. F. (1980) Reproductive cycles of *Miniopterus schreibersii natalensis* (Kuhl, 1819) and *Miniopterus fraterculus* (Thomas & Schwann, 1906). *Annals of the Transvaal Museum*, **32**, 55–64.

Bhatnagar, K. P., Wible, J. R. & Karim, K. B. (1996) Development of the vomeronasal organ in *Rousettus leschenaulti* (Megachiroptera, Pteropodidae). *Journal of Anatomy*, **188**, 129–35.

Bhiwgade, D. A. (1990) Comparative electron microscopy of chorio-allantoic placental barrier in some Indian Chiroptera. *Acta Anatomica*, **138**, 302–17.

Bhiwgade, D. A., Singh, A. B., Manekar, A. P. & Menon, S. N. (1992) Ultrastructural development of chorioallantoic placenta in the Indian *Miniopterus* (sic) bat *Miniopterus schreibersii fuliginosus* (Hodgson). *Acta Anatomica*, **145**, 248–64.

Bjorkman, N. H. & Wimsatt, W. A. (1968) Allantoic placenta of the vampire bat *(Desmodus rotundus murinus)*: a reinterpretation of its structure based on electron microscopic observations. *Anatomical Record*, **162**, 83–98.

Blandau, R. J. (1961) Biology of eggs and implantation. In *Sex and Internal Secretions*, vol. 2, 3rd edn, ed. W. C. Young & G. W. Corner, pp. 797–882. Baltimore: Williams & Wilkins.

Bradshaw, G. V. R. (1962) Reproductive cycles of the California leaf-nosed bat, *Macrotus californicus*. *Science*, **136**, 645–6.

de Bonilla, H. & Rasweiler, J. J. IV (1974) Breeding activity, preimplantation development, and oviduct histology of the short-tailed fruit bat, *Carollia*, in captivity. *Anatomical Record*, **179**, 385–404.

Enders, A. C. (1965) A comparative study of the fine structure of the trophoblast in several hemochorial placentas. *American Journal of Anatomy*, **116**, 29–68.

Enders, A. C. & Wimsatt, W. A. (1968) Formation and structure of the hemodichorial chorio-allantoic placenta of the bat, *Myotis lucifugus lucifugus*. *American Journal of Anatomy*, **122**, 453–89.

Forman, G. L., Smith, J. D. & Hood, C. S. (1989) Exceptional size and unusual morphology of spermatozoa in *Noctilio albiventris* (Noctilionidae). *Journal of Mammalogy*, **70**, 179–84.

Gardner, R. L. (1975) Analysis of determination and differentiation in the early mammalian embryo using intra- and interspecific chimeras. In *The Developmental Biology of Reproduction*, ed. C. L. Markert & J. Papaconstantinov, pp. 207–36. New York: Academic Press.

Gardner, R. L. (1990) Location and orientation of implantation. In *Establishing a Successful Human Pregnancy*, ed. R. G. Edwards, pp. 225–38. New York: Raven.

Gopalakrishna, A. (1949) Studies on the embryology of Microchiroptera. Part III. The histological changes in the genital organs and the accessory reproductive structure during the sex cycle of the vespertilionid bat *Scotophilus wroughtoni* (Thomas) *Proceedings of the Indian Academy of Sciences*, **30**, 17–46.

Gopalakrishna, A. & Badwaik, N. (1990) Vascularization of placenta in some bats. *Proceedings of the Indian Academy of Sciences*, **99**, 284–303.

Gopalakrishna, A. & Chari, G. C. (1983) A review of the taxonomic position of *Miniopterus* based on embryological characters. *Current Science*, **50**, 152–4.

Gopalakrishna, A. & Chari, G. C. (1985) Early development, implantation of the blastocyst and amniogenesis in the bat, *Miniopterus schreibersii fulginosus* (Hodgson). *Proceedings of the National Academy of Sciences, India*, **55**, 1–9.

Gopalakrishna, A. & Karim, K. B. (1973) Blastocyst-uterus relationship in the Indian leaf-nosed bat, *Hipposideros fulvus fulvus* (Gray). *Current Science*, **42**, 860–2.

Gopalakrishna, A. & Karim, K. B. (1979) Fetal membranes and placentation in Chiroptera. *Journal of Reproduction and Fertility*, **56**, 417–29.

Gopalakrishna, A. & Karim, K. B. (1980) Female genital anatomy and the morphogenesis of fetal membranes of Chiroptera and their bearing on the phylogenetic relationships of the group. Golden Jubilee volume of the *National Academy of Sciences*, Allahabad, pp. 379–428.

Gopalakrishna, A. & Khaparde, M. S. (1972) Variable orientation of the embryonic mass during the implantation of the blastocyst in the false vampire bat, *Megaderma lyra lyra* (Geoffroy) *Current Science*, **41**, 738–9.

Gopalakrishna, A. & Khaparde, M. S. (1978) Early development, implantation and amniogenesis in the Indian false vampire bat, *Megaderma lyra lyra* (Geoffroy). *Proceedings of the Indian Academy of Sciences*, **87B**, 91–104.

Gopalakrishna, A. & Moghe, M. A. (1960) Development of the foetal membranes in the Indian leaf-nosed bat, *Hipposideros bicolor pallidus*. *Zeitschrift für Anatomie und Entwicklungsgeschichte*, **122**, 137–49.

Gopalakrishna, A., Karim, K. B. & Rajgopal, G. (1974) Post-ovulatory changes in the eggs of some Indian bats. *Current Science*, **43**, 454–5.

Gopalakrishna, A., Bernard, R. T. F., Rasweiler, J. J. IV & Badwaik, N. (1992) The anatomy of the female genitalia and the structure of the foetal membranes in Mormoopidae. *Bat Research News*, **31**, 17 (abstract).

Gopalakrishna, A., Bernard, R. T. F., Rasweiler, J. J. IV & Badwaik, N. K. (1995) Implantation development of the fetal membranes and placentation in the slit faced bat, *Nycteris thebiaca* (Family Nycteridae). *Bat Research News*, **36**, 67 (abstract).

Grosser, O. (1909) *Eihäute und der Placenta*. Vienna: Wilhelm Braumüller.

Grosser, O. (1927) *Frühentwicklung, Eihautbildung und Placentation des Menschen und der Saügetiere*. Munich: Bergman.

Hamlett, G. W. D. (1935) Notes on the embryology of a phyllostomid bat. *American Journal of Anatomy*, **56**, 327–53.

Hayssen, V., Tienhoven, Ari V. & Tienhoven, Ans V. (1993) *Asdell's Patterns of Mammalian Reproduction*, revised 2nd edn. Ithaca: Cornell University Press.

Heideman, P. D. (1989) Delayed development in Fischer's pygmy fruit bats, *Haplonycteris fischeri*, in the Philippines. *Journal of Reproduction and Fertility*, **85**, 363–82.

Heideman, P. D. & Powell, K. S. (1998) Age-specific reproductive strategies and delayed embryonic development in an old world fruit bat, *Ptenochirus jargori*. *Journal of Mammalogy*, **79**, 295–311.

Heideman, P. D., Cummings, J. A. & Heaney, L. R. (1993) Reproductive timing and early embryonic development in the Old World fruit bat, *Otopteropus cartilagonodus* (Megachiroptera). *Journal of Mammalogy*, **74**, 621–30.

Hughes, R. L. (1989) Observations on the placenta of the flying foxes *Pteropus scapulatus* and *Pteropus poliocephalus*. *Macroderma*, **5**, 13.

Inamdar, S. M. (1986) Embryology of a bat, *Hipposideros ater ater* (Templeton). PhD dissertation, Nagpur University, India.

Karim, K. B. (1971) Early development, development of the fetal membranes, and placentation in the Indian fruit bat, *Rousettus leschenaulti* (Desmarest). PhD dissertation, Nagpur University, India.

Karim, K. B. (1972) Foetal membranes and placentation in the Indian leaf-nosed bat, *Hipposideros fulvus fulvus* (Gray). *Proceedings of the Indian Academy of Sciences*, **76B**, 71–8.

Karim, K. B. (1973) Giant cells in the placenta of two species of Indian bats. *Current Science*, **42**, 282–4.

Karim, K. B. (1975) Early development of the embryo and implantation in the Indian vespertilionid bat, *Pipistrellus mimus mimus* (Wroughton). *Journal of Zoological Society of India*, **27**, 119–36.

Karim, K. B. (1976) Embryology of some Indian Chiroptera. DSc dissertation, Nagpur University, India.

Karim, K. B. (1986) Blastocyst–uterus relationship in the Indian rhinopomatid, *Rhinopoma hardwickei hardwickei*. In *Progress in Developmental Biology*, Part A, ed. H. C. Slavkin, pp.427–34. New York: Alan Liss.

Karim, K. B. & Bhatnagar, K. P. (1996) Observations on the chorioallantoic placenta of the Indian flying fox, *Pteropus giganteus giganteus*. *Annals of Anatomy*, **178**, 523–30.

Karim, K. B. & Fazil, M. (1985) Post-implantation development of the Indian rhinopomatid, *Rhinopoma hardwickei* (Gray). *Myotis*, **23–24**, 63–9.

Karim, K. B. & Fazil, M. (1987) Early embryonic development and preimplantation changes in the uterus of *Rhinopoma hardwickei hardwickei*, Gray (Rhinopomatidae). *American Journal of Anatomy*, **178**, 341–51.

Karim, K. B., Wimsatt, W. A. & Gopalakrishna, A. (1978) Structure of the definitive placenta in the Indian bat, *Rousettus leschenaulti* (Pteropidae). *Anatomical Record*, **190**, 438.

Khan, N. A. (1996) Embryological studies correlated with histochemistry and biochemistry of ovary, uterus, and placenta with role of progesterone during gestation in *Hipposideros lankadiva lankadiva* (Kelaart). PhD dissertation, Nagpur University, India.

Kirby, D. R. S. (1971) Blastocyst–uterine relationship before and during implantation. In *The Biology of the Blastocyst*, ed. R. J. Blandau, pp. 393–411. Chicago: University of Chicago Press.

Koopman, K. F. (1994) Chiroptera: Systematics. In *Handbook of Zoology*, vol. 8, *Mammalia*, ed. J. Niethammer, H. Schliemann & D. Starck, pp. 1–217. Berlin: Walter de Gruyter.

Kothari, A. & Bhiwgade, D. A. (1992) Ultrastructural studies of interhemal membrane in three species of hipposiderid bats. *Proceedings of the Ninth International Bat Research Conference*, **54**, 134.

Krutzsch, P. H. & Crichton, E. G. (1991) Fertilization in bats. In *A Comparative Overview of Mammalian Fertilization*, ed. B. S. Dunbar & M. O'Rand, pp.137–49. New York: Plenum Press.

Luckett, W. P. (1980) The use of fetal membrane data in assessing chiropteran phylogeny. In *Proceedings of the Fifth International Bat Research Conference*, ed. D. E. Wilson & A. L. Gardner, pp. 245–66, Lubbock: Texas Tech University.

Luckett, W. P. (1993) Developmental evidence from the fetal membranes for assessing archontan relationships. In *Primates and their Relatives in Phylogenetic Perspective*, ed. R. D. E. MacPhee, pp. 149–86. New York: Plenum Press.

Martin, L., Towers, P. A., McGuckin, M. A., Little, L., Luckoff, H. & Blackshaw, A. W. (1987) Reproductive biology of flying foxes (Chiroptera: Pteropodidae). *Australian Mammalogy*, **10**, 115–18.

Miller, G. S. Jr. (1907) *The Families and Genera of Bats*. Bulletin 57, United States National Museum. Washington: Smithsonian Institution Press.

Moghe, M. A. (1956) On the development and placentation of the megachiropteran bat, *Cynopterus sphinx gangeticus*. *Proceedings of the National Institute of Sciences, India*, B, **22**, 48–55.

Mori, T. & Uchida, T. A. (1981) Ultrastructural observations of fertilization in the Japanese long-fingered bat *Miniopterus schreibersii fuliginosus*. *Journal of Reproduction and Fertility*, **63**, 231–5.

Mossman, H. W. (1987) *Vertebrate Fetal Membranes: Comparative Ontogeny and Morphology; Evolution; Phylogenetic Significance; Basic Functions; Research Opportunities*. New Brunswick: Rutgers University Press.

Perry, J. S. (1981) The mammalian fetal membranes. *Journal of Reproduction and Fertility*, **62**, 321–35.

Peyre, A. & Malassine, A. (1969) L'équipement stéroidodeshydrogénasique et la fonction endocrine du placenta de Minioptère (Chiroptère). *Comptes rendus des séances de la Société de biologie et de ses filiales* (Paris), **163**, 914–17.

Potts, D. M. & Racey, P. A. (1971) A light and electron microscope study of early development in the bat, *Pipistrellus pipistrellus*. *Micron*, **2**, 322–48.

Pow, C. S. T. (1992) Studies on the vasculature of the female reproductive tract of *Pteropus* and other Megachiroptera. PhD dissertation, University of Queensland, Brisbane.

Pow, C. S. T. & Martin, L. (1994) The ovarian–uterine vasculature in relation to unilateral endometrial growth in flying foxes (genus *Pteropus*, suborder Megachiroptera, order Chiroptera). *Journal of Reproduction and Fertility*, **101**, 247–55.

Quintero, F. & Rasweiler, J. J. IV (1974) Ovulation and early embryonic development in the captive vampire bat, *Desmodus rotundus*. *Journal of Reproduction and Fertility*, **41**, 265–73.

Ramakrishna, P. A. (1977) Parturition in the Indian rufous bat *Rhinolophus rouxi* (Temminck). *Journal of the Bombay Natural History Society*, **75**, 473–5.

Rasweiler, J. J. IV (1977) Preimplantation development, fate of zona pellucida, and observations on the glycogen-rich oviduct of the little bulldog bat, *Noctilio albiventris*. *American Journal of Anatomy*, **150**, 269–300.

Rasweiler, J. J. IV (1978) Unilateral oviducal and uterine reactions in the little bulldog bat, *Noctilio albiventris*. *Biology of Reproduction*, **19**, 467–92.

Rasweiler, J. J. IV (1979) Early embryonic development and implantation in bats. *Journal of Reproduction and Fertility*, **56**, 403–16.

Rasweiler, J. J. IV (1982) The contribution of observations on early pregnancy in the little sac-winged bat, *Peropteryx kappleri*, to an understanding of the evolution of reproductive mechanisms in monoovular bats. *Biology of Reproduction*, **27**, 681–702.

Rasweiler, J. J. IV (1990) Implantation, development of the fetal membranes and placentation in the captive mastiff bat, *Molossus ater*. *American Journal of Anatomy*, **187**, 109–36.

Rasweiler, J. J. IV (1991a) Spontaneous decidual reactions and menstruation in the black mastiff bat, Molossus ater. American Journal of Anatomy, 191, 1–22.

Rasweiler, J. J. IV (1991b) Development of the discoidal hemochorial placenta in the black mastiff bat, Molossus ater: evidence for a role of maternal endothelial cells in the control of trophoblastic growth. American Journal of Anatomy, 191, 185–207.

Rasweiler, J. J. IV (1993) Pregnancy in Chiroptera. Journal of Experimental Zoology, 266, 495–513.

Rasweiler, J. J. IV & Badwaik, N. (1996) Unusual aspects of inner cell mass formation, endoderm differentiation, Reichert's membrane development, and amniogenesis in the lesser bulldog bat, Noctilio albiventris. Anatomical Record, 246, 293–304.

Rasweiler, J. J. IV & Badwaik, N. K. (1999) Discoidal placenta. In Encyclopedia of Reproduction, vol. 1, ed. E. Knobil & J. D. Neil, pp. 890–902. San Diego: Academic Press.

Reichert, C. B. (1861) Beitrage zur Entwicklungsgeschichte des Meerschweinchens. Abhandlungen der deutschen Akademie der Wissenschaften zu Berlin, pp. 98–216. (From Royal Society Catalogue of Scientific Papers, vol. 5, 1871, p.144. Scarecrow Reprint Corp.: Metuchen, NJ, 1968.)

Richardson, E. G. (1977) The biology and evolution of the reproductive cycle of Miniopterus schreibersii and M. australis (Chiroptera: Vespertilionidae). Journal of Zoology, London, 183, 353–75.

Rohatgi, L., Bhiwgade, D. A. & Menon, S. N. (1992) Electron microscopic studies on the chorioallantoic placenta of the emballonurid bat, Taphozous melanopogon. Bat Research News, 34, 39.

Sandhu, S. K. (1986) Studies on the embryology of some Indian Chiroptera. PhD dissertation, Nagpur University, India.

Sapkal, V. M. (1981) Orientation of the embryonic mass in the blastocysts of the emballonurid bat, Taphozous melanopogon. Current Science, 30, 100.

Simpson, G. G. (1945) The principles of classification and a classification of mammals. Bulletin of the American Museum of Natural History, 85, 1–350.

Stephens, R. J. (1969) The development and the fine structure of the allantoic placental barrier in the bat, Tadarida brasiliensis cynocephala. Journal of Ultrastructural Research, 28, 371–98.

Uchida, T. (1953) Studies on the embryology of the Japanese house bat, Pipistrellus tralatitius abramus (Temminck) II. From the maturation of the ova to the fertilization, especially on the behaviour of the follicle cells at the period of fertilization. Science Bulletin, Faculty of Agriculture, Kyushu University, 14, 153–68.

van der Merwe, M. (1982) Histological study of implantation in the natal clinging bat, Miniopterus schreibersii natalensis. Journal of Reproduction and Fertility, 65, 319–23.

van der Stricht, O. (1909) La structure de l'oeuf des mammifères (chauve-souris), Vesperugo noctula. 3e part.: L'oöcyte de la fin du stade d'accroissement, au stade de la maturation, au stade de la fécondation et au début de la segmentation. Mémoires de l'académie royale de medicine de Belgique, vol. 2.

Wible, J. R. & Bhatnagar, K. P. (1996) Chiropteran vomeronasal complex and the interfamilial relationship of bats. Journal of Mammalian Evolution, 3, 285–314.

Wimsatt, W. A. (1944a) Further studies on the survival of spermatozoa in the female reproductive tract of the bat. Anatomical Record, 88, 193–204.

Wimsatt, W. A. (1944b) An analysis of implantation in the bat, Myotis lucifugus. American Journal of Anatomy, 74, 355–411.

Wimsatt, W. A. (1945) Notes on breeding behavior, pregnancy and parturition in some vespertilionid bats of eastern United States. Journal of Mammalogy, 26, 23–33.

Wimsatt, W. A. (1954) The fetal membranes and placentation of the tropical American vampire bat, *Desmodus murinus murinus* with notes on the histochemistry of the placenta. *Acta Anatomica*, **21**, 285–341.

Wimsatt, W. A. (1958) The allantoic placental barrier in Chiroptera: a new concept of its organization and histochemistry. *Acta Anatomica*, **32**, 141–86.

Wimsatt, W. A. (1975) Some comparative aspects of implantation. *Biology of Reproduction*, **12**, 1–40.

Wimsatt, W. A. (1980) Personal communication on *Natalus*. See Mossman (1987).

Wimsatt, W. A. & Enders, A. C. (1980) Structure and morphogenesis of the uterus, placenta, and preplacental organs of the neotropical disc-winged bat, *Thyroptera tricolor spix* (Microchiroptera-Thyropteridae). *American Journal of Anatomy*, **159**, 209–43.

4

Brain ontogeny and ecomorphology in bats

INTRODUCTION

From a neuroethological perspective, the spectacular diversity in behavioral ecology among chiropterans is suggestive of a similar richness in brain organization. Dramatic evidence that this is the case is exemplified by the evolution of expanded auditory processing centers subserving echolocation in microchiropterans, specialized visual pathways in some megachiropterans, and novel development of olfactory structures in the Phyllostomidae and Desmodontinae (Henson 1970; McDaniel 1976; Baron *et al.* 1996c). The overwhelming majority of investigations concerning chiropteran neurobiology have been carried out on adult animals. However, the large number of extant chiropteran taxa represents fruitful ground for studies aimed at understanding the developmental processes underlying mammalian brain evolution. As discussed in this chapter, such data are of great significance with respect to two major issues: bat monophyly and the differential (mosaic) evolution of brain regions in response to co-evolutionary selection pressures.

In this chapter, we first examine chiropteran brains from a regional perspective. We then discuss the role that patterns of neuronal connections have played in discussions concerning bat evolution, with an emphasis on the importance of developmental data to this debate. New data are presented on layer VII of the cerebral cortex, as an example of an adult trait whose variation has specific developmental connotations. Finally, we present a brief account of the relationship of brain traits to behavioral ecology.

CHIROPTERAN BRAIN DEVELOPMENT

The wide range of diversity among chiropteran brains may be appreciated most simply by reference to surface views illustrating the range of development of sulci and fissures in the cerebrum and cerebellum, and the relative sizes of various cranial nerves (Fig. 4.1A–S). In addition to allometric effects of brain size, these variations also convey some idea of the range of specializations in neural machinery suited to particular niches, and remind us that adult morphology represents the endpoint of dynamic developmental processes.

Several chiropteran brain atlases and monographs concerning adult bat brains exist (Schneider 1957, 1966, 1972; Mann 1963; Henson 1970; McDaniel 1976). By far, the most comprehensive and carefully executed study of adult bat brains (372 species) has been performed by Baron *et al.* (1996*a*, *b*, *c*). However, there have been very few developmental studies of the bat central nervous system. An early general study of the fetal brain of *Vesperugo (Nyctalus) noctula* (15 embryos) was made by Werkman (1913). Koike (1924) described several major organ systems during development including the brain, hypophysis, and the vomeronasal organ in the vespertilionid *Scotophilus temmincki* (23 embryos, 1.25–15.2 mm crown–rump length), and the time course of appearance of brain features is summarized in Table 4.1. The development of the vomeronasal organ system was also described in *Rousettus leschenaulti* by Bhatnagar *et al.* (1996). Brown's pioneering studies (1962, 1966, 1967, 1968, 1980, 1987) involved the development of the hippocampal formation and amygdaloid complex in *Tadarida brasiliensis (= mexicana)*, and the development of nervus terminalis in *T. brasiliensis* and *Myotis lucifugus*. Humphrey (1936, 1966) and Crosby & Humphrey (1939) reported in detail on the telencephalon of *Tadarida*. Misek (1989, 1990, 1991) studied the developing neocortex of *Myotis myotis*, while Sterba (1990) described surface features in the embryos, newborns, and juveniles of *Myotis myotis* and *Miniopterus schreibersii*. Pirlot & Bernier (1974, 1991) reported on the development of ten brain components in the Jamaican fruit bat (*Artibeus jamaicensis*; also see Table 4.5). Aside from these, there are only fragmentary reports on adult bat brains in the older literature, and these are not pertinent to the present discussion. In the subsections below we present developmental data for particular brain regions.

Nervus terminalis

The nervus terminalis is a system of nerves sometimes designated Cranial Nerve 0. They are associated with the olfactory nerves (CN I) and are considered to be entirely sensory in function with recent evidence suggesting a role in pheromonal detection in some mammals (Oelschläger 1988; Schwanzel-Fukuda *et al.* 1992). Brown (1980, 1987)

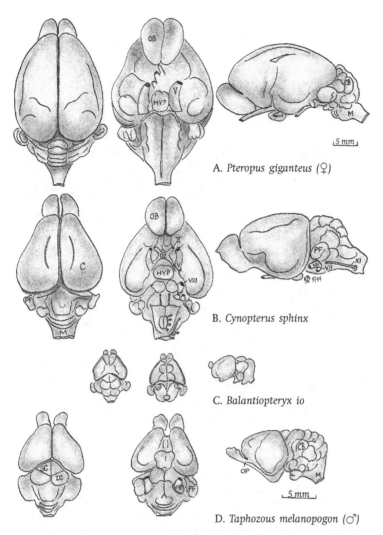

A. *Pteropus giganteus* (♀)

B. *Cynopterus sphinx*

C. *Balantiopteryx io*

D. *Taphozous melanopogon* (♂)

Figure 4.1. Camera lucida tracings of adult brains drawn from carefully dissected specimens. Gender is indicated when known. Except for *Pteropus giganteus*, brains are drawn at the same scale, magnified about 4.5× (scale bar = 5 mm). Dorsal, ventral, and lateral views are shown, depicting various structures for comparison. Abbreviations: C – cerebrum; CB – cerebellum; HYP – hypophysis; IC – inferior colliculus (tectum); M – medulla oblongata; OB – olfactory bulb; OC – optic chiasma; OP – olfactory peduncle; PF – paraflocculus (lobus petrosus); SC – superior colliculus (tectum); II – optic nerve; V – trigeminal nerve; VIII – vestibulo-cochlear nerve; XI – spinal accessory nerve. Anatomical material is from the same species as those published in Bhatnagar & Kallen (1974).

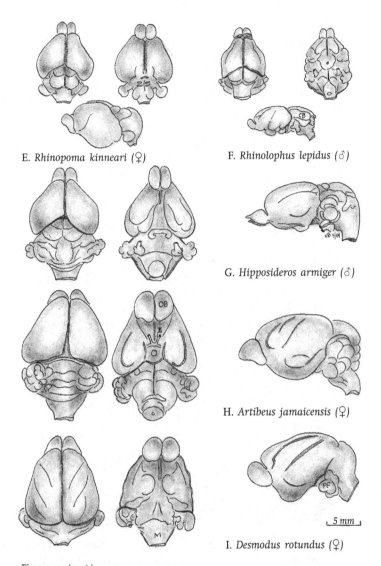

E. *Rhinopoma kinneari* (♀)

F. *Rhinolophus lepidus* (♂)

G. *Hipposideros armiger* (♂)

H. *Artibeus jamaicensis* (♀)

5 mm

I. *Desmodus rotundus* (♀)

Figure 4.1 (*cont.*)
A) *Pteropus giganteus* (female) (courtesy of Dr. K. B. Karim, Nagpur), B) *Cynopterus sphinx* (female), C) *Balantiopteryx io*, D) *Taphozous melanopogon* (male), E) *Rhinopoma kinneari* (female), F) *Rhinolophus lepidus* (male), G) *Hippposideros armiger* (male), H) *Artibeus jamaicensis* (female), I) *Desmodus rotundus* (female)

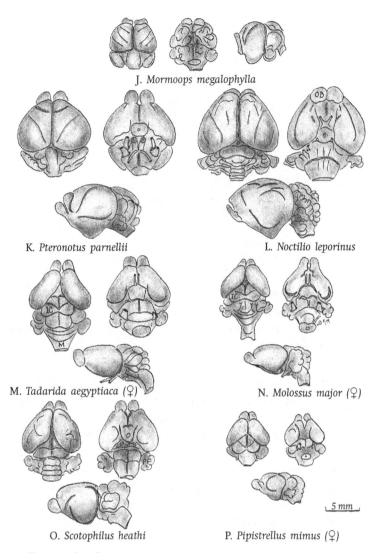

J. *Mormoops megalophylla*

K. *Pteronotus parnellii*

L. *Noctilio leporinus*

M. *Tadarida aegyptiaca* (♀)

N. *Molossus major* (♀)

5 mm

O. *Scotophilus heathi*

P. *Pipistrellus mimus* (♀)

Figure 4.1 (*cont.*)

J) *Mormoops megalophylla*, K) *Pteronotus parnellii*, L) *Noctilio leporinus*, M) *Tadarida aegyptiaca* (female), N) *Molossus major* (female), O) *Scotophilus heathi*, P) *Pipistrellus mimus* (female)

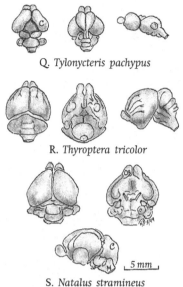

Q. *Tylonycteris pachypus*

R. *Thyroptera tricolor*

S. *Natalus stramineus*

Figure 4.1 (*cont.*)

Q) *Tylonycteris pachypus* (courtesy of Dr. D. A. Bhiwgade, Bombay), R) *Thyroptera tricolor*, S) *Natalus stramineus*.

reported that in the insectivorous bats *Myotis lucifugus* and *Tadarida brasiliensis* the nervus terminalis was clearly present before birth, but could not be identified in adults. However, the use of immunocytochemical techniques in recent years (Oelschläger 1988; Oelschläger & Northcutt 1992) has demonstrated the persistence of the nervus terminalis in both embryos and adult microchiropterans (*Myotis myotis*, *M. bocagei*, *Miniopterus* spp., *Eptesicus fuscus*, *Rhinolophus mehelyi*, *R. blasii*, *Hipposideros caffer*), and reportedly in one species of megabat (*Pteropus* spp.) (unpublished observations cited in Oelschläger & Northcutt 1992).

Vomeronasal organ

The vomeronasal organ is a chemosensory structure located in the anteroventral nasal septum of phyllostomid bats (for exceptions, see Bhatnagar & Meisami 1998). Its functions are not known in bats, but it has been shown to be an organ contributing to the mediation of sexual behavior in hamsters (Winans & Powers 1974). Of interest, epithelium of the rudimentary vomeronasal organ is found in many chiropteran and primate lineages. However, its presence or absence (Fig. 4.2) does not necessarily match the taxonomic patterns drawn from existing cladistic

Table 4.1. *Developing brain components in embryos of* Scotophilus temmincki *(Vespertilionidae)*

Embryo crown–rump length, mm	Event/First appearance of structure
1.2	Endoderm stage
2.0	Ectoderm stage
2.4	Medullary tube
2.7	Medullary plate
2.8	Neuromere, fourth ventricle, spinal ganglion
3.2	Glossopharyngeal, vagal ganglia
3.4	Brain begins to form
3.4	Five neuromeres, hypophysis
3.6	Ventricle forms; trigeminal and acousticofacial ganglia
4.6	Brain divisions, semilunar ganglion
4.6	22 spinal ganglia, 13 spinal nerves
5.2	Cerebellar commissure, hypophyseal infundibulum
5.4	Branchial groove organ
6.8	Sympathetic cord
7.2	Choroid plexus
7.2	Vomeronasal organ (Jacobson's organ) absent
7.4	Epiphyseal anlagen absent, vomeronasal organ present
9.0	Ventricles well developed
9.0	Vomeronasal organ and sympathetic cord well developed
9.4	Epiphysis present
10.2	Choroid plexus well developed
13.0	Epiphysis is solid, posterior and habenular commissures present
13.0	Olfactory lobe of vomeronasal organ well developed
15.2	Epiphysis well developed

Source: After Koike 1924.

analyses (Figs. 4.3, 4.4; Wible & Bhatnagar 1996; Bhatnagar & Meisami 1998; Meisami & Bhatnagar 1998).

A curious phenomenon concerning the vomeronasal organ in bats is of note. The central terminal of the vomeronasal organ, the accessory olfactory bulb, develops transiently but then disappears in some megadermatid and hipposiderid species (personal observations). In other bat species, where a rudimentary vomeronasal organ persists into adulthood, an accessory olfactory bulb has not been recorded (Cooper & Bhatnagar 1976; Bhatnagar 1980; Bhatnagar & Meisami 1998). An exception to this

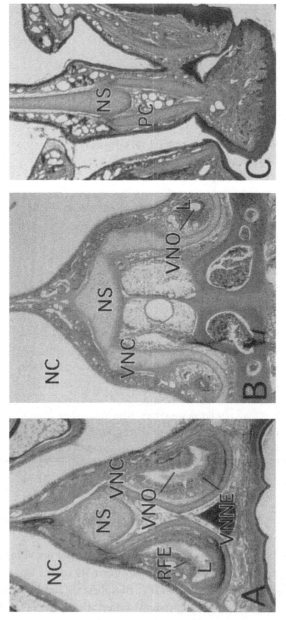

Figure 4.2. Coronal sections through the anteroventral nasal septal region in three representative species of adult bats. Note the well-developed vomeronasal organ (VNO) in *Diaemus youngii*, male (A), rudimentary stage in *Hipposideros lankadiva*, female (B), and the complete absence of the VNO in *Eptesicus fuscus*, female (C). L – lumen of the VNO, NC – nasal cavity, NS – nasal septum, PC – paraseptal cartilage, RFE – receptor-free epithelium, VNC – vomeronasal cartilage, VNNE – vomeronasal neuroepithelium.

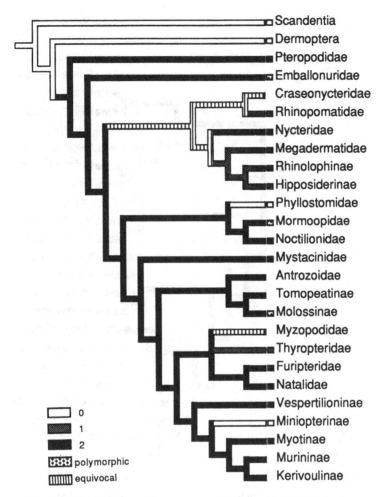

Figure 4.3. Distribution of the vomeronasal organ epithelial tube in comparison with a recent interfamilial classification tree for bats (Simmons 1998). Character states: 0 – tube is well developed, i.e., exhibits neuroepithelium; 1 – tube is rudimentary, i.e., does not exhibit neuroepithelium; 2 – tube is absent. The box to the immediate left of each taxon indicates the character state of that group. States are unknown for taxa without boxes. Scandentia and Dermoptera are presented as outgroups. The consensus tree (after Simmons 1998) is from Wible & Bhatnagar (1996) with permission.

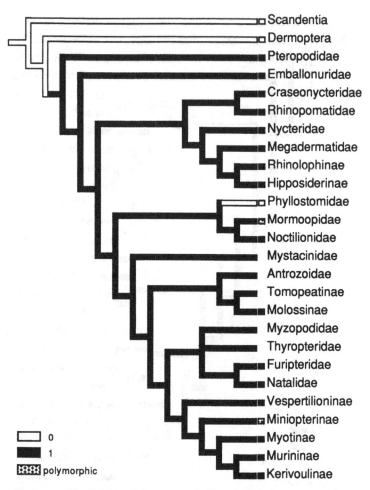

Figure 4.4. Distribution of the accessory olfactory bulb (0 – presence of the accessory olfactory bulb, 1 – absence of the accessory olfactory bulb). The consensus tree (after Simmons 1998) is from Wible & Bhatnagar (1996) with permission.

generalization was reported by Koike (1924) who described the major organ systems in the vespertilionid *Scotophilus temmincki* (23 embryos, 1.25–15.20 mm crown–rump length). In early embryonic stages (1.25–7.20 mm crown–rump length), the vomeronasal organ was not observed; however, the organ was reported as being well-developed in later stages (7.40–15.20 mm crown–rump length; Table 4.1).

The occurrence of accessory olfactory bulbs only in the family Phyllostomidae remains an enigma (Bhatnagar & Meisami 1998; Meisami &

Bhatnagar 1998), but may indicate a comparatively greater use of pheromonal communication in this group. Without the benefit of direct physiological analysis, many comparative brain studies proceed on the assumption that the size of a brain sub-component correlates directly with its physiological capacity. Therefore, statements such as 'olfaction, as indicated by the size of the main olfactory bulbs and the paleocortex, is the most variable sensory system in Chiroptera' and 'the finest nose of all bats is not that of a pteropodid, but that of the spear-nosed bat *Phylloderma stenops*' (Baron *et al.* 1996c: 1083), must be viewed with caution. Bigger may be better, but it is also possible that in many cases size reductions driven by selection pressures may leave information processing capacity unaffected. There are virtually no data which address this issue directly for any taxa.

Amygdala

The amygdala is a subcortical component of the telencephalon, positioned between the basal ganglia and cerebral cortex. It modulates visceral functions, emotional behavior, fear and rage reactions, sexual activity, and memory. Damage to the amygdala produces tameness, hypersexuality, and tendencies in experimental subjects to put all objects encountered in their mouth. The amygdala is reciprocally connected to the hypothalamus, thalamus, and portions of the cerebral cortex and brainstem. It receives input from several cortical areas related to the limbic system: the orbitofrontal, infralimbic, prelimbic, anterior cingulate, insular, entorhinal, perirhinal, and subicular areas (Aggleton *et al.* 1980; Russchen 1982). The amygdala sends two main efferent projections to other brain regions. The stria terminalis projection innervates the bed nucleus of the stria terminalis, ventral striatum, hypothalamus, and septal area. The ventral amygdalofugal pathway projects to the hypothalamus, medial nucleus of the thalamus, and striatum. Cortical projections from the amygdala reach widespread areas, including frontal, cingulate, orbitofrontal, insular, and parahippocampal fields (Krettek &Price 1977).

The amygdala receives major afferent inputs from the main olfactory and accessory olfactory (vomeronasal) systems, and lesser inputs from other sensory systems. Each of the component nuclei of the amygdala has a specific set of neuronal connections and functional attributes. For example, the accessory olfactory bulb projects to the medial and posterior cortical nuclei, whereas the main olfactory bulb projects to the anterior cortical and posterolateral cortical nuclei (Winans & Scalia 1970; Broadwell 1975; Scalia & Winans 1975). The amygdaloid nuclei related to the main olfactory and vomeronasal systems have different efferent projections, thereby maintaining the segregated nature of information related to these two chemosensory systems. A principal output of the vomeronasal part of the amygdala is to hypothalamic areas which regulate reproductive behavior (Kevetter & Winans 1981). These findings raise interesting questions regarding the organization of the amygdala in bats lacking vomeronasal organs and accessory olfactory bulbs (suborder Megachiroptera), compared to those having them (family Phyllostomidae) (see Figs. 4.3, 4.4). For example, are the medial and posterior cortical nuclei reduced in size in the megachiropterans, reflecting a lack of vomeronasal input, or do they have expanded connections with non-vomeronasal systems? Although Baron *et al.* (1996*a*) reported amygdala volumes for a wide variety of bats, the relative sizes of the subnuclei and patterns of fiber connections are not known for chiropterans.

The central amygdaloid nucleus receives gustatory and visceral sensory input from the solitary and parabrachial nuclei of the brainstem, and is involved in visceral and cardiovascular regulation. The basolateral amygdala is primarily interconnected with the cerebral cortex and projects to the striatum. It receives highly processed input from the auditory, visual, and somatic sensory systems, and is involved in creating an affective dimension to non-olfactory sensory experience and motor behavior (LeDoux 1996). Nothing is known about these connections in bats.

A descriptive account of the development of the amygdaloid complex in fetal insectivorous bats (*Tadarida brasiliensis* and *Myotis lucifugus*) was given by Brown (1967).

Cerebral cortex

Due to its range of variation, the bat telencephalon is a valuable resource for studying neocortical evolution. The neocortex, together with olfactory bulb, hippocampus, schizocortex, and septum pellucidum are more elaborate in frugivorous bats than in insectivorous microbats. As discussed in a later section below, this may be related to spatial processing and memory functions related to feeding strategy (Eisenberg & Wilson 1978). Additionally, there are major differences between mega- and microchiropteran brains, particularly in reference to the relatively larger telencephalon of megachiropterans, and the greater development of the medulla, cerebellum, and midbrain in microchiropterans (Tables 4.2–4.4).

Misek (1989, 1990) described prenatal neocortical development in *Myotis myotis* using the following sequence: 1) beginning of neocortex (27 days in gestation), 2) cortical plate layers I–VI (36 days), 3) increase of thickness of cortical plate (43 days), and 4) differentiation of layers (50

Table 4.2. *Average percentages for the five main regions of the brain of 47 megachiropteran and 225 microchiropteran species; the values for human brain are provided for comparison*

	Megachiroptera		Microchiroptera		Human
	%	Range	%	Range	%
Medulla oblongata	7.1	(5.3–8.9)	13.1	(7.8–19.7)	0.8
Cerebellum	13.5	(11.5–16.0)	19.5	(15.3–28.9)	11.0
Mesencephalon	5.4	(4.2–6.9)	9.8	(5.5–14.6)	0.6
Diencephalon	8.6	(7.9–9.4)	8.0	(6.7–10.6)	2.7
Telencephalon	65.3	(60.1–70.2)	49.6	(37.7–63.4)	85.0

Source: After Baron *et al.* 1996*a*.

days of gestation through 20 days ante-partum). During this period only layers I, V, and VI appear, whereas layers II–IV are not differentiated. Considering the general six-layered organization of mammalian cerebral cortex, Misek asked whether in fact there are fewer layers in the neocortex of *Myotis*; perhaps certain layers merge later in development. It was determined that certain layers of the neocortex simply do not appear until quite late in development and continue to differentiate after birth in *Myotis myotis* (Misek 1990). Therefore, based on the poor and late development of layer II in a variety of vespertilionid species (*Myotis myotis, Nyctalus leisleri, Pipistrellus pipistrellus,* and *Plecotus auritus*), Misek and others consider the chiropteran neocortex to be primitive (Kurepina 1966). Sterba (1990) described surface features in the embryos, newborns, and juveniles of *M. myotis* and *Miniopterus schreibersii,* noting the slow rate of prenatal brain development. Pirlot & Bernier (1974) described embryonic brain growth in the frugivore *Artibeus jamaicensis* (Table 4.5; Pirlot & Bernier 1974, 1991), noting that in contrast to an aerial insectivore (*Molossus major*), *Artibeus* has a proportionately larger cerebrum and a smaller cerebellum and brainstem.

The hippocampal formation develops from the medial wall of the cerebral cortex, and is involved in spatial functions and memory. It is quite large in frugivorous and nectarivorous megachiropteran and microchiropteran bats (Baron *et al.* 1996*a, b*). The hippocampus is much smaller in species with lesser degrees of overall brain size, and generally parallels the trends for telencephalon as a whole (compare Figs. 138, 154 of Baron *et al.* 1996*a*). Reporting on the early development of hippocampal formation of *Tadarida brasiliensis,* Humphrey (1966) concluded that the 'sequence of

Table 4.3. *Brain characteristics of Megachiroptera and Microchiroptera*

	Suborder Megachiroptera (Fig. 4.1A, B)	Suborder Microchiroptera (Fig. 4.1C–S)
Absolute size	Large brains, 480–10 450 mg	Small brains, 68–2587 mg
Relative size (encephalization quotient)	0.95–1.37	0.52–1.23
Olfactory ventricle	Variably developed	Traces in a few species
Olfactory bulb	Large, consistent with frugivory	Generally small, but larger in frugi-nectarivorous species
Accessory olfactory bulb	Absent	Found primarily in Phyllostomidae
Cranial nerves	Twelve plus nervus terminalis	Twelve plus nervus terminalis; vomeronasal nerve in Phyllostomidae and a few other species
I Olfactory	Large number of nerves; entire cribriform plate perforate	Small number of nerves generally; cribriform plate perforate dorsally
II Optic	Visual system well developed; large nerves diverge widely (except *Pteropus, Dobsonia*); chiasm is x-shaped	Insectivorous microchiropteran optic nerves are thin, small and remain uncrossed; the optic chiasm with the optic tracts form a straight transverse band and with the optic nerves these three structures appear as ⊥⊥-shaped (see Fig. 4.1H)
III Oculomotor	Largest of those innervating eye muscles	Relatively large
IV Trochlear	Difficult to trace	Very small

V Trigeminal	Well developed	Well developed
VI Abducens	Difficult to trace	Very small
VII Facial	Small	Very small
VIII Vestibulocochlear	Small; no echolocatory abilities (except tongue-clicking in *Rousettus*)	Large; echolocation highly developed, cochlear division is extramedullary (*Hipposideros*); cochlear nuclei often reach the internal auditory meatus; ventral cochlear nuclei project over the medulla; trapezoid body is large
IX Glossopharyngeal	Not large	Not large; some peculiarities (see Henson 1970)
X Vagus	Large; recurrent laryngeal nerve is lacking in *Hypsignathus* (but this needs confirmation)	Large
XI Accessory	Considerable in size	Considerable in size
XII Hypoglossal	Moderate size	Moderate size; huge in *Desmodus* (consistent with control of fine tongue movements)
Spinal cord	Variable; all three funiculi are well developed; dorsal roots enter the cord laterally	Henson (1970) considered the microchiropteran spinal cord the most primitive among mammals; substantia gelatinosa is massive; posterior funiculus is very small; dorsal roots enter medially; spinal ganglia enormous
Medulla oblongata	Narrow; the medulla–spinal cord transition is gradual	Broad (supporting auditory structures); medulla–spinal cord transition is abrupt
Pons	Well developed	Small; compressed ventrally by the hypophysis and laterally by the trigeminal nerve in some species

Table 4.3 (cont.)

	Suborder Megachiroptera (Fig. 4.1A, B)	Suborder Microchiroptera (Fig. 4.1C–S)
Cerebellum	Complex (Henson 1970) with deep fissures; paraflocculus dorsalis not apparent	Smooth surface; paraflocculus dorsalis is prominent
Cerebral peduncles	Massive	Poorly developed
Midbrain tectum	Covered by cerebral hemispheres and cerebellum; superior and inferior colliculi well developed; dorsal cochlear nuclei are larger than ventral nuclei	Desmodontines and some phyllostomatids are similar to the megachiropteran condition; in others the colliculi are exposed; superior colliculi poorly developed in insectivorous taxa (except for *Saccopteryx*); inferior colliculi massive, ventral cochlear nuclei are larger than dorsal
Subcommissural organ	Well developed	Poorly developed
Pineal gland	Generally large; type AB only (Bhatnagar et al. 1986, 1990)	Generally small; mostly type AB, but other types seen as well (Bhatnagar et al. 1986, 1990)
Hypophyseal fossa	Variable in shape and depth	Variable
Hypothalamus	Well developed	Very complicated
Telencephalon	Large, fissured hemispheres; large corpus callosum	Small, smooth hemispheres; correspondingly small caudate putamen; medium-size corpus callosum

Paleocortex (olfactory)	Occupies most of the basal telencephalon	Occupies the basal telencephalon posterior and posterolateral hemispheric cortices
Neocortex (isocortex)	Well developed; visual cortex covers entire posterior pole	Poorly devloped in insectivores but better in *Artibeus* and *Desmodus*; (echolocation ability is independent of auditory cortex: Suga 1964)
Pyramidal system	Well developed	Poorly developed
Falx cerebri	Well developed; does not extend to tentorium cerebelli	Absent (Schneider 1957) but this needs confirmation

Table 4.4. *Comparison of microchiropteran brain features. Classification follows Koopman (1993) amended by Simmons (1998)*

	Cerebrum	Colliculi (tectum)	Paraflocculus (lobus petrosus)	Pineal	Size of olfactory bulb	Accessory olfactory bulb	Optic nerves	Comments
Emballonuridae		exposed	in few spp. only	*	small, without sulcus cinguli		small	Fig. 4.1C, D. *Epiphysis is not visible in *Taphozous*
Rhinopomatidae	low hemispheres	exposed	not obvious		medium			Fig. 4.1E. Pons covered by hypophysis, large dorsal cochlear nucleus
Craseonycteridae							small, divergent	Similar to non-*Taphozous* emballonurids
Nycteridae		exposed	exposed, very large		small, disc-like, no lumen		small, parallel	
Megadermatidae	large neocortex*	exposed	exposed, large**				small, parallel	*Largest neocortex of all carnivorous bats; ** Except *Cardioderma*/dorsal cochlear nucleus is enormous
Rhinolophinae		exposed	exposed, large*	visible**	long, medium**		small, divergent	Fig. 4.1F. *Except *Ascellisus*, **Rhinolophus luctus, R. trifoliatus*
Hipposiderinae		exposed	exposed, large	visible*	long, medium**		small, divergent	Fig. 4.1G. *Hipposideros commersoni, H. lankadiwa, Asellia tridens*; similar to Rhinolophinae

Mystacinidae								No data
Phyllostomidae	highly variable*		present**		large	present	small, divergent	Fig. 4.1H. Sagittal, anterior rhinal and hippocampal sulci well developed. *Glossophagines have long flat brains **Except *Brachyphylla* (McDaniel 1976)
Desmodontinae	long, high	covered	not exposed		medium*	present	medium	Fig. 4.1I. *Contradicts Baron *et al.* (1996a)
Mormoopidae	very short	partly covered	not exposed		round or flat	present*	small, divergent	Figs. 4.1J, K. *Present in *Pteronotus parnellii*, *P. gymnonotus*
Noctilionidae	short, wide, high	exposed	partly visible		small, flat		small, divergent	Fig. 4.1L. Medial geniculate well developed
Antrozoidae								No data
Molossidae	short, wide, flat	exposed		often visible	thick crista galli, lack lumen		small, parallel	Figs. 4.1M, N.
Vespertilioninae	highly variable							Figs. 4.1O–Q. Highly variable in structure
Miniopterinae	short, high		not exposed		small, no lumen	present	small, divergent	Rostral and sylvian sulci prominent, separate Miniopterines from other vespertilionids
Murininae	highly variable	exposed			small			

Table 4.4. (cont.)

	Cerebrum	Colliculi (tectum)	Paraflocculus (lobus petrosus)	Pineal	Size of olfactory bulb	Accessory olfactory bulb	Optic nerves	Comments
Myotinae			finger-like projection*		large, lack lumen			*Finger-like caudal projection large in *Harpiocephalus*
Kerivoulinae	short, high	partly covered*	projection**		small, lack lumen		very small	*Fully exposed in *Kerivoula* **Finger-like caudo-medial projection
Myzopodidae								No data
Thyropteridae								Fig. 4.1R. No data.
Furipteridae		mostly exposed		large	small, disc-like		very small	
Natalidae	short	mostly covered	not exposed, large flocculus				very small	Fig. 4.1S. Small hemispheres without fissures

Table 4.5. *Quantitative data on brain growth in* Artibeus *and* Molossus

	Artibeus jamaicensis	*Molossus major*
Family	Phyllostomidae	Molossidae
Diet	Fruit	Insects
Gestation	120 days without delay; 210 days with delay	89 days (*M. ater*)
Number of stages examined	7 + adult	5 + adult
Embryo weights as percent of birth weight	10.9, 15.5, 31.4, 58.8, 60.0, 71.6, 100 (birth), 375 (adult)	3.3, 5.2, 5.4, 7.5, 100 (birth), 255 (adult)
Adult body weight[a]	41 gm	13 gm
Adult brain weight[a]	1016 mg	267 mg
Relative size of brain parts as a percentage of whole brain	Embryos A1–A7, Adult	Embryos M1–M5, Adult
Telencephalon + Neocortex	48.0–63.0, 56.0	26.0–43.0, 38.0
Neocortex alone	23.0–34.0, 28.0	11.0–23.0, 18.5
Diencephalon	9.0–12.0, 9.0	8.0–15.0, 8.3
Cerebellum	7.0–18.0, 18.0	16.0–32.0, 19.0
Brain stem	8.0–34.0, 18.0	30.0–42.0, 35.0

Note: [a] Baron *et al.* (1996a).
Source: After Pirlot & Bernier (1991).

development of the hippocampal formation in insectivorous bats and in humans is similar; e.g. in both forms the cornu ammonis, after appearance, develops more rapidly in the early stages and gyrus dentatus appears later.' However, humans and bats differ in that the initial cell layer does not form in the primordial gyrus dentatus in bats, rather, a cell concentration appears within the pyramidal cell layer of the cornu ammonis. Brown (1966, 1967, 1980, 1987) found that all layers of the cornu ammonis were identifiable in a 12-mm embryo of *T. brasiliensis*.

Midbrain – corpora quadrigemina

The midbrain lies between the pons and the diencephalon. Its posterior surface (tectum) presents four elevations – the corpora quadrigemina.

The rostral pair of elevations are the superior colliculi (SC), and the more caudal pair are the inferior colliculi (IC). The SC are laminated structures and show considerable morphological variation among bats. Retinal input from the eye projects to the SC and is integrated with a variety of inputs from the cerebral cortex and other regions to generate head and eye orientation movements. The IC are the largest subcortical auditory structures in bats. They receive projections from all auditory nuclei except the medial nucleus of the trapezoid body and medial geniculate body (Zook & Casseday 1982, 1985, 1987). The IC in bats is unique in that all ascending auditory fibers synapse on neurons within the IC, thus making it an obligatory synaptic relay in the auditory system unlike other mammals where it is non-obligatory. This obligatory nature of the IC of Chiroptera reflects a highly integrated response to auditory stimuli while bats are flying and foraging at night.

Consistent with its known functions, the SC is proportionately larger in the highly visual megachiropterans (27 species) than in the echolocating microchiropterans (123 species), whether its volume is considered as a percentage of the midbrain as a whole or of the midbrain tectum. Likewise, the IC is proportionately larger in the microchiropterans, with very little overlap between microchiropterans and megachiropterans (Baron et al. 1996a). Larsen et al. (1984) compared the IC in three species of bats and found distinct differences which they could correlate with echolocating behavior. Cytoarchitectonically, the IC of Eptesicus fuscus is densely packed with small neurons. Rousettus leschenaulti has large, widely distributed neurons, whereas the IC of Cynopterus sphinx is sparsely populated with both small and large neurons.

Cerebellum

The cerebellum is chiefly concerned with the maintenance of muscle tone and the fine control of muscular activity during locomotion. It is composed of several sub-components including a central vermis that is flanked by paired cerebellar hemispheres. Posterior and ventral to these hemispheres are small appendages that have distinct gross morphologies that are of some taxonomic and perhaps functional significance to chiropteran evolution – the flocculus and paraflocculus dorsalis. The paraflocculus dorsalis has afferent connections from the cerebellar cortex and is involved in the coordination of somatic motor activity, balance, and muscle tone. In contrast to the Insectivora, in bats the paraflocculus is stalked. Ecoethological correlates to its development and morphology are worthy of future investigation.

Cerebellar size indices overlap considerably among the micro-chiropterans and megachiropterans (Baron et al. 1996a). However, the largest relative sizes are found in nectarivorous and frugivorus micro-chiropterans, and the lowest indices in aerial insectivores, particularly the vespertilionids.

Spinal cord

The microchiropteran spinal cord is one of the shortest among mammals, with the conus medullaris (terminus of cord) located between the eighth and twelfth thoracic vertebrae (Artibeus jamaicensis and Pteronotus parnellii, respectively). By comparison, the cord terminates at the level of the first lumbar vertebra in humans. Accordingly, the chiropteran cauda equina is relatively long. Consistent with adaptations for flight (wing membranes, musculature), the spinal cord exhibits a large cervical enlargement and the dorsal cervical roots have remarkably large ganglia (Pettigrew et al. 1989). In contrast, the lumbar enlargement is unimpressive, and the dorsal funiculus is small and deeply situated.

Comparative brain development

Neuroethological observations of neonatal and juvenile bats could be used to map out the patterns in which fiber tracts, nuclear groups, and other supporting nerve tissue have evolved (see Pirlot 1987). However, at present we are limited to the comparative studies of quantitative and qualitative brain growth and differentiation made by Pirlot & Bernier (1974, 1991; see Table 4.5). These studies utilized developmental series of Artibeus jamaicensis (7 embryos, 1 adult) and Molossus major (5 embryos, 1 adult). From their qualitative data the authors concluded that 'the telencephalon, diencephalon, and cerebellum grow faster in Artibeus, whereas the brainstem maintains a greater relative volume throughout' in Molossus. It should be emphasized that these data are drawn from small series; preserved brains are highly shrinkable, and there is considerable variation among individuals with regards to brain weight or volume (Baron et al. 1996a). Considering these limitations, the conclusions drawn by Pirlot & Bernier warrant further research. However, conclusions derived from morphology in their study suggest that after an initial delay in development of the cerebellum, mitral cell layer of the olfactory bulb, caudate nucleus and layer II of the neocortex, growth rates advance later on in development in Artibeus. As expected, fewer mitral cells were reported (no numbers are provided) in Molossus than in the frugivorous and larger *Artibeus*. The overall density of the neocortical neurons in the hippocampal region exhibited a greater density in *Molossus* than in *Artibeus*, especially in neocortical layer II.

To date, the best examples of comparative neural development among chiropterans come from the echolocation literature and detail the development of the cochlea and cochlear nerves (Poljak 1926; see Vater, this volume). Comparative analyses of bat brains are further complicated when differences in gestation length are taken into account (44–240 days, *Pipistrellus pipistrellus* and *Desmodus rotundus* respectively). Since newborn bats may be altricial, intermediate, or precocial, one can expect a full range of neonatal brain development. Certainly, comparisons of neonatal brains among Insectivora, Chiroptera, Primates, and Dermoptera would be of great evolutionary interest; however, little work has been done on this topic even within a single group.

NEURONAL CONNECTIONS AND THE CHIROPTERAN MONOPHYLY DEBATE

Pettigrew's (1986) report of primate-like retinotectal connections in megachiropterans, based upon data from *Pteropus poliocephalus*, *P. alecto*, and *P. scapulatus*, and his further development of the 'flying primate' hypothesis using other neural and non-neural characters (Pettigrew *et al.* 1989) renewed the debate on bat monophyly, which soon revolved around familiar issues of character independence, character weighting, synapomorphies vs. homoplasies, and the likelihood of reversals (Baker *et al.* 1991; Pettigrew 1991). The startling nature of Pettigrew's initial report on retinotectal connections was tempered somewhat by the findings of Thiele *et al.* (1991), who discovered that *Rousettus egyptiacus* exhibits the general mammalian condition whereby some, but not all, of the temporal retinal ganglion cells project to the ipsilateral superior colliculus. Together, these data may be taken to suggest that the genus *Pteropus* has independently evolved the primate pattern (convergence). Alternatively (though less parsimonious) it is possible that all megachiropterans initially had the primate pattern, but that *Rousettus* has secondarily developed the general mammalian arrangement (Thiele *et al.* 1991). Baker *et al.* (1991) favor an even stronger version of this hypothesis, in which an original homology of visual brain traits among Chiroptera and primates was followed by reversal in Microchiroptera (and, presumably, *Rousettus*) – Pettigrew's 'blind cave bat' scenario. However, as Pettigrew (1991) has argued, this scenario does not explain the divergence between megachiropterans and microchiropterans for a substantial number of other non-visual nervous system traits.

Because the discourse has focused so heavily upon the use of adult characters and the issue of chiropteran monophyly, there has been little attention paid to the significance of developmental data (see Adams & Thibault, this volume). It seems probable that many of the visual system characters identified by Pettigrew *et al.* (1989) are not independent. Currently, outgroup analysis is the only method available to decide this issue when adult characters are used. In contrast, developmental studies offer the possibility of illuminating the processes involved, rather than simply focusing on the endpoints of those processes – the adult characters. The advantage of studying dynamic processes is that a much richer basis is obtained upon which to make determinations of character independence. For example, one may distinguish between characters which are linked in causal chains versus being members of separate quasi-independent chains (Raff 1996). However, developmental studies have their own set of pitfalls and limitations, and are tedious to perform.

The retinotectal system represents perhaps the most tractable opportunity to investigate developmental events underlying neural characters which have a central bearing on the bat phylogeny debate. If different developmental events are operative in establishing ipsilateral projections from the temporal retina to the superior colliculus (e.g., sorting of retinal cells destined to project ipsilaterally) in pteropodids than in primates, one could argue that the difference in developmental processes was of greater significance than the similarity between their outcomes (the adult character). In the usual parlance, it would constitute evidence in favor of convergent or parallel homoplasy rather than homology. If similar developmental patterns emerged, the issue would remain unresolved, because it could represent homology or parallel homoplasy via activation of a similar developmental chain of events selected for the same local optimum on a fitness landscape (Kauffman 1993; Striedter 1998). The use of developmental events as characters solves one problem and creates at least one other. On the one hand, we potentially gain a tremendous amount of resolution in considering dynamic events (ontogenetic trajectories) as characters rather than simply relying on morphological endpoints. However, in another sense the issues of character weighting get worse because of this added complexity. There is also a familiar difficulty – that of evaluating the relative weight of developmental characters. For example, do we give more weight to the identities of genes or the trajectories of axon bundles?

By similar reasoning, it is important to know whether the pattern seen in *Rousettus egyptiacus* develops in a manner comparable to that in most non-primate mammals. Studies such as these go to the heart of the

possibility that because of developmental constraints in tightly integrated gene complexes, the same transformations are likely to occur
more than once in phylogeny, as in vertebrate electroreception (New
1997; Northcutt 1997). Investigations of wing development in megachiropterans and microchiropterans will be similarly informative in the
debate over whether chiropteran wings evolved once or twice.

Do retinotectal and other visual pathways represent an example of
phyletic constraint (Raff 1996), wherein the megabat genome was pre-
adapted to respond to visual selection pressures by re-inventing the same
or similar visual pathways as those that evolved in primates? Such constraints would represent a bias that leads to rediscovery of the same local
optimum. By applying experimental techniques to the development of
megabat visual pathways, it should be possible to determine the degree to
which the traits enumerated by Pettigrew et al. (1989) are linked to each
other.

Bats exhibit a unique specialization of the somatosensory system
compared to all other mammalian taxa examined. In most mammals the
cortical representation of the body surface is organized as a topographic
map with the head region located laterally, the hindlimb medially, and
the ventral surface and digits pointing rostrally. However, in the three
chiropterans examined so far (*Pteropus poliocephalus*, *P. scapulatus*, and
Macroderma gigas), the forelimb digits are oriented caudally (Calford et al.
1985; Wise et al. 1986; Krubitzer & Calford 1992). Is this pattern related to
a quirk of phylogenetic history or does it have functional significance
with regard to the evolution of powered flight? The authors of these
studies favored the latter view, and suggested the possibility that use of
the forelimb during development influences the resulting topography
(Wise et al. 1986). They also pointed to the potential of developmental
studies for comparing the formation of topography and neural connections of this unique forelimb representation in the cerebral cortex with
that of other taxa, especially primates.

LAYER VII OF CEREBRAL CORTEX

One neural trait which varies among chiropteran taxa and has an
intriguing developmental history is layer VII of the cerebral cortex.
Among mammals as a whole, layer VII is most robust in rodents, where it
constitutes the deepest layer of the cortex, located directly superficial to
the underlying white matter (Reep & Goodwin 1988). Neurons in layer VII
participate in local and long-distance intrahemispheric corticocortical
connections (Divac et al. 1987; Vandevelde et al. 1996; Clancy & Cauller

1999), and the cell-sparse stratum immediately superficial to layer VII contains many of the corticocortical axons originating in layers III, V, and VII (Reep & Goodwin 1988; Vandevelde *et al.* 1996). The developmental history of layer VII is of significance because of its relationship to the subplate, an early-developing structure which helps establish orderly topography in thalamocortical and corticothalamic connections (Allendoerfer & Shatz 1994). Most subplate cells are transient and die before birth. However, in rats and mice, layer VII represents that portion of the subplate cells which persists throughout life (Valverde *et al.* 1995; Price *et al.* 1997). In cats and primates the subplate survivors become the interstitial cells of Cajal, scattered throughout the white matter (Chun & Shatz 1989; Kostovic & Rakic 1990), and these participate in connections with other neurons (Shering & Lowenstein 1994).

These observations raise the issue of how widespread persisting subplate cells are among various mammalian taxa, and why in some taxa they are coalesced into a distinct layer VII, whereas in others they are found in the white matter. Important variables would appear to be the time of onset of subplate development, duration of subplate existence prior to programmed cell death, and onset and duration of white matter development in relation to the subplate. One might predict that in taxa with a prolonged period of neurogenesis there is greater overlap in time between the subplate and the developing white matter. Thus, the subplate cells would become spread out as the white matter increases in thickness, and survivors would form an interstitial population rather than a distinct lamina. In contrast, in taxa with a brief time-course of cortical development the period of cotemporaneity between the pre-cell-death subplate and developing white matter could be so brief that subplate survivors are confined to a distinct lamina, layer VII.

The presence or absence of layer VII is readily identifiable on stained brain sections (Fig. 4.5). Adult brains in the Comparative Mammalian Brain Collection at Michigan State University and the University of Wisconsin were examined for this trait. Two specimens were assessed from the brain atlases in Baron *et al.* (1996a). A score of 3 was given to species having a well-defined layer VII, appearing as a thin, densely packed cell layer located immediately superficial to the subcortical white matter, and separated from layer VI by a distinct cell-sparse zone. A score of 2 was given in cases exhibiting a well-defined layer VII that was restricted in rostrocaudal extent, usually also having a less distinct overlying cell-sparse zone. A score of 1 represented cases with an indistinct or intermittent layer VII, often lacking a cell-sparse zone. A score of 0 was given in cases with no visible layer VII.

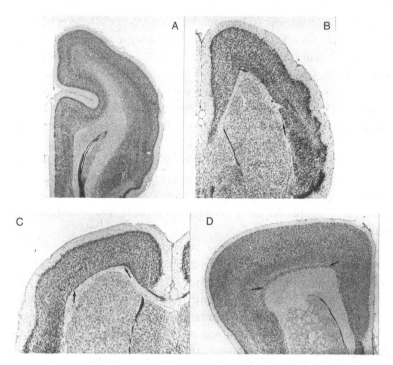

Figure 4.5. The four character states of cerebral cortical layer VII, as seen in coronal stained sections of selected taxa. A) State o, layer VII absent, *Pteropus giganteus*. B) State 1, layer VII indistinct or intermittent, *Rhinolophus hipposideros*. C) State 2, layer VII well-defined but restricted in extent, *Myotis lucifugus*. D) State 3, layer VII robust and extensive; cell-sparse zone present superficially, *Sciurus carolinensis* (gray squirrel). Arrows indicate layer VII.

Of 144 species from 24 orders, 43 species in 6 orders exhibit a layer VII. These include: among Rodentia, the suborder Sciurognathi but not the suborder Hystricognathi; all Insectivora examined; Paucituberculata (shrew opossums); Paramelemorphia (bandicoots); some Xenarthra, and some Chiroptera (classification scheme of Wilson & Reeder 1993). The sciurognath rodents examined all possess a prominent layer VII and distinct cell-sparse stratum, whereas members of the remaining five orders exhibit less consistency. The observed distribution of layer VII in this sample of mammalian taxa suggests that layer VII is a typical feature in some taxa, but is not present in most orders. It is known that primates and carnivores have interstitial cells rather than a layer VII; other taxa (e.g., ungulates) may as well.

The heterogeneous occurrence of layer VII seen in Chiroptera (Table 4.6) may indicate species-level differences in the timing of cortical devel-

Table 4.6. *Cortical layer VII in 18 species of the order Chiroptera (taxonomic scheme of Koopman, 1993)*

Family	Genus, species	Layer VII condition[a]
Megachiroptera		
Pteropodidae	*Pteropus giganteus*	0
	Rousettus amplexicaudatus	1
Microchiroptera		
Emballonuridae	*Saccopteryx bilineata*	0
Molossidae	*Molossus major*	0
Natalidae	*Natalus micropus*	0
Noctilionidae	*Noctilio labialis*	0
Phyllostomidae	*Brachyphylla cavernarum*	0
	Carollia perspicillata	0
	Monophyllus redmani	0
	Trachops cirrhosus	1
	Vampyrum spectrum	0
Rhinolophidae	*Rhinolophus ferrumequinum*	2
	Rhinolophus hipposideros	1
	Hipposideros armiger	0
Vespertilionidae	*Eptesicus fuscus*	2
	Myotis lucifugus	2
	Myotis myotis	2
	Myotis montivagus	2

Notes:
[a] 0 – layer VII absent
1 – layer VII indistinct or intermittent
2 – layer VII well defined but restricted in extent
3 – layer VII robust and extensive; cell sparse zone superficially

opment. These differences, in turn, are likely to be related to a number of variables including altriciality vs. precociality and associated early behavioral repertoires. In particular, it would be instructive to know how the dynamics of subplate development differ in phyllostomids (no layer VII) and vespertilionids (layer VII present). These two families have overlapping body size ranges (Nowak 1994) but the phyllostomids have larger absolute brain sizes, much higher encephalization quotients, and larger telencephalic and neocortical size indices than the vespertilionids (Baron *et al.* 1996a). In general they also have more prolonged gestation times than vespertilionids (~ 4 months vs. 2 months; Hayssen *et al.* 1993). These facts suggest that there is a more prolonged period of neurogenesis in the phyllostomids, resulting in the development of an interstitial population of subplate survivors rather than a subplate. Alternatively, there could be more extensive programmed cell death in this group, or a different time course to the establishment of neuronal connections. Answers to questions such as these will get at the heart of the processes represented by the adult character, layer VII, and will provide a richer understanding of its significance.

ONTOGENY AND PHYLOGENY

Few would now argue with Garstang's (1922) view that 'Ontogeny does not recapitulate phylogeny: it creates it' (cited in Northcutt 1990). However, central issues remain: to discover how ontogenetic changes are related to phylogenetic changes, to ascertain whether a single set of mechanisms is responsible for both types of change (Northcutt 1990), and to determine exactly how the dynamics of developmental processes can be integrated into systematic formulations (Alberch 1985). This involves examination of changes in the molecular dynamics of developmental events. Heterochrony, which includes changes in the timing or rate of developmental events, represents one means by which a mosaic of tightly integrated (constrained) developmental programs can be altered enough to produce a significantly different outcome without disrupting the integrity of the system as a whole (Gould 1992). Alberch (1985) highlighted some of the difficulties of interpreting ontogenetic sequences in a phylogenetic context. Of particular concern is the concept of sequences, in which developmental morphologies replace adult features as characters. He focused attention instead on the importance of initial conditions and dynamic modes of interaction; processes rather than resultant morphologies. McKinney & McNamara (1991) emphasized the role of dissociated heterochronies, those specific to particular features rather than to the whole body. Raff (1996) has likewise emphasized the view that development involves multiple independent chains of causally linked processes, providing a priori conditions for dissociation and thus, mosaic evolution. He also gives examples of processes which can lead to heterochronic outcomes without requiring timing changes. Together, these considerations suggest that specific systems (e.g., visual pathways) will exhibit a combination of developmental constraints and plasticity that defines the range of phenotypic space available to that taxon for that system, and that subtle changes in the timing of gene activity or the amount of product produced can significantly affect morphological outcomes. Striedter (1998) has outlined many of these issues with reference to neural development, in the context of epigenetic landscapes and attractors in state space, an approach elaborated by Kauffman (1993). The major issue is to identify what genetic or developmental changes occur in derived vs. ancestral phenotypes, and to discover how readily such transitions occur (Rice 1997).

An instructive example is the use of developmental data in the ongoing debate concerning the hypothesis that birds evolved from therapod dinosaurs. Burke & Feduccia (1997) addressed the issue of digit identity in birds by examining the positions of embryological precursors of the digits. They concluded that birds have the generic amniote mode of development, resulting in an adult digit identity of II–III–IV in contrast to the I–II–III pattern seen in dinosaurs. While this finding has not resolved the bird phylogeny issue, it does illustrate that a complete understanding of the significance of adult characters, and their usefulness in systematics, ultimately rests on an ontogenetic foundation.

Pettigrew *et al.* (1989) analyzed 24 neural characters, 18 of which were related to the visual system. Because it is quite possible that there are multiple pleiotropic effects among these characters during development of the visual system, it is probably misleading to consider them as independent characters in a phylogenetic analysis. Similar reasoning can also be applied to the use by others (Simmons 1998) of a variety of non-neural morphological characters. One way out of this dilemma is the use of empirical data derived from careful developmental studies. As these efforts proceed, more emphasis is likely to be placed on patterns of dynamic processes rather than on static endpoints (characters). We are sure to gain a richer appreciation of the range of possible ways different lineages have appropriated and altered developmental programs, and the ways in which these may interact. Then, species may come to be viewed as assemblages of multiple dynamic, interacting developmental programs as much as lineages of organisms.

BRAIN TRAITS AND BEHAVIORAL ECOLOGY

A relationship between brain size and feeding strategies among chiropterans was first established by Stephan & Pirlot (1970). In addition to noting that different brain regions appear to evolve at different rates, they found a general correlation between the relative size of the neocortex and feeding habits. Among 18 species from eight families, the nectarivorous and frugivorous phyllostomids had higher neocorticalization indices than insectivores (rhinolophids, hipposiderids, vespertilionids, and molossids), and three pteropids were higher still. This approach was later refined by Eisenberg & Wilson (1978) and applied to 225 species of bats from 14 families. Using Wilson's (1973) classification for feeding strategies, they found that frugivores of the family Pteropodidae and phyllostomid subfamily Stenoderminae, and nectarivores of the Pteropodidae and the phyllostomid subfamilies Glossophaginae and Phyllostominae have consistently larger relative brain sizes than aerial insectivores in the families Mormoopidae, Emballonuridae, Vespertilionidae, and Molossidae. Chiropteran taxa classified as foliage gleaners, carnivores, piscivores, or sanguivores generally had cranial capacities intermediate between these extremes.

Eisenberg & Wilson (1978) suggested that large relative brain size is associated with chiropteran taxa whose food resources are distributed in isolated pockets of space, often unpredictable with regard to their temporal distribution. This relationship is typified by the subfamily Stenoderminae within the Phyllostomidae. The stenodermines were characterized by Bonaccorso (1979) as canopy frugivores because they feed on fruit which grows on trees, vines, and epiphytes in the upper level of the canopy. This requires intensive exploration, which is likely to involve computationally intensive brain processes such as spatial memory. As a group, the stenodermines have the largest overall brain size of any subfamily within the Phyllostomidae (Baron et al. 1996c). Eisenberg & Wilson (1978) suggested that small relative brain size in aerial insectivores may have evolved in response to selection pressures to reduce weight to facilitate active flight, or because echolocation does not require large neocortical volume. In addition to feeding strategies, other behavioral ecological variables which may influence relative brain size include flight performance and wing loading, and roosting preferences (Findley & Wilson 1982).

The correlation of large relative brain size with feeding strategies that involve intensified spatial memory is not novel to the order Chiroptera. Birds that engage in food-storing behavior have been shown to possess extremely well-developed spatial memory (Vander Wall 1982; Balda & Kamil 1989), and this is correlated with relatively large size of the hippocampal formation (Sherry & Healy 1998), a major component of the cerebral hemispheres. Furthermore, experimental lesions of the hippocampus impair food retrieval (Sherry & Healy 1998). Because the cerebral cortex comprises the bulk of brain mass in mammals (Hofman 1989; Finlay & Darlington 1995), the findings of Eisenberg & Wilson (1978) using overall brain size are congruent with those of Stephan & Pirlot (1970), for which neocortical size was used. The bird findings further imply that an analysis like that of Eisenberg & Wilson (1978) on whole brain size in relation to feeding strategies may be profitably extended to a consideration of the sizes of various brain components in relation to a suite of behavioral, ecological, and life history variables. We suspect that this will be done most fruitfully when brain component structures are grouped by functional systems (e.g., auditory, visual, spatial, or motor) rather than as major divisions (e.g., medulla, pons, cerebellum).

Baron *et al.* (1996*a*, *b*, *c*) have provided an extensive account of the relative sizes of whole brains and numerous brain regions in Chiroptera, and this is also the most thorough such compilation for any mammalian order. They identified several broad trends associating behavioral ecology with brain traits, as summarized below. However, these are perhaps best viewed as qualified speculations requiring multivariate quantitative analysis as suggested above.

Aerial insectivory

Aerial insectivores are relatively fast fliers but have relatively small brains, i.e., neocortex, striatum, hippocampus, and the vestibular nuclei. In rhinopomatids and molossids, the nucleus fasicularis gracilis that concerns proprioreceptive information is very large.

Moderate and slow-flying bats

Great variation in brain size and the relative size of the various brain components reflects the behavioral heterogeneity in this somewhat artificial group. In most cases, the brains, vestibular complex, and acoustic nuclei are small when compared to generalized microchiropteran values (e.g., *Pipistrellus*, *Miniopterus*, and *Saccopteryx*).

Constant frequency (CF) bats

The inferior colliculus is large and the main olfactory bulb is deviated in both the Rhinolophidae and Hipposideridae. The neocortex is relatively small in both of these families. The hippocampus is large in rhinolophids and average in hipposiderids.

Trawlers

Fishing bats (*Noctilio*) fly close to the ground over water surfaces. They have large brains and a small hippocampus. The nucleus gracilis is large and receives somatosensory input from the hind body, wing membranes, and joints.

Gleaners

Together, gleaners exhibit well-developed visual (superior colliculi) and olfactory centers. The inferior colliculi are enormous as is the sensory nucleus of the trigeminal nucleus (e.g., *Antrozous, Euderma, Craseonycteris, Nycteris, Macroderma, Trachops, Vampyrum*, and several *Myotis*).

Sanguinivores

These have short brains, with long and high hemispheres, small rostral sulcus, and olfactory bulbs of medium to large size; there is a well-developed vomeronasal organ (e.g., *Desmodus, Diaemus*, and *Diphylla*).

Nectarivores

Given that many of these highly specialized animals are able to hover, it is no surprise that their cerebellar development is the greatest among the Chiroptera (e.g., *Leptonycteris, Choeronycteris, Anoura*, and *Glossophaga*).

Frugivores

Frugivores have large olfactory bulbs and well-developed visual systems. Whereas megachiropteran frugivores lack both sonar and an accessory olfactory system, the microchiropteran frugivores (Phyllostomidae) possess both.

Tight-crevice roosting species

Several chiropteran species that roost in rock crevices or other narrow restricted spaces (e.g., bamboo internodal segments) exhibit flattened skulls with correspondingly 'flattened' brains (e.g., *Platymops setiger, Rhinopoma* spp., *Pipistrellus mimus, Eudiscopus denticulatus, Taphozous* spp., *Tylonycteris* spp.). Certainly, brain evolution and brain development did not drive the observed gross morphological distortion of these skulls, rather, skull form has constrained the shape of the brain (see Pedersen, this volume). It would be of some interest to investigate how the brain has accommodated these spatial restrictions in the extreme forms, i.e., *Platymops* or *Tylonycteris*.

In the principal method of analysis used by Baron *et al.* (1996a) the volumes of specified chiropteran brain regions are expressed as 'progression indices,' in reference to 'basal insectivores' (tenrecs) as representative of a presumed unspecialized ancestral condition. However, it is unlikely that any species can be considered unspecialized, and some tenrecs are known to employ a form of echolocation (Eisenberg 1981), surely an exemplary neurobiological specialization. This approach was originally formulated out of a desire to assess 'primitive' versus 'advanced' brain traits among Chiroptera, Insectivora, and Primates (see Baron *et al.* 1996*a*: 27). However, the most recent accounts by this group of investigators (Baron *et al.* 1996*a, b, c*) place more emphasis on variations due to ecological specializations like echolocation and feeding habits, though they continue to make reference to progression indices. On the other hand, raw volumetric data are provided in this exhaustive account, so it is possible for anyone who wishes to perform other kinds of analyses on this data set to do so.

Relative overall brain size is influenced by many variables including gestation length, maternal investment, metabolic rate, and behavioral ecology (Harvey & Krebs 1990; Barton *et al.* 1995). Therefore, in searching for meaningful brain–behavior correlates, many investigators have focused on systems within the brain whose functional characteristics are known; for example, the various sensory systems. As Baron *et al.* (1996*c*: 1196) put it, 'The strong interrelation of morphologic traits, echolocating abilities and behavior with environmental factors implies that brain structure is essentially an ecological variable.' This viewpoint naturally raises the question of the possible range of differential expansion of particular functional systems and brain regions, and to what degree developmental constraints oppose differential expansions. Finlay & Darlington (1995) and Finlay *et al.* (1998) pointed to the importance of understanding developmental dynamics, and the possibility that developmental/evolutionary changes in one brain region might necessitate changes in other regions simply because of linkages (constraints) in the developmental programs. They found that anatomically defined brain subregions (e.g., medulla, diencephalon, etc.) scale with ordinal-specific allometry according to changes in overall brain size, in primates, bats, and insectivores. In addition, they provided experimental evidence that structures which become disproportionately large have longer periods of neurogenesis. Based upon these findings, they suggested that selection for any specific behavioral capacity might necessitate volumetric increase not only in the neural systems subserving that function, but also in the brain as a whole. Interestingly, an exception was the olfactory bulb, which scaled more independently of overall brain size, suggesting that different results may be obtained when regional brain volumetrics are analyzed by functional systems rather than by major anatomical divisions.

Just such an approach was taken by Barton *et al.* (1995) using a different analytic approach to essentially the same data set (see also van Essen *et al.* 1992). They analyzed the visual and olfactory systems of insectivores, bats, and primates from the perspective that differences in brain structure are related to ecological selection pressures. They concluded that diet, habitat, and diurnality each had an effect on the evolutionary radiation of mammalian sensory systems. Specifically, after eliminating brain size allometry, they found that the relationship between visual and olfactory centers was negatively correlated in insectivores and primates, but positively correlated in Chiroptera. They suggested that the negative correlations in primates may be indicative of a trade-off whereby reduction in the olfactory system was necessary to achieve gains in the visual system. Associations with diet were also noted; nocturnal frugivorous primates possess enlarged olfactory structures, whereas diurnal frugivorous primates have larger geniculostriate visual systems. In Chiroptera, correlations between olfactory and visual components are positive, and these sensory systems are relatively larger (compared to the rest of the brain) in frugivorous lineages within the Pteropodidae and Phyllostomidae. Barton *et al.* (1995) suggested that in Chiroptera the evolutionary trade-off may have been between vision plus olfaction, versus echolocation.

Another approach to the general issue of differential expansion of sensory specializations was taken by Johnson & Kirsch (1993). In order to obtain an easily measured index of auditory versus visual specialization, they compared the volumes of the inferior colliculus (IC) (auditory) and superior colliculus (SC) (visual) in a wide variety of mammalian taxa, thereby elaborating upon an approach first used by Pettigrew *et al.* (1989). They found that IC/(IC + SC) ratios range from 0.15–0.79 and correlate well with known behavioral specializations, including cetacean echolocation and primate vision. As mentioned in an earlier section above, Baron *et al.* (1996a) noted that as a group, the visually specialized megachiropterans have proportionately larger superior colliculi than do the echolocating microchiropterans, and that the microchiropterans have relatively expanded inferior colliculi. More recently, Johnson and co-workers (1998) reported IC/(IC + SC) ratios in 161 chiropteran species, including 152 for which raw data were obtained from Table 25 of Baron *et al.* (1996a). Among Megachiroptera, the 28 species examined have ratios of 0.22–0.42, consistent with their role as visually dominant frugivores. Sixteen species of frugivorous Microchiroptera have ratios of 0.54–0.67, suggesting a balance between the auditory and visual modalities. Insectivorous

Microchiroptera (88 species) have ratios of 0.63–0.78, indicative of special-izations related to echolocation. Although it is known that the superior colliculus is an integrative center not limited strictly to the visual modal-ity (Dean *et al.* 1989; Thiele *et al.* 1996), this analysis, like that of Barton *et al.* (1995), represents a significant advance over whole-brain correlations of brain–behavior relationships. It should be noted that the range of IC/(IC + SC) ratios (0.15–0.79) reported by Johnson *et al.* (1998) is greater than the factor of 2.5 mentioned by Finlay & Darlington (1995) as encompassing the range of variation explained by their two major factors on principal component analysis. This gives quantitative weight to the suggestion that differential expansion of structures may occur independent of changes in overall brain size, and thus that different functional systems can evolve somewhat independently.

All of the above approaches are limited by the fact that the signifi-cance of volumetric change is quite unclear. It is true that neuron number generally increases as volume enlarges (see Finlay *et al.* 1998), but this is not always the case, as exemplified by the findings of Campos & Welker (1976) in cerebral cortex. The extensiveness of dendritic branching and concomi-tant changes in synaptic density are probably more pertinent outcomes of volumetric increase, from the viewpoint of information processing capac-ity. However, these traits are tedious to document, in the extreme.

The most carefully investigated sensory system of bats is clearly that of echolocation, with many detailed studies having been made of the ontogeny and timing of echolocation (see treatments by Vater and by Jones, this volume). However, very little is known concerning the ontog-eny of brain regions and neural pathways involved in sonar production and reception (Brown & Grinell 1980). Neonatal bats range from altricial to precocial levels of maturation (Gould 1974, 1975; Eisenberg 1981). We would predict that the relative degree of maturation in brain regions, nuclear groups, laryngeal musculature, etc. concerned with sonar devel-opment would follow in kind, thereby affecting the timing of the first appearance of structural calls made by the neonate.

SUMMARY

This chapter has emphasized the need for systematic investigations of neuroontogeny and neuroethology in bats. Data from such studies per-tain to issues involving the relationship of development to evolutionary change. The large number and broad diversity of chiropteran taxa offers considerable unexplored territory in this regard. Consideration of devel-opmental events will expand our appreciation of the meaning of adult characters by focusing on the dynamics of their construction rather than viewing them simply as static endpoints, and will also clarify the extent to which epigenetic processes are operative in creating convergent adult characters.

A second major area of interest is the relationship between behavioral ecology and brain traits. The approach we envision is the integration of information on behavior and life history traits with a functionally oriented analysis of brain component volumetry. The diversity of behavioral ecology and life history traits among the Chiroptera leads us to expect that such an analysis would represent the most wide-ranging synthesis of brain–behavior data for any mammalian order, and would thereby serve to delineate the range of evolutionary potential in mammalian nervous system organization. Developmental studies could reveal the dynamics operative in differential expansion of particular functional systems, as well as illuminating the developmental constraints involved. Much remains to be learned.

ACKNOWLEDGMENTS

The authors greatly appreciate the contributions of Scott Pedersen to the development of this chapter. We also have a debt of gratitude to Susan Hodge for carefully typesetting the numerous initial drafts of this manuscript. Thanks go to Jack Johnson for encouragement and suggestions, and for the use of the Pirlot brain collection, now housed at Michigan State University. Our gratitude goes to Wally Welker for kindly allowing usage of the Comparative Mammalian Brain Collection at the University of Wisconsin. We also appreciate the hospitality of John Seyjagat, Director of The Lubee Foundation, Inc. The comments of two anonymous reviewers were helpful.

REFERENCES

Aggleton, J. P., Burton, M. J. & Passingham, R. E. (1980) Cortical and subcortical afferents to the amygdala of the rhesus monkey (Macaca mulatta). Brain Research, 190, 347–68.
Alberch, P. (1985) Problems with the interpretation of developmental sequences. Systematic Zoology, 34, 46–58.
Allendoerfer, K. L. & Shatz, C. J. (1994) The subplate, a transient neocortical structure: its role in the development of connections between thalamus and cortex. Annual Review of Neuroscience, 17, 185–218.
Baker, R. J., Novacek, M. J. & Simmons, N. B. (1991) On the monophyly of bats. Systematic Zoology, 40, 216–31.

Balda, R. P. & Kamil, A. C. (1989) A comparative study of cache recovery by three corvid species. *Animal Behavior*, **38**, 486–95.

Baron, G., Stephan, H. & Frahm, H. D. (1996a) *Comparative Neurobiology in Chiroptera*, vol. 1, *Macromorphology, Brain Structures, Tables and Atlases*. Basel: Birkhäuser Verlag.

Baron, G., Stephan, H. & Frahm, H. D. (1996b) *Comparative Neurobiology in Chiroptera*, vol. 2, *Brain Characteristics in Taxonomic Units*. Basel: Birkhäuser Verlag.

Baron, G., Stephan, H. & Frahm, H. D. (1996c) *Comparative Neurobiology in Chiroptera*, vol. 3, *Brain Characteristics in Functional Systems, Ecoethological Adaptation, Adaptive Radiation and Evolution*. Basel: Birkhäuser Verlag.

Barton, R. A., Purvis, A. & Harvey, P. H. (1995) Evolutionary radiation of visual and olfactory brain systems in primates, bats, and insectivores. *Philosophical Transactions of the Royal Society of London B*, **348**, 381–92.

Bhatnagar, K. P. (1980) The chiropteran vomeronasal organ: its relevance to the phylogeny of bats. In *Proceedings of the Fifth International Bat Research Conference*, ed. D. E. Wilson & A. L. Gardner, pp. 289–315. Lubbock: Texas Tech Press.

Bhatnagar, K. P. & Kallen, F. C. (1974) Cribriform plate of ethmoid, olfactory bulb and olfactory acuity in forty species of bats. *Journal of Morphology*, **142**, 71–89.

Bhatnagar, K. P. & Meisami, E. (1998) The vomeronasal organ in bats and primates: extremes of structural variability and its phylogenetic implications. *Microscopy Research and Technique*, **43**, 465–75.

Bhatnagar, K. P., Frahm, H. D. & Stephan, H. (1986) The pineal organ of bats: a comparative morphological and volumetric investigation. *Journal of Anatomy*, **147**, 143–61.

Bhatnagar, K. P., Frahm, H. D. & Stephan, H. (1990) The megachiropteran pineal organ: a comparative morphological and volumetric investigation with special emphasis on the remarkably large pineal of *Dobsonia praedatrix*. *Journal of Anatomy*, **168**, 143–66.

Bhatnagar, K. P., Wible, J. R. & Karim, K. B. (1996) Development of the vomeronasal organ in *Rousettus leschenaulti*. *Journal of Anatomy*, **188**, 129–35.

Bonaccorso, F. J. (1979) Foraging and reproductive ecology in a Panamanian bat community. *Bulletin of Florida State Museum, Biological Science*, **24**, 359–408.

Broadwell, R. D. (1975) Olfactory relationships of the telencephalon and diencephalon in the rabbit. I. An autoradiographic study of the efferent connections of the main and accessory olfactory bulbs. *Journal of Comparative Neurology*, **163**, 329–45.

Brown, J. W. (1962) The embryonic development of the olfactory formation in the bat. *Anatomical Record*, **145**, 211–12 (abstract).

Brown, J. W. (1966) Some aspects of the early development of the hippocampal formation in certain insectivorous bats. In *Evolution of the Forebrain*, ed. R. Hassler & H. Stephan, pp. 92–103. Stuttgart: Thieme Press.

Brown, J. W. (1967) The development of the amygdaloid complex in insectivorous bat embryos. *Alabama Journal of Medical Sciences*, **4**, 399–415.

Brown, J. W. (1968) Some features of certain fiber pathways of the telencephalon of insectivorous bats. *Anatomical Record*, **160**, 321–2.

Brown, J. W. (1980) Developmental history of nervus terminalis in embryos of insectivorous bats. *Anatomical Record*, **196**, 23–35.

Brown, J. W. (1987) The nervus terminalis in insectivorous bat embryos and notes on its presence during human ontogeny. *Annals of New York Academy of Sciences*, **519**, 184–200.

Brown, P. E. & Grinell, A. D. (1980) Echolocation ontogeny in bats. In *Animal Sonar Systems, Proceedings NATO Advanced Study Institute*, ed. R. G. Busnel & J. F. Fish, pp. 355–77. New York: Plenum Press.

Burke, A. C. & Feduccia, A. (1997) Developmental patterns and the identification of homologies in the avian hand. *Science*, **278**, 666-8.

Calford, M. B., Graydon, M. L., Huerta, M. F., Kaas, J. H. & Pettigrew, J. D. (1985) A variant of the mammalian somatotopic map in a bat. *Nature*, **313**, 477-9.

Campos, G. B. & Welker, W. I. (1976) Comparisons between brains of a large and a small hystricomorph rodent: capybara, *Hydrochoerus* and guinea pig, *Cavia*; neocortical projectrion regions and measurements of brain subdivisions. *Brain, Behavior and Evolution*, **13**, 243-66.

Chun, J. J. M. & Shatz, C. J. (1989) Interstitial cells of the adult neocortical white matter are the remnant of the early generated subplate neuron population. *Journal of Comparative Neurology*, **282**, 555-69.

Clancy, B. & Cauller, L. J. (1999) Widespread projections from subgriseal neurons (layer VII) to layer I in adult rat cortex. *Journal of Comparative Neurology*, **407**, 275-86.

Cooper, J. G. & Bhatnagar, K. P. (1976) Comparative anatomy of the vomeronasal organ complex in bats. *Journal of Anatomy*, **122**, 571-601.

Crosby, E. C. & Humphrey, T. (1939) Studies of the vertebrate telencephalon I. The nuclear configuration of the olfactory and accessory olfactory formations and of the nucleus olfactorius anterior of certain reptiles, birds, and mammals. *Journal of Comparative Neurology*, **71**, 121-213.

Dean, P., Redgrave, P. & Westby, G. W. M. (1989) Event or emergency? Two response systems in the mammalian superior colliculus. *Trends in Neuroscience*, **12**, 137-47.

Divac I., Marinkovic, S., Mogensen, J., Schwerdtfeger, W. & Regidor, J. (1987) Vertical ascending connections in the isocortex. *Anatomy and Embryology*, **175**, 443-55.

Eisenberg, J. F. (1981) *The Mammalian Radiations*. Chicago: University of Chicago Press.

Eisenberg, J. F. & Wilson, D. E. (1978) Relative brain size and feeding strategies in the Chiroptera. *Evolution*, **32**, 740-51.

Findley, J. S. & Wilson, D. E. (1982) Ecological significance of chiropteran morphology. In *Ecology of Bats*, ed. T. H. Kunz, pp. 243-60. New York: Plenum Press.

Finlay, B. L. & Darlington, R. B. (1995) Linked regularities in the development and evolution of mammalian brains. *Science*, **268**, 1578-84.

Finlay, B. L., Hersman, M. N. & Darlington, R. B. (1998) Patterns of vertebrate neurogenesis and the paths of vertebrate evolution. *Brain, Behavior and Evolution*, **52**, 232-42.

Garstang, W. (1922) The theory of recapitulation: a critical restatement of the biogenic law. *Zoological Journal of the Linnaean Society of London*, **35**, 81-101.

Gould, E. (1974) Experimental studies of the ontogeny of ultrasonic vocalizations in bats. *Developmental Psychobiology*, **8**, 333-46.

Gould, E. (1975) Neonatal vocalizations in bats of eight genera. *Journal of Mammalogy*, **56**, 15-29.

Gould, S. J. (1992) Ontogeny and phylogeny - revisited and reunited. *BioEssays*, **14**, 275-9.

Harvey, P. H. & Krebs, J. R. (1990) Comparing brains. *Science*, **249**, 140-9.

Hayssen V., van Tienhoven, A. & van Tienhoven, A. (1993) *Asdell's Patterns of Mammalian Reproduction*. Ithaca: Cornell University Press.

Henson, O. W. (1970) The central nervous system. In *Biology of Bats*, vol. 2, ed. W. A. Wimsatt, pp. 57-152. New York: Academic Press.

Hofman, M. A. (1989) On the evolution and geometry of the brain in mammals. *Progress in Neurobiology*, **32**, 137-58.

Humphrey, T. (1936) The telencephalon of the bat. I. The non-cortical nuclear masses and certain pertinent fiber connections. *Journal of Comparative Neurology*, **65**, 603-711.

Humphrey, T. (1966) The development of the human hippocampal formation correlated with some aspects of its phylogenetic history. In *Evolution of the Forebrain*, ed. R. Hassler & H. Stephan, pp. 104–16. Stuttgart: Thieme Press.

Johnson, J. I. & Kirsch, J. A. W. (1993) Phylogeny through brain traits: interordinal relationships among mammals including primates and chiroptera. In *Primates and their Relatives in Phylogenetic Perspective*, ed. R. D. E. MacPhee, pp. 293–331. New York: Plenum Press.

Johnson, J. I., Osborn, C. E., Morris, J. A., Sheppard, S. E. & Gorayski, P. M. (1998) A neuroanatomical measure of visual and auditory dominance in behavioral guidance. *American Zoologist*, **38**, 58A.

Kauffman, S. A. (1993) *The Origins of Order: Self-Organization and Selection in Evolution*. Oxford: Oxford University Press.

Kevetter, G. A. & Winans, S. S. (1981) Connections of the corticomedial amygdala in the golden hamster. I. Efferents of the 'vomeronasal amygdala.' *Journal of Comparative Neurology*, **197**, 81–98.

Koike, K. (1924) Die Herausbildung der äusseren Körperform und der Entwicklungsgrad der Organe bei einer javanischen Kleinfledermaus (*Scotophilus temmincki*, Hoesfield). *Zeitschrift für Anatomie Entwicklungsgeschichte*, **72**, 510–41.

Koopman, K. F. (1993) Order Chiroptera. In *Mammal Species of the World: A Taxonomic and Geographical Reference*, 2nd edn, ed. D. E. Wilson & D. M. Reeder, pp. 137–241. Washington: Smithsonian Institution Press.

Kostovic, I. & Rakic, P. (1990) Developmental history of the transient subplate zone in the visual and somatosensory cortex of the macaque monkey and human brain. *Journal of Comparative Neurology*, **297**, 441–70.

Krettek, J. E. & Price, J. L. (1977) Projections from the amygdaloid complex to the cerebral cortex and thalamus in the rat and cat. *Journal of Comparative Neurology*, **172**, 687–722.

Krubitzer, L. A. & Calford, M. B. (1992) Five topographically organized fields in the somatosensory cortex of the flying fox: microelectrode maps, myeloarchitecture, and cortical modules. *Journal of Comparative Neurology*, **317**, 1–30.

Kurepina, M. M. (1966) Anatomical peculiarities of the cerebral cortex in ecologically different bats. *Vsesojuz, s'jezda anatomov, gistologov I embryologov*, **6**, 136.

Larsen, S. A., Bhatnagar, K. P. & Cox, R. L. (1984) Comparative cytoarchitectonic investigation of the inferior colliculus of *Eptesicus*, *Rousettus*, and *Cynopterus*. *Bat Research News*, **25**, 46.

LeDoux, J. (1996) *The Emotional Brain*. New York: Simon & Schuster.

Mann, G. (1963) Phylogeny and cortical evolution in the Chiroptera. *Evolution*, **17**, 589–91.

McDaniel, V. R. (1976) Brain Anatomy. In *Biology of Bats of the New World Family Phyllostomatidae*, Special Publications of the Texas Tech University Museum No. 13, ed. R. J. Baker, J. K. Jones & D. C. Carter, pp. 147–200. Lubbock: Texas Tech Press.

McKinney, M. L. & McNamara, K. J. (1991) *Heterochrony: The Evolution of Ontogeny*. New York: Plenum Press.

Meisami, E. & Bhatnagar, K. P. (1998) Structure and diversity in mammalian accessory olfactory bulb. *Microscopy, Research and Technique*, **43**, 476–99.

Misek, I. (1989) Prenatal development of the neocortex of the large mouse-eared bat (*Myotis myotis*, Microchiroptera). *Folia Zoologica*, **38**, 339–47.

Misek, I. (1990) Morphogenesis of the mammalian neocortex. II. Prenatal development of the neocortex in selected mammals. *Acta Scientiarum Naturalium Academiae Scientarium Bohemoslovacae Brno*, **24**, 1–40.

Misek, I. (1991) Neocortical lamination in area postcentralis communis in selected small mammals. *Folia Zoologica*, **40**, 367–75.

New, J. G. (1997) The evolution of vertebrate electrosensory systems. *Brain, Behavior and Evolution*, **50**, 244–52.

Northcutt, R. G. (1990) Ontogeny and phylogeny: a re-evaluation of conceptual relationships and some applications. *Brain, Behavior and Evolution*, **36**, 116–40.

Northcutt, R. G. (1997) Evolution of gnathostome lateral line ontogenies. *Brain, Behavior and Evolution*, **50**, 25–37.

Nowak, R. M. (1994) *Walker's Bats of the World*. Baltimore: Johns Hopkins University Press.

Oelschläger, H. A. (1988) Persistence of the nervus terminalis in adult bats: A morphological and phylogenetical approach. *Brain, Behavior and Evolution*, **32**, 330–9.

Oelschläger, H. A. & Northcutt, R. G. (1992) Immunocytochemical localization of luteinizing hormone-releasing hormone (LHRH) in the nervus terminalis and brain of the big brown bat, *Eptesicus fuscus*. *Journal of Comparative Neurology*, **315**, 433–63.

Pettigrew, J. D. (1986) Flying primates? Megabats have the advanced pathway from eye to midbrain. *Science*, **231**, 1304–6.

Pettigrew, J. D. (1991) Wings or brain? Convergent evolution in the origins of bats. *Systematic Zoology*, **40**, 199–216.

Pettigrew, J. D., Jamieson, B. G. M., Robson, S. K., Hall, L. S., McNally, K. I. & Cooper, H. M. (1989) Phylogenetic relations between microbats, megabats and primates (Mammalia: Chiroptera and Primates). *Philosophical Transactions of the Royal Society of London B*, **325**, 489–559.

Pirlot, P. (1987) Contemporary brain morphology in ecological and ethological perspectives. *Journal für Hirnforschung*, **28**, 145–211.

Pirlot, P. & Bernier, R. (1974) Embryonic brain growth in a fruit bat. *Anatomy and Embryology*, **146**, 193–208.

Pirlot, P. & Bernier, R. (1991) Brain growth and differentiation in two fetal bats: qualitative and quantitative aspects. *American Journal of Anatomy*, **190**, 167–81.

Poljak, S. (1926) The connections of the acoustic nerve. *Journal of Anatomy*, **60**, 465–9.

Price, D. J., Aslam, S., Tasker, L. & Gillies, K. (1997) Fates of the earliest generated cells in the developing murine neocortex. *Journal of Comparative Neurology*, **377**, 414–22.

Raff, R. A. (1996) *The Shape of Life*. Chicago: University of Chicago Press.

Reep, R. L. & Goodwin, G. S. (1988) Layer VII of rodent cerebral cortex. *Neuroscience Letters*, **90**, 15–20.

Rice, S. H. (1997) The analysis of ontogenetic trajectories: when a change in size or shape is not heterochrony. *Proceedings of the National Academy of Sciences, U.S.A.*, **94**, 907–12.

Russchen, F. T. (1982) Amygdalopetal projections in the cat. I. Cortical afferent connections. A study with retrograde and anterograde tracing techniques. *Journal of Comparative Neurology*, **206**, 159–79.

Scalia, F. & Winans, S. S. (1975) The differential projections of the olfactory bulb and accessory olfactory bulb in mammals. *Journal of Comparative Neurology*, **161**, 31–55.

Schneider, R. (1957) Morphologische Untersuchungen am Gehirn der Chiropteren (Mammalia). *Abhandlungen der senckenbergischen Naturforschenden Gesellschaft*, **495**, 1–92.

Schneider, R. (1966) Das Gehirn von *Rousettus aegyptiacus* (E. Geoffroy 1810) (Megachiroptera, Chiroptera, Mammalia). Ein mit Hilfe mehrerer schnittserien erstellter Atlas. *Abhandlungen der senckenbergischen Naturforschenden Gesellschaft*, **513**, 1–166.

Schneider, R. (1972) Zur quantitativen Morphologie und Evolution des Fledermausgehirns. *Anatomischer Anzeiger*, **130**, 332–46.

Schwanzel-Fukuda, M., Zheng, L. M., Bergen, H., Weesner, G. & Pfaff, D. W. (1992) LHRH-neurons: functions and development. *Progress in Brain Research*, **93**, 189–203.

Shering, A. F. & Lowenstein, P. R. (1994) Neocortex provides direct synaptic input to interstitial neurons of the intermediate zone of kittens and white matter of adult cats: a light and electron microscopic study. *Journal of Comparative Neurology*, **347**, 433–43.

Sherry, D. & Healy, S. (1998) *Spatial Representation in Animals*. Oxford: Oxford University Press.

Simmons, N. B. (1998) A reappraisal of interfamily relationships of bats. In *Bat Biology and Conservation*, ed. T. H. Kunz & P. A. Racey, pp. 3–26. Washington: Smithsonian Institution Press.

Stephan, H. & Pirlot, P. (1970) Volumetric comparisons of brain structures in bats. *Zeitschrift für zoologische Systematik und Evolutionsforschung*, **8**, 200–36.

Sterba, O. (1990) Prenatal development of *Myotis myotis* and *Miniopterus schreibersii*. *Folia Zoologica*, **39**, 73–83.

Striedter, G. F. (1998) Stepping into the same river twice: homologues as recurring attractors in epigenetic landscapes. *Brain, Behavior and Evolution*, **52**, 218–31.

Suga, N. (1964) Single unit activity in cochlear nucleus and inferior colliculus of echo-locating bats. *Journal of Physiology*, **130**, 332–46.

Thiele, A., Vogelsang, M. & Hoffmann, K.-P. (1991) Pattern of retinotectal projection in the megachiropteran bat *Rousettus aegyptiacus*. *Journal of Comparative Neurology*, **314**, 671–83.

Thiele, A., Rübsamen, R. & Hoffmann, K.-P. (1996) Anatomical and physiological investigation of auditory input to the superior colliculus of the echolocating megachiropteran bat *Rousettus aegyptiacus*. *Experimental Brain Research*, **112**, 223–36.

Valverde, F., López-Mascaraque, L., Santacana, M. & De Carlos, J. A. (1995) Persistence of early-generated neurons in the rodent subplate: assessment of cell death in neocortex during the early postnatal period. *Journal of Neuroscience*, **15**, 5014–24.

Vandevelde, I., Duckworth, E. & Reep, R. L. (1996) Layer VII and the gray matter trajectory of corticocortical axons in rats. *Anatomy and Embryology*, **194**, 581–93.

Vander Wall, S. B. (1982) An experimental analysis of cache recovery in Clark's Nutcracker. *Animal Behavior*, **30**, 84–94.

van Essen, D. C., Anderson, C. H. & Felleman, D. J. (1992) Information processing in the primate visual system: an integrated systems perspective. *Science*, **255**, 419–23.

Werkman, H. (1913) L'évolution ontogénique de la paroi antérieure du cerveau intermédiaire et des commissures du cerveau antérieur chez les mammifères inférieurs. *Archives neerlandaises des sciences exactes et naturelles, Série III B*, **2**, 1–90.

Wible, J. R. & Bhatnagar, K. P. (1996) Chiropteran vomeronasal complex and the interfamilial relationships of bats. *Journal of Mammalian Evolution*, **3**, 285–314.

Wilson, D. E. (1973) Bat faunas: A trophic comparison. *Systematic Zoology*, **22**, 14–29.

Wilson, D. E. & Reeder, D. M. (1993) *Mammal Species of the World: A Taxonomic and Geographical Reference*, 2nd edn. Washington: Smithsonian Institution Press.

Winans, S. S. & Powers, J. B. (1974) Neonatal and two-stage olfactory bulbectomy; effects on male hamster sexual behavior. *Behavioral Biology*, **10**, 461–71.

Winans, S. S. & Scalia, F. (1970) Amygdaloid nuclei: new afferent input from the vomeronasal organ. *Science*, **170**, 330–2.

Wise, L. Z., Pettigrew, J. D. & Calford, M. B. (1986) Somatosensory cortical representation in the Australian ghost bat, *Macroderma gigas*. *Journal of Comparative Neurology*, **248**, 257–62.

Zook, J. M. & Casseday, J. H. (1982) Origin of ascending projections to inferior colliculus in the mustache bat, *Pteronotus parnellii. Journal of Comparative Neurology*, **207**, 14–28.

Zook, J. M. & Casseday, J. H. (1985) Projections from the cochlear nuclei in the mustache bat, *Pteronotus parnellii. Journal of Comparative Neurology*, **237**, 307–24.

Zook, J. M. & Casseday, J. H. (1987) Convergence of ascending pathways at the inferior colliculus in the mustache bat, *Pteronotus parnellii. Journal of Comparative Neurology*, **261**, 347–61

5

Evolutionary plasticity and ontogeny of the bat cochlea

INTRODUCTION

One of the central questions in zoology concerns the adaptations of sensory filtering mechanisms to species-specific environments. Echo-locating bats represent ideal models in auditory neurobiology to understand the underlying principles from both phylogenetic and ontogenetic points of view. They have adapted basic mammalian features of sound transmission by external and middle ears, and of frequency analysis by the inner ear to the working range of their biosonar, which in certain species operates more than three octaves above the high frequency limit of hearing in humans. Furthermore, for matched filtering of the dominant components of their echolocation signals, certain species have evolved the most sharply tuned auditory receptor systems of the animal kingdom.

Following a brief introduction into sonar systems, the first part of this chapter describes comparative anatomical studies of the cochlea in bats. Special emphasis is devoted to an analysis of the structural variations on the common mammalian cochlear *Bauplan* as adaptations to ecological constraints and specialized acoustic performance. Diversity of cochlear mechanisms and the adaptive value of specialized features in cochlear morphology will be judged by a comparative analysis of 1) non-related bats who evolved Doppler-sensitive sonar systems via convergent evolution (*Rhinolophus rouxi* – *Pteronotus parnellii*), and 2) related bats that employ different types of sonar (*P. parnellii* – *Pteronotus quadridens*). The second part of the chapter addresses the question of how Doppler-sensitive sonar and specialized frequency analysis by the cochlea develops during ontogeny in horseshoe bats.

A brief note on biosonar

Among the more than 800 species of Chiroptera, different degrees of auditory specializations are found, ranging from the use of audition in passive listening tasks only (most Megachiroptera) to its incorporation into the active orientation system of biosonar (Microchiroptera) (for review: Schnitzler & Henson 1980; Neuweiler 1990; Schnitzler & Kalkow 1998). Within the Megachiroptera, only a few species of the genus *Rousettus* echolocate. They use tongue-generated clicks during obstacle avoidance. All Microchiroptera echolocate with ultrasonic calls that are produced by the larynx and use the returning echoes for detection, localization, and classification of the reflecting objects. Within the same taxonomic group, there is often a variety of different auditory adaptations and echolocation strategies. Insectivorous bats rely most heavily on echolocation, whereas those species that feed on nectar, fruit, small vertebrates, or blood do not rely exclusively on their sonar while foraging, but use other sensory systems to find and classify their food, or rely on prey-generated acoustic clues.

Echolocation signals

In adaptation to the physical constraints of species-specific ecological niches, microchiropteran bats have evolved different types of biosonar (reviews: Pye 1980; Schnitzler & Henson 1980; Simmons & Stein 1980; Neuweiler 1990; Fenton 1995; Schnitzler & Kalkow 1998) based upon three basic signal elements which differ in their information content. Brief (1–5 ms) downward frequency-modulated (FM) signals (e.g., Fig. 5.1B, right) give precise time marks and are ideally suited for measurement of target distance based on the temporal delay of the echo relative to the emitted call. Furthermore, the wide bandwidth (larger than ½ octave) provides detailed information about target texture based on spectral clues. The other two types of signals are of narrow bandwidth. Quasi-constant-frequency signals (QCF) cover a frequency range of only a few kHz and last between 3 ms and 10 ms and serve as a detection signal due to high energy focused in a narrow frequency band. Long constant-frequency (CF) signals (10–100 ms) (e.g., Fig. 5.1C, right) are also well suited for detection, and as a velocity-sensitive signal form are employed by certain species to analyze relative motion. These basic signals forms are incorporated into species-characteristic echolocation patterns and depending on species or echolocation situations differ in frequency range, harmonic content, sound pressure level, and duration.

Those species that predominantly rely on broadband FM-based sonar, whether or not they also employ QCF signals for detection, are often denoted as FM bats. FM bats represent the majority of Microchiroptera. In contrast, a minority of species routinely use a bimodal signal form by combining CF and FM components into a single CF–FM call. These CF–FM bats are represented by about 130 species of the Old World genera *Rhinolophus* and *Hipposideros* and by one species of the family Mormoopidae, *Pteronotus parnellii*. Their orientation signal consists of a long CF component followed by a downward FM sweep of about 15 kHz bandwidth and contains two harmonics (*Rhinolophus* and *Hipposideros*) or up to four harmonics (*Pteronotus parnellii*) (e.g., Schnitzler & Henson 1980). The second harmonic CF signal component (CF2) is loudest. The CF2 frequency range differs among species, and different bats from the same species have their individual CF2 frequency, with females typically emitting at a frequency 3–5 kHz higher than males (e.g., Kössl & Vater 1985a; Rübsamen & Schäfer 1990a).

CF–FM bats are characterized by the behavior of Doppler-shift compensation (for review: Schnitzler & Henson 1980). When confronted with echoes that are shifted upward in frequency, they lower the frequency of the emitted call proportionally to the Doppler shift, so that returning echoes are kept at a fixed reference frequency with a precision of 50 Hz at an absolute frequency of about 80 kHz in *Rhinolophus ferrumequinum*. This performance is made possible by the evolution of unique specializations in sender and receiver systems for hunting in densely cluttered spaces (review: Neuweiler 1990). These specializations are optimized to evaluate the small periodic changes in frequency and amplitude of the echoes that encode the wing beats of prey insects (Schnitzler 1987).

Hearing characteristics

Hearing at frequencies well above 10 kHz is a basic evolutionary achievement of the mammalian ear (review: Fay 1988; Echteler *et al.* 1994), and the audiograms of many non-echolocating mammals reach into the ultrasonic range (Fig. 5.1A). Bats exploit frequency ranges that partly overlap with the middle and high frequency range of the audiogram of many non-echolocating mammalian species and also use frequencies exceeding the upper limit of hearing in non-echolocating mammals. Thus the capacity for hearing at ultrasonic frequencies alone does not define an echolocator. Echolocation is an ability that requires matched performance of sensory, motor, and attentional systems of the brain. The extension of the hearing range to frequencies up to 200 kHz in certain bats (Schuller 1980; Rübsamen *et al.* 1988) is at the cost of sensitivity to frequencies below 10 kHz.

A detailed comparison of the hearing curves of FM and CF–FM bats is shown in Figs. 5.1B, C. The audiograms of different species of FM

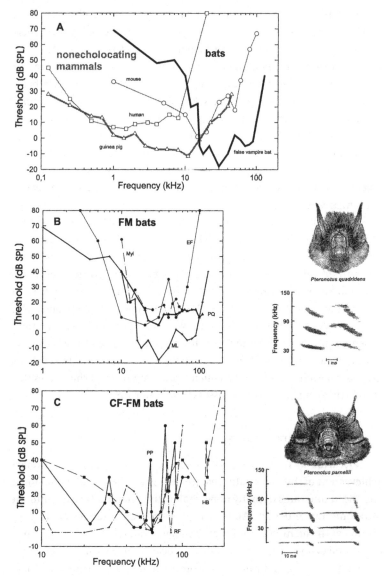

Figure 5.1. A) Comparison of audiograms of non-echolocating mammals and a bat (guinea pig, human, mouse: after Fay 1988). B) Left: Audiograms of FM bats: EF – *Eptesicus fuscus* and Myl – *Myotis lucifugus* (after Dalland 1965); PQ – *Pteronotus quadridens* (after Kössl *et al.* 1997); ML – *Megaderma lyra* (after Schmidt *et al.* 1984). Right: Portrait and orientation calls of *Pteronotus quadridens* (after Kössl *et al.* 1997). C) Left: Audiograms of CF–FM bats. PP – *Pteronotus parnellii* (after Kössl *et al.* 1997); RF – *Rhinolophus ferrumequinum* (after Long & Schnitzler 1975); HB – *Hipposideros bicolor* (after Schuller 1980). Right: Portrait and orientation calls of *Pteronotus parnellii* (after Kössl *et al.* 1997).

bats differ in upper and lower cut-off frequencies but typically show rather broad frequency regions of maximal sensitivity. These sensitivity maxima are either matched to the dominant frequencies of the echolocation signals or are found in frequency bands important for passive listening tasks. The exquisite threshold sensitivity found in *Megaderma lyra* is due to amplification by the specialized pinnae (e.g., Rübsamen *et al.* 1988).

The match of spectral content of call with sensitivity and filter properties of the auditory system is most strikingly revealed in those species that emit long CF–FM signals. The audiograms of CF–FM bats show sharply tuned minima and maxima (Fig. 5.1C). In both *Rhinolophus* and *Pteronotus parnellii*, a prominent and sharply tuned sensitivity maximum is located slightly above the frequency of the CF2 component emitted by the non-flying bat. *P. parnellii* features a further sensitivity maximum at the third harmonic CF component. Within these narrow frequency ranges, tuning sharpness of single auditory neurons is enhanced by a factor of at least 20 beyond that normally encountered in mammals or in other frequency ranges. Furthermore, the frequency range encompassing the CF2 signal component is vastly over-represented within the central auditory system (review: Kössl & Vater 1995). These characteristics are due to specialized hydromechanical processing in the cochlea (Pollak *et al.* 1972; Bruns 1976a, b; Suga & Jen 1977; Vater *et al.* 1985; Kössl & Vater 1985a, b; Henson & Henson 1991; Kössl & Russell 1995; review: Kössl & Vater 1995).

COMPARATIVE FUNCTIONAL ANATOMY AND EVOLUTION OF THE BAT COCHLEA

The mammalian cochlea functions as a hydromechanical frequency analyzer by processing different frequencies at different locations of the receptor surface and relays this information in topographically ordered channels to the central auditory system (Békésy 1960). The bat cochlea is built from the common mammalian set of sensory-neural and supporting structures and has adapted basic mammalian mechanisms for sound transduction and frequency analysis to the working range far above the human hearing range. Rather than introducing new anatomical structures, the design principles used are 1) modifications of the structural quality of components of the micro- and macromechanical frequency analysis system, and 2) modifications of the longitudinal gradients in dimensions of tectorial membrane, basilar membrane and organ of Corti that are the basis of the cochlear frequency map. Certain anatomical features can be interpreted as general adaptations for high frequency

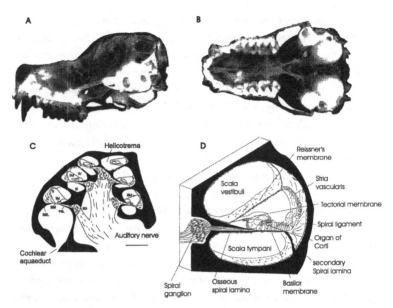

Figure 5.2. A) Lateral and B) ventral view of the skull of *Pteronotus parnellii*. Note the large size of the cochlea in relation to basicranial width and middle ear. C) Schematic drawing of a midmodiolar section through the cochlea of an FM bat, *Trachops cirrhosus* (after Bruns *et al.* 1988). D) Enlarged schematic drawing of a cross-section through the upper basal turn of an FM bat. Abbreviations: BM – basilar membrane, RM – Meissner's membrane, PSL – primary (osseous) spiral lamina, SG – spiral ganglion, SV – scala vestibuli, SM – scala media, ST – scala tympani, SL – spiral ligament, SSL – secondary spiral lamina, TM – tectorial membrane.

hearing, and are commonly found with only little variation across different taxa independent of the type of sonar used. Other features show clear correlations with the use of Doppler-sensitive sonar.

Gross anatomy of the bat cochlea

The cochlea of some microchiropteran bats occupies nearly half the width of the basicranium. Its volume is typically larger than that of the middle ear cavity (Henson 1970) (Fig. 5.2A, B). The osseous labyrinth of the cochlea forms a helical coil whose basal turn has two membraneous windows facing the middle ear cavity: the oval window and the round window. The footplate of the stapes is attached to the oval window and transmits the movement of the tympanic membrane that is mediated by the chain of the three middle ear ossicles (malleus, incus, stapes) to the

perilymphatic space of the scala vestibuli. At the helicotrema of the cochlear apex, the scala vestibuli communicates with the scala tympani (Fig.
5.2C). The round window of the scala tympani is covered by a thin membrane and serves as a pressure-release device. In some bats of the genus
Macrotus and in *P. parnellii*, the round window is exceptionally large.
Another opening situated posterior to the round window and facing the
cranial cavity, the ductus perilymphaticus or cochlear aqueduct, is
enlarged in certain bats (Henson 1970; Henson *et al.* 1977).

The modiolus or central axis of the cochlea has a hollow center that
houses the auditory nerve (Fig. 5. 2C). The spiral ganglion is either located
within the internal acoustic meatus (e.g., *Trachops cirrhosus* [Fig. 5.2C],
Eptesicus fuscus, *Myotis lucifugus*), or within spiral cavities of the modiolar
bone (e.g., basal turn of *P. parnellii*, *Rhinolophus*, *Hipposideros* [Fig. 5.3]). The
peripheral processes of the spiral ganglion cells radiate outwards towards
the organ of Corti via channels within the osseous spiral lamina (Fig.
5.2C, D).

The endolymphatic space of the cochlea, the cochlear duct or scala
media, is delineated by 1) Reissner's membrane, 2) the organ of Corti
which is situated on the basilar membrane, and 3) the stria vascularis (Fig.
5.2C, D) The stria is attached to the spiral ligament. The attachment of the
spiral ligament to the bone is mediated by a specialized cell type – anchoring cells or tension fibroblasts – which are thought to exert radial tension
on the spiral ligament/basilar membrane complex (e.g., Henson & Henson
1988) and thereby influence hydromechanical processing. The spiral ligament is especially large and well developed in the basal cochlear turn. The
region of the spiral ligament that borders the scala tympani is supported
by a prominent osseous secondary spiral lamina (Fig. 5. 2C, D) in both
extant Microchiroptera and Eocene bats (*Hassianycteris messelenis*,
Palaeochiropteryx tupaiadon: Habersetzer & Storch 1992). This feature likely
represents an adaptation for high frequency hearing, since a secondary
spiral lamina is only found at the very base of the cochlea in mammals
with pronounced low frequency hearing. In *Rhinolophus* (Bruns 1980) and
Hipposideros (Dannhof & Bruns 1991), the secondary spiral lamina of the
basal turn is of specialized shape, and likely contributes to specialized resonance properties.

The number of cochlear turns varies from 1.75 in *Pteropus* to 3.5 in
Rhinolophus with most species having 2.5 to 3 turns (e.g., Pye 1966; review:
Henson 1970). The importance of the acoustic sense in behavior is mir-
rored by considerable differences in cochlear size among the Chiroptera.
In Microchiroptera, the basal cochlear turn is larger than in Megachirop-
tera and its most extreme dimensions are found in bats emitting long

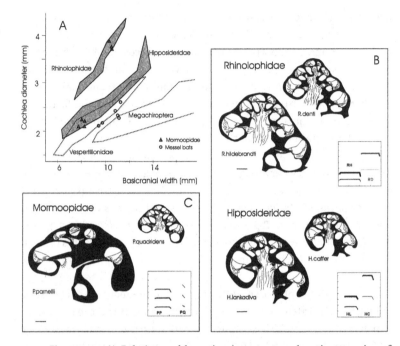

Figure 5.3. A) Relative cochlear size in extant and extinct species of Chiroptera. Cochlear size in CF–FM bats (rhinolophids, hipposiderids, shaded areas) is significantly larger than in FM bats of the family vespertilionids. Differences in cochlear size among mormoopids (open triangles) rerflect specializations to CF–FM sonar and FM sonar. Relative cochlear size in Megachiroptera is significantly smaller than in Microchiroptera. Open circles: extinct Eocene Microchiroptera from Messel (after Habersetzer & Storch 1992). B) Midmodiolar sections through the cochlea of related Old World genera *Rhinolophus* and *Hipposideros* together with sonagrams of the orientation calls (ordinate: full scale 150 kHz, abscissa: full scale 50 ms for *Hipposideros*, 100 ms for *Rhinolophus*). C) Midmodiolar sections through the cochlea of mormoopids together with sonagrams of orientation calls (ordinate: 150 kHz, abscissa: 50 ms).

CF–FM signals (Henson 1970). According to Habersetzer & Storch (1992), cochlear size not only reflects taxonomic relations but also different stages of acoustical specializations: 1) small cochleae (diameter of basal turn as compared to basicranial width) are found in all Megachiroptera regardless of whether they echolocate or not (Fig. 5.3A), and in Microchiroptera belonging to a wide systematic spectrum which are not typical aerial insect feeders but feed on nectar, pollen, blood, or small vertebrates, and do not rely exclusively on echolocation (e.g., *Phyllostomus has-*

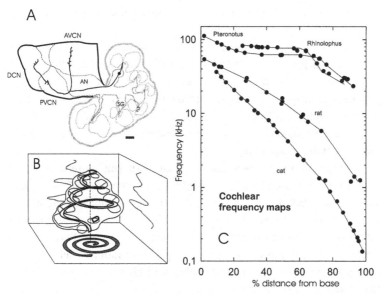

Figure 5.4. A) Frequency-specific connectivity of type I spiral ganglion cells of the cochlea with the three subnuclei of the cochlear nucleus: anteroventral (ACVN), posteroventral (PCVN), and dorsal (DCN). Tracer injections into physiologically characterized regions of the cochlear nucleus allow the construction of the cochlear frequency map by analysis of retrograde transport and three-dimensional reconstruction of the cochlea (B). C) Cochlear frequency maps obtained by tracing techniques in non-echolocating mammals and CF–FM bats (cat: after Liberman 1982; rat: after Müller 1991a; *Rhinolophus rouxi*: after Vater *et al.* 1985; *Pteronotus parnellii*: after Kössl & Vater 1985).

tatus, Desmodus rotundus, Carollia perspicillata, Megaderma lyra), 2) medium-sized cochleae are found in all vespertilionids as typical aerial hunters most of which employ broadband FM sonar (Fig. 5. 3A), 3) enlarged cochleae characterize the genus *Hipposideros* which echolocate with short CF–FM signals (Figs. 5.3A, 5.4). Extremely enlarged cochleae are found in the Old World genus *Rhinolophus* and in one species of the New World Mormoopidae, *Pteronotus parnellii* (Fig. 5.3A). These bats employ long CF–FM signals and have developed Doppler-sensitive sonar in an example of convergent evolution. Cochlear size in extinct Microchiroptera from the Eocene (*Hassianycteris* and *Palaeochiropteryx*) overlaps with the lower extreme of the vespertilionid range, and cochlear size in the even more plesiomorphic family *Archaeonycteris* overlaps with the Megachiroptera (Fig. 5.3A). This could indicate that echolocation performance was less advanced in 'Eochiroptera' during the Eocene, representing one possible reason for extinction and replacement by modern families (Habersetzer & Storch 1992).

The gross overall appearance of the cochlea, as shown by midmodiolar sections (Fig. 5.3B), differs between the related genera *Rhinolophus* and *Hipposideros*. Within a genus, taxon-specific external and internal morphological appearance is conserved but scaled according to the dominant frequency of biosonar signals; larger species that emit lower-frequency CF2 signals possess a larger cochlea (Vater & Duifhuis 1986; Francis & Habersetzer 1998).

These allometric relations contrast with the pattern observed within the Mormoopidae (Fig. 5.3C). This family comprises FM bats such as *Pteronotus quadridens*, *P. macleayi* and *Mormops blainvillii* and one species of CF–FM bats, *P. parnellii*. The cochlea of the CF–FM bat *P. parnellii* not only differs considerably in size from the cochlea of related FM bats, but also exhibits different internal and external specializations (see below). Thus, within the mormoopids, the adaptability of biosonar type to cochlear structure is most strikingly revealed.

Cochlear frequency maps

Cochlear enlargement in CF–FM bats is due to the implementation of an auditory fovea, i.e., an expanded representation of a narrow frequency band that includes the CF2 signal range (Bruns 1976b; Vater et al. 1985; Kössl & Vater 1985b; for review: Neuweiler 1990; Kössl & Vater 1995; Vater 1998).

The frequency map of the cochlea in *Rhinolophus rouxi* and *P. parnellii* was constructed by analysis of retrograde transport of horseradish peroxidase to the cochlea following extracellular tracer injections into physiologically characterized regions of the cochlear nucleus, the central termination site of all auditory nerve fibers (Fig. 5.4A; Feng & Vater 1985; Vater et al. 1985; Kössl & Vater 1985b; Zook & Leake 1989). The exact locus of labeled spiral ganglion cells and their peripheral processes were obtained in three-dimensional reconstructions of the cochlea (Fig. 5.4B).

The cochlear frequency map of acoustically non-specialized mammals such as cats (Liberman 1982) and rats (Müller 1991a) has an exponential course with maximal mapping coefficients of 2.4 mm per octave (Fig. 5.4C). Both species of CF–FM bats deviate from this pattern. There is a significant over-representation of frequencies between 54 to 70 kHz in *Pteronotus* which includes the CF2 signal at 61 kHz, and from 74 to at least 85 kHz in *Rhinolophus* which includes the CF2 signal frequency of

78 kHz with mapping coefficients amounting to 70 mm per octave (Vater *et al.* 1985; Kössl & Vater 1985*b*). Within the auditory fovea, those auditory nerve fibers that are tuned to CF2 ±1 kHz possess exceptionally narrow frequency threshold curves (for a more detailed review see Kössl & Vater [1995]). In *Pteronotus*, there is a second area of increased tuning sharpness located at CF3 ±1 kHz within a region of the basilar membrane where the frequency map is non-specialized.

Since the first description of an auditory fovea in CF–FM bats (Bruns 1976*b*; Schuller & Pollak 1979; Bruns & Schmieszek 1980), foveal representations of certain frequency ranges have been described in one FM bat (*Tadarida brasiliensis*; Vater & Siefer 1995), and in non-echolocators such as mole rat (Müller *et al.* 1992) and the barn owl (Köppl *et al.* 1993). In mammals, auditory foveae are generated by specializations of the passive hydromechanical system of the cochlea (see below).

As a common feature, auditory foveae occur in frequency ranges of highest biological importance and produce an over-representation of the respective frequency ranges in the central auditory pathway for parallel processing of different information aspects. But only in CF–FM bats are parts of the auditory fovea characterized by exceptional sharp tuning (for detailed discussion see Kössl & Vater 1995; Vater 1998).

General organization of the organ of Corti

A detailed description of cochlear fine structure in adult bats is a prerequisite for understanding the development of the major morphological features during ontogeny and their functional relevance. The composition of the organ of Corti in adult bats is shown schematically in Fig. 5.5, and in semithin sections and scanning electron microscopy in Figs. 5.6A, B. As in other mammals, there is a dual system of mechanoreceptor cells, the inner and outer hair cells which distinctly differ in cellular organization, innervation pattern, and placement relative to basilar membrane and tectorial membrane. Sensory transduction involves deflection of the stereocilia by shearing motion between the tectorial membrane and the reticular lamina of the organ of Corti that is generated by displacement of the basilar membrane. The similarities in ultrastructure of bat hair cells with those in other mammals suggests the same division of work; inner hair cells are commonly regarded as classical passive receptor cells whose function is sensory transduction and relay of information to central auditory system, whereas the outer hair cells are mainly involved in mechanical preprocessing in the cochlea (reviews: Lim 1986; Dallos 1992).

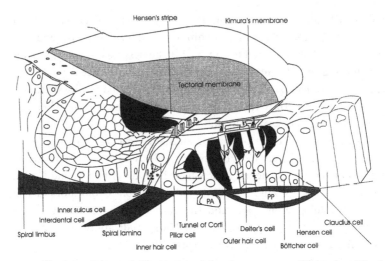

Figure 5.5. Schematic illustration of the ultrastructure of the organ of Corti in a generalized bat. PA – pars arcuata of the basilar membrane, PP – pars pectinata of the basilar membrane.

Sensory cells

Inner hair cells are situated on the edge of the osseous spiral lamina, and thus not directly affected by motion of the basilar membrane. Furthermore, they are completely encased by supporting cells (Fig. 5.5). As in other mammals, the inner hair cells receive the bulk of afferent endings (90–95%; Bruns & Schmieszek 1980; Zook & Leake 1989) and thus are the main relay for information ascending the auditory pathway. The bottom of the pear-shaped cell body is contacted by numerous afferent dendrites of type I spiral ganglion cells whose central processes form the auditory nerve (Vater *et al.* 1992). In FM bats, the number of afferent fibers on one inner hair cell can be as large as 70 (Ramprashad *et al.* 1978) whereas in CF–FM bats, the maximum number of afferents per inner hair cell agrees with that found in other mammals (24–37; Bruns & Schmieszek 1980; Zook & Leake 1989). As in other mammals, efferent endings only rarely contact the inner hair cell bodies directly; their main targets are the afferent dendrites of type I ganglion cells.

The three rows of outer hair cells are strategically located at the region of maximal basilar membrane displacement (Fig. 5.5). The cylindrical cell body of the outer hair cell is tightly attached only at its apical and basal pole. The apical attachment within the reticular lamina consists of tight junctions between the cuticular plate and supporting cell phalanges. The basal attachment is formed by a large cup-like formation of the Deiter's cells (Vater *et al.* 1992; Kuhn & Vater 1995). In between these

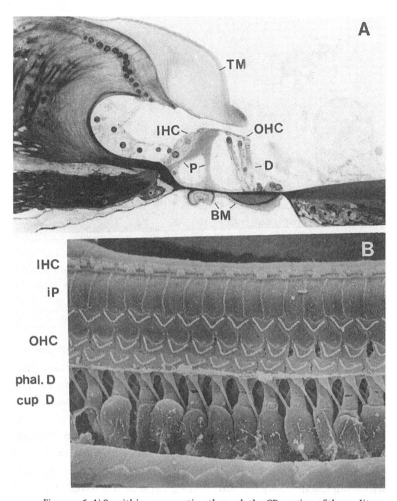

Figure 5.6. A) Semithin cross-section through the CF2 region of the auditory fovea in *Pteronotus parnellii* (after Vater & Kössl 1996). B) Scanning electron microscopic picture of the receptor surface in the upper basal turn of *Rhinolophus rouxi* (Vater & Lenoir, unpublished). Abbreviations: BM – basilar membrane, D – Deiters cell, IHC – inner hair cell, iP – inner pillar cell, OHC – outer hair cell, phal. of D – phalangeal process of D, P – pillar cells, TM – tectorial membrane.

attachment zones, the lateral wall of outer hair cells is surrounded by the fluid-filled spaces of Nuel. The ultrastructural composition of the lateral wall of the outer hair cells is similar to that found in other mammals being composed of an intricate submembraneous cytoskeleton linked to the membrane via electron-dense pillars and a single layer of subsurface cisternae (Vater *et al.* 1992). In analogy to other mammals, this region is

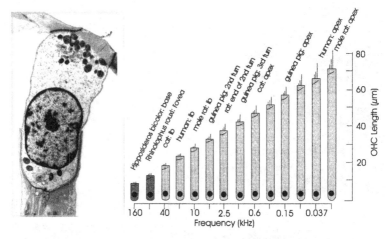

Figure 5.7. Left: Fine structure of bat outer hair cell (after Vater *et al.* 1992). Right: Outer hair cell (OHC) length is related to frequency range across different mammalian species. Note that bats possess the shortest outer hair cells (after Pujol *et al.* 1992).

assumed to contain the cellular machinery responsible for fast motility of the outer hair cell body. This capacity likely represents the basis of the 'cochlear amplifier' which is assumed to boost sensitivity and tuning at low sound pressure levels (reviews: Dallos 1992; Holley 1996; Nobili *et al.* 1998).

The afferent supply of outer hair cells is derived from type II ganglion cells and contacts the bottom of the outer hair cell within the Deiter's cup. There are maximally four afferent fibers per outer hair cell which implies that only 5–10% of afferent fibers belong to the outer hair cell system (for more detailed review: Vater & Kössl 1996). The efferent supply of outer hair cells is less abundant in bats than in the cochlea of large non-echolocating mammals. In many bat species but also in mice there is typically only one single large efferent ending per outer hair cell (*P. parnellii*: Bishop & Henson 1988; Xie *et al.* 1993; *E. fuscus, T. brasiliensis*: Vater, unpublished). In horseshoe bats (Bruns & Schmieszek 1980; Vater *et al.* 1992) and *Hipposideros lankadiva* (Bishop & Henson 1988) there is a complete lack of efferents to the outer hair cells.

As an adaptation to the high frequency range of hearing, the cell bodies and stereocilia of bat outer hair cells are miniatuarized (Dannhof & Bruns 1991; Dannhof *et al.* 1991; Pujol *et al.* 1992; Vater & Lenoir 1992; Vater *et al.* 1992; Vater & Siefer 1995; Vater & Kössl 1996) (Figs. 5.6, 5.7). The length of the cell body ranges from 8 μm in the basal turn of *H. bicolor* to a maximal length of 15 μm reported for the apex. These gradations are significantly smaller than those observed in mammals with good hearing capacities in both low and high frequency ranges, such as the guinea pig (Pujol *et al.* 1992). A stereocilia length of only 0.8 μm, as reported for the basal turn of *Rhinolophus* (Bruns & Goldbach 1980; Vater & Lenoir 1992), likely represents the lower limit in stereocilia dimensions of the mammalian cochlea.

Supporting cells

The supporting cells (pillar and Deiter's cells) play an integral role in the micromechanics of the organ of Corti (e.g., Lim 1986; Slepecky 1996). The inner and outer pillar cells form the arches of the tunnel of Corti and provide rigid support in the radial direction. Their heads are interlocked and form portions of the reticular lamina (Fig. 5.5). They are reinforced by F-actin and microtubules. The stalks of pillar cells are mechanically strengthened by densely packed bundles of microtubules and microfilaments. The pillar cells are unusually thick and sturdy throughout most of the cochlea of bats, thereby increasing mechanical stiffness which can be seen as an adaptation for processing high frequencies (Vater *et al.* 1992; Kuhn & Vater 1995).

The Deiter's cells are the specialized supporting cells of the outer hair cells (Fig. 5.5). Their basal pole is attached to the basilar membrane by a wide hemidesmosome-like structure from which a cytoskeletal bundle ascends into the phalangeal process and the cup of the Deiter's cell. The cup completely envelopes the basal portion of the outer hair cell and is reinforced with an F-actin cytoskeleton. The function of this intricate cellular architecture is to provide a stiff support for the outer hair cell and to optimize transfer of mechanical energy between outer hair cell and basilar membrane (Vater *et al.* 1992). The adaptive value of this feature for processing of high frequency signals is shown by comparative anatomical data; prominent, cytoskeletally reinforced Deiter's cups are present throughout the bat cochlea whereas in other mammals such arrangements are only found in basal high frequency regions of the cochlea (Kuhn & Vater 1995). Additionally, Deiter's cells provide mechanical coupling in the longitudinal direction of the cochlea via their phalangeal processes that arise laterally just below the Deiter's cup and ascend in an apical direction to participate in formation of the reticular lamina (Vater & Lenoir 1992).

In conclusion, the cellular organization of the organ of Corti in bats emphasizes and exaggerates features of micromechanical relevance that are typically only encountered in the most basal regions of the cochlea in non-echolocating mammals. They thus likely represent general adaptations of the cochlear micromechanical system for processing high frequencies. Although species- or genus-specific characters are present in the organization of the reticular lamina and efferent innervation pattern (e.g., Bruns & Goldbach 1980; Bishop & Henson 1988), these do not clearly relate to the use of different sonar types. As described below, such clearcut correlations are seen in the organization of the basilar membrane and tectorial membrane which coincide with region-specific maxima and minima in afferent innervation density.

Basilar membrane

The basilar membrane is anchored on the osseous spiral lamina and the spiral ligament and composed of filamentous layers and ground substance (review: Echteler *et al.* 1994). The density of filaments is the main factor that provides its stiffness. The mediolateral extent of the basilar membrane can be divided into two regions: the pars tecta (arcuata), situated below the tunnel of Corti, and the more lateral region of pars pectinata (Fig. 5.5). In non-echolocating mammals, the pars tecta contains a single layer of filaments embedded in a small amount of ground-substance, the pars pectinata can be thickened due to a thick layer of ground-substance sandwiched between two layers of filaments, and the undersurface of the basilar membrane is bordered by a thin layer of mesothelial cells (the tympanic cover layer). These basic arrangements are also observed in echolocating bats, but in several species the morphology of the basilar membrane is specialized. In many bats (e.g., genus *Rhinolophus*: Pye 1966; Bruns 1976a; Vater *et al.* 1985; *Hipposideros*: Dannhof & Bruns 1991; *Eptesicus fuscus*: Vater, unpublished; *Pteronotus parnellii*: Kössl & Vater 1985b; Vater & Kössl 1996; *Tadarida brasiliensis*: Vater & Siefer 1995), the pars tecta reveals a thickening that consists of homogenous ground-substance with few filaments along the tympanic margin. This thickening is typically present in the basal turn, and up to at least the middle turn, in both CF–FM bats and FM bats, suggesting that it is a feature related to processing the high to middle frequencies of the bats hearing range independent of the type of sonar used.

Ultrastructural specializations of the pars pectinata are, however, clearly correlated with type of sonar. In bats using Doppler-sensitive sonar, there are prominent thickenings (i.e., thickness >15 μm) of the pars pectinata of the basilar membrane in the basal turn. In the genus *Rhinolophus* (Bruns 1976a; Vater *et al.* 1985), the pars pectinata, throughout the basal

25% of cochlea length, is composed of two thickened portions situated both on the tympanic and vestibular side of the basilar membrane. Both thickenings are due to the incorporation of filaments into the ground-substance. The tympanic thickening contains an abundance of interwoven filaments that are predominantly oriented in the radial direction, whereas the vestibular thickening predominantly contains longitudinally directed filaments. A comparable specialization is found between about 25 and 45% distance from base in the mustached bat, *P. parnellii* (Vater & Kössl 1996). Incorporation of filaments will lead to an increase in basilar membrane stiffness. Furthermore, the presence of the specialized vestibular thickening is expected to create or increase longitudinal coupling over long stretches of the basilar membrane. In the genus *Hipposideros* (Dannhof & Bruns 1991), the pars pectinata in the basal-most 10% of basilar membrane is also prominently thickened (27 μm), but in contrast to the species emitting long CF–FM signals, the thickening only contains radially directed filaments and is confined to the tympanic side. Apart from these ultrastructural specializations, the baso-apical gradients in width and thickness of the basilar membrane are specialized in CF–FM bats (see below). A further type of basilar membrane specialization is found in the Molossidae. Within the foveal region of the cochlea of *Tadarida brasiliensis*, the basilar membrane is thickened throughout its modiolar to abmodiolar extent due to the presence of a thick, homogeneous ground-substance which likely represents a mass load (Vater & Siefer 1995).

Tectorial membrane

The tectorial membrane of the mammalian cochlea is an acellular flap or beam-like mass that forms the roof of the organ of Corti within the scala media (Figs. 5.5, 5.6A). It is anchored to the spiral limbus and its subsurface contacts the stereocilia of hair cells. Its body is composed of ground-substance and fibrillar elements (Lim 1987). The upper surface is formed by the cover net, and the undersurface consists of the Hensen's stripe above the inner hair cell region with an amorphous layer (Kimura's membrane) above the outer hair cell region. This basic organization is conserved in echolocating bats. In *P. parnellii*, more subregions of the tectorial membrane can be defined than in non-echolocating mammals according to the packing density and trajectories of protofibrils. Furthermore the packing density of filaments in certain subregions is greater than in other mammals (Vater & Kössl 1996). This indicates that the resonance properties are more complex and tectorial membrane stiffness is increased.

On the subsurface of the tectorial membrane of Horseshoe bats (Vater & Lenoir 1992; Vater *et al*. 1992), imprints of the tallest stereocilia of both inner and outer hair cells are observed throughout the cochlea. In non-echolocating mammals, imprints of outer hair cell stereocilia are commonly observed, but imprints of inner hair cell stereocilia are not found at all or are confined to the very basal cochlea (Lenoir *et al*. 1987; Lim 1987). Imprints of inner hair cell stereocilia typically are seen on the pronounced Hensen's stripe that is present on the subsurface of the tectorial membrane in all bat species that have been studied. This suggests that inner hair cells are excited at high frequencies of hearing by a direct displacement of stereocilia due to shear movement of the tectorial membrane (Vater & Lenoir 1992), rather than by fluid motion in the subtectorial space.

Diversity of cochlear mechanics in bats

According to Zwislocki (1986), there are two resonator systems in the cochlea, one consisting of the distributed mass of the organ of Corti supported by the stiffness of the basilar membrane, and the other consisting of the tectorial membrane mass and its elastic attachment to the spiral limbus. These two systems appear to be the main targets of evolutionary change that underly specialized hearing capabilities in CF–FM bats.

In non-echolocating mammals, the width of the basilar membrane systematically increases from base to apex, and in many species, thickness of the basilar membrane decreases from base to apex (review: Echteler *et al*. 1994). The systematic longitudinal change in basilar membrane dimensions and thus stiffness are the basis of the cochlear frequency map (Békésy 1960). A similar pattern is observed in FM bats (review: Kössl & Vater 1995) and illustrated for *Pteronotus quadridens* in Fig. 5.8. Cross-sectional area of the tectorial membrane increases gradually from base to apex in non-echolocating mammals and in FM bats (Fig. 5.8). In contrast, the non-related CF–FM bats *Rhinolophus rouxi* and *Pteronotus parnellii* feature highly specialized longitudinal gradients in both basilar and tectorial membrane morphology (Fig. 5.8). The morphological specializations coincide with subregions of the auditory fovea. In the sparsely innervated zone (SI) that represents the frequency range slightly above the CF2 frequency (62–80 kHz in *Pteronotus*; above 83 kHz in *Rhinolophus*), basilar membrane thickness is increased, but decreases abruptly at the transition to the CF2 region, which is densely innervated. The CF2 region is characterized by a plateau in basilar membrane

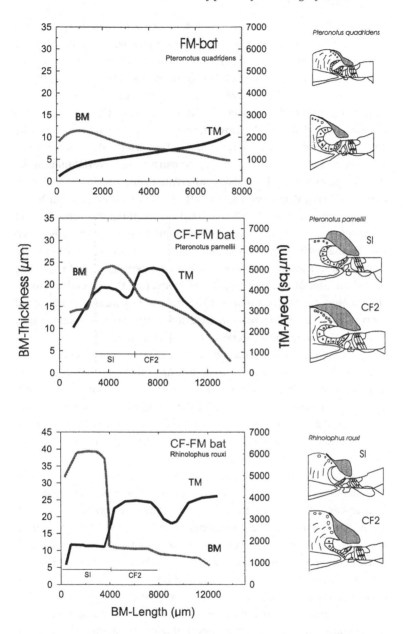

Figure 5.8. Left: Comparison of longitudinal profiles in basilar membrane (BM) thickness and cross-sectional area of the tectorial membrane (TM) in the related FM bats and CF–FM bats (*Pteronotus quadridens* and *P. parnellii*); and in non-related CF–FM bats (*P. parnellii* and *Rhinolophus rouxi*). SI – sparsely innervated zone, CF2 – second harmonic frequency. Right: Cross-sections of the organ of Corti (after Vater 1997).

thickness (Fig. 5.8) and width (Vater & Kössl 1996; Vater 1997). Plateau regions of basilar membrane dimensions typically house the auditory foveae in mammals (e.g., Bruns 1976a; Kössl & Vater 1985b; Vater et al. 1985; Müller et al. 1992). In the CF2 region, cross-sectional area of the tectorial membrane reaches an absolute maximum in Pteronotus, and a relative maximum in Rhinolophus. Furthermore, in the CF2 region of both CF–FM bats both tectorial membrane area and its attachment site to the spiral limbus are enlarged (Figs. 5.6B, C). The latter feature may restrict freedom of motion. There are sharp transitions in tectorial membrane morphology towards the sparsely innervated zone: cross-sectional area, and thus mass, decreases and the attachment to the spiral limbus is reduced considerably. For Pteronotus, it is assumed that the distinct change in shape endows the tectorial membrane of the sparsely innervated zone with unique resonance and motion properties (Henson & Henson 1991; Kössl & Vater 1996a, b; Vater & Kössl 1996; Kössl 1997). In a model calculation (Steele 1997), it was shown that this specialized tectorial membrane, which is shaped like a hammer, strongly resonates and moves up and down in a vertical direction, rather than in a radial direction as is typical for non-specialized tectorial membrane.

The specializations of tectorial and basilar membranes observed in both CF–FM bats likely represent salient components of a passive mechanical mechanism which enhances tuning beyond values normally encountered in mammals. Furthermore, the sharp transitions in morphology of the basilar membrane are thought to create reflection points for traveling waves (review: Kössl & Vater 1995) which lead to frequency-specific summation and cancellation effects.

In terms of the relationship between the size of specialized tectorial membrane and the thickened basilar membrane, Pteronotus and Rhinolophus differ significantly (Fig. 5.8). These differences may well underlie known differences in damping of cochlear resonance mechanisms evidenced by recordings of the cochlear microphonic and otoacoustic emissions (Henson et al. 1985; Kössl 1994a, b). But it is striking that the basic modifications in design are similar in both species which have developed Doppler-sensitive sonar in convergent evolution. This is indicative of a common algorithm in developmental programs that leads to a specialization of longitudinal morphological gradients. A plausible assumption on the mechanism responsible for creation of specialized gradients in basilar and tectorial membrane morphology are regional-specific differences in timing and duration of the secretory activity of cells that produce these extracellular structures during embryonic development. The inverse relationship of basilar membrane and tectorial membrane dimensions (thin basilar in areas of large tectorial membrane) that is not only found in specialized areas of the CF–FM bat cochlea, but also typifies baso-apical gradients in non-specialized cochleae, suggests the involvement of at least partly interdependent regulatory mechanisms. Such changes in the spatiotemporal sequence during development of the extracellular components of the organ of Corti must be well balanced so as not to lead to complete impairment of hearing. In CF–FM bats, the specialized interplay of tectorial membrane and basilar membrane resonator systems of the cochlea not only gives rise to enhanced tuning in certain frequency bands, but also creates frequency-specific insensitivities (see Fig. 5.1). An elegant way to cope with non-uniform peripheral filter properties and threshold distributions is a regulatory adjustment of the pitch of the dominant echolocation signal component to match peripheral sensory capacities. Viewed that way, Doppler-sensitive sonar may have emerged as a phylogenetic learning process in motor and attentional systems within the central auditory pathway in response to a profoundly changed sensory filter. These peripheral changes heightened the ability of the hearing system to perceive relative motion as an integral quality of a sound source.

ONTOGENY OF ECHOLOCATION (HEARING AND VOCALIZATION).

Functional biosonar requires the concerted action of two systems: the vocal motor system and the hearing system. The bats' abilities to hunt on the wing, under guidance of sonar, develop during the first few postnatal weeks (e.g., Rübsamen 1987; Rübsamen & Schäfer 1990a) Bats thus offer the unique opportunity to study developmental principles that integrate hearing and vocal motor systems into an active sensory system. The development of hearing and biosonar calls is most thoroughly studied in horseshoe bats whose specialized Doppler-sensitive sonar system involves highly tuned cochlear mechanics and a sophisticated auditory–vocal feedback loop.

Development of vocalization and hearing capabilities

Horseshoe bats are born blind and deaf but are capable of vocalization. When separated from their mothers, newborn bats emit long sequences of loud isolation calls through the mouth. Individual calls are brief (<20 ms), harmonically structured, and contain energy in frequency bands that are audible to the human ear (Fig. 5.9). Echolocation

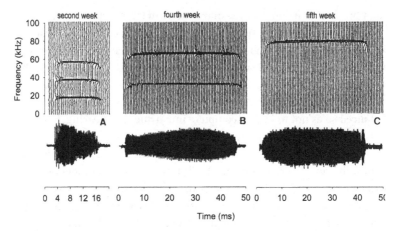

Figure 5.9. Sonograms (top) and spectra (bottom) of an isolation call (A) and echolocation calls (B, C) of juvenile horseshoe bats. (After Rübsamen 1987.)

calls which are emitted through the nostrils and first appear within the third postnatal week. The CF2 component is, however, up to 15 kHz lower than in the adult, and the first and third harmonics are more intense (Fig. 5.9). The CF2 frequency gradually increases at a rate of about 1 kHz per day, until in the fifth postnatal week when the adult frequency range is achieved (Rübsamen 1987). During development, as in adults, there is a clear sexual dimorphism in the frequency of echolocation calls. The CF2 frequencies of female bats are on the average 2 kHz higher than in the males; however, no such difference is noted in the spectral content of isolation calls (Rübsamen 1987; Rübsamen & Schäfer 1990a).

Recordings of evoked potentials from the auditory midbrain of the horseshoe bat revealed a dynamic maturation of hearing capabilities up to the fifth postnatal week (Rübsamen 1987). The first auditory responses emerged during the second postnatal week. They were insensitive and confined to the low frequency region of the adults audiogram (i.e., below about 50 kHz). The first sharply tuned responses emerged within the third postnatal week, and were tuned to frequencies between 57–60 kHz matching the individual's CF2 emission frequency at that age. Parallel with the development of echolocation calls, the frequency of sharply tuned responses shifted upward with increasing age.

These pioneering studies laid the basis for several sets of experiments which addressed two basic questions: 1) Which mechanisms underlie the postnatal extension of the hearing range and the developmental shift in absolute frequency range of sharp tuning? 2) Do the bats behaviorally monitor ontogenetic shifts in responses characteristic of the

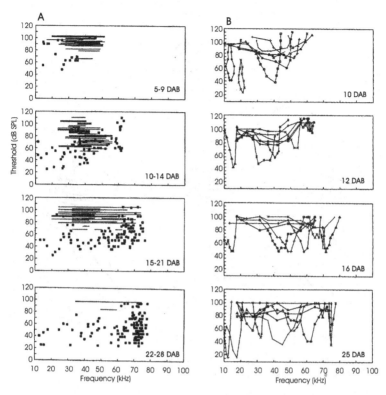

Figure 5.10. Development of threshold sensitivity (A) and tuning curves (B) of single cochlear nucleus neurons in *Rhinolophus rouxi*. DAB – days after birth. (Vater & Rübsamen, unpublished.)

receptor organ by adjusting their vocal output, or does the maturing vocalization system influence the development of audition?

Single unit recordings employing stereotactic localization of recording sites in the auditory midbrain (Rübsamen & Schäfer 1990a) and in the cochlear nucleus (Fig. 5.10; Vater & Rübsamen 1992) also show an improvement of threshold sensitivity and tuning properties within the first few postnatal weeks. Already, between 5 to 10 days after birth, tuned responses can be recorded, but only in frequency ranges below 50 kHz (indicated by squares in Fig. 5.10A), whereas neurons of the central representation site of the auditory fovea are broadly tuned and insensitive (indicated by lines in Fig 5.10A). Around 12 days after birth, the hearing range has widened, and the first sharply tuned responses can be recorded from the foveal region. These responses are characterized by insensitive tip regions and pronounced tail regions of the tuning curve (Fig. 5.10B). The tip is tuned to the individuals CF2 frequency which is about 12 kHz

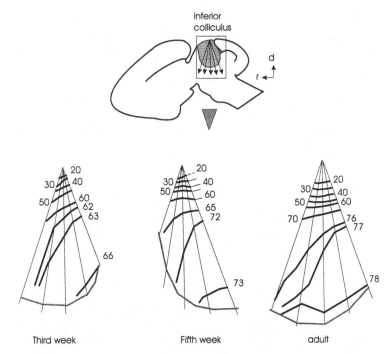

Inferior
colliculus

Third week   Fifth week   adult

Figure 5.11. Development of tonotopy in the auditory midbrain of the horseshoe bat. (After Rübsamen 1992.)

lower than in the adult. At that age, there are still numerous broadly
tuned neurons. During the following weeks, threshold sensitivity within
the fovea improves and its frequency range shifts upward into the adult
range, but there are no significant changes in specialized tuning sharpness during this time period. These data compare well with observations
in other mammals. A restriction of the hearing range to the low frequency regions of the adult audiogram at the onset of hearing with a progressive recruitment of higher frequencies is a shared trait in the
physiological development of mammalian audition (review: Romand
1983; Rubel 1984; Rübsamen 1992). The exact dates of the onset of hearing
vary widely among mammals, ranging from prenatal stages up to several
weeks after birth (review: Romand 1983).

Mapping studies of the tonotopic arrangements in the auditory midbrain (Rübsamen & Schäfer 1990a) and the cochlear nucleus (Vater &
Rübsamen 1992), show that during postnatal development, the central representation sites of the auditory fovea dramatically change their frequency
response characteristics during the first postnatal month, whereas the frequency representation of the lower frequencies remained stable (Fig. 5.11).

Since the major limiting factor for the establishment of central tonotopy is the maturation state of the receptor organ (reviews: Romand 1983; Rubel 1984; Pujol & Uziel 1988), these shifts in frequency mapping most likely reflect developmental changes in spectral response properties of the cochlea.

Development of the cochlear frequency maps

The development of the cochlear frequency map in horseshoe bats was studied using the same techniques as employed in the adult, namely by single and multiunit recordings in the cochlear nucleus, followed by tracing the origin of auditory nerve fibers with retrograde transport of horseradish peroxidase (Vater & Rübsamen 1992).

The development of the cochlear frequency representation is schematically illustrated by plotting characteristic frequency on the corresponding locus of normalized basilar membrane length (Fig. 5.12). This is justified, since there are no significant changes in basilar membrane dimensions after birth (Vater 1988) and the communication lines between cochlea and central auditory pathway are established (Vater & Rübsamen 1992). The data from tracing experiments show that basilar membrane sites containing a foveal representation of the CF2 signal range, and a few kHz below, change their frequency response characteristics with maturation. Starting from broadly tuned insensitive responses to the low frequency range, there is an upward shift in characteristic frequency of sharply tuned tips by about 1/3 octave towards the adult value that takes place between about postnatal day 14 and the fifth postnatal week.

These ontogenetic shifts are much larger than temperature-dependent shifts reported in adult bats (Huffman & Henson 1993). At each age, the center frequency of the auditory fovea matches the CF frequency of the individual's echolocation signal (Rübsamen 1987; Vater & Rübsamen 1988, 1992; Rübsamen & Schäfer 1990a). Apical cochlear regions start to respond with a slight time delay (Rübsamen & Schäfer 1990a), but gain adult-like frequency response characteristics earlier than basal cochlear regions. Thus their frequency representation remains stable during the time period of pronounced ontogenetic mapping shifts in the basal turn.

Shifts in the cochlear frequency place map have also been demonstrated in non-echolocating mammals (Harris & Dallos 1984; Arjmand et al. 1988; Echteler et al. 1989; Müller 1991b) Significantly, in the gerbil cochlea, ontogenetic upward shifts in the frequency representation are likewise confined to basal, high frequency sites, and the frequency representation in the apical cochlea remains stable.

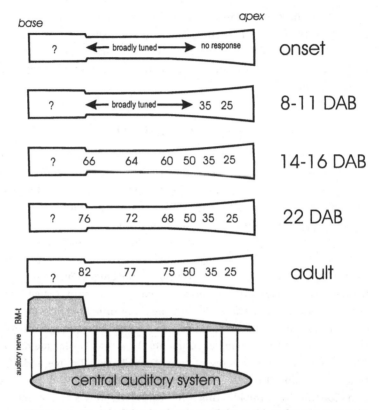

Figure 5.12. Model of the development of the cochlear frequency map in *Rhinolophus rouxi* based on cochlear frequency map (Vater & Rübsamen 1988), and recordings from auditory midbrain (Rübsamen & Schäfer 1990a). Given loci on basilar membrane and corresponding central representation sites change frequency response with age. Top: Numbers give the frequency in kHz at their respective representation place on basilar membrane. Question marks denote regions for which the frequency representation is unknown. DAB – days after birth. Bottom: Throughout development, basilar membrane dimensions and connectivity of the cochlea with the central auditory system remain stable.

The first phenomenological model of functional development of the cochlea was based on results derived from acoustic overstimulation of the avian ear (Rubel 1984; Rubel *et al.* 1984), and incorporates the anatomical finding of a baso-apically directed gradient in cochlear maturation. It was proposed that at onset of hearing, only basal cochlear sites, which are the most mature, transduce acoustic stimuli, but their frequency tuning is lower than in the adult. Later, these regions are tuned to progressively higher frequencies, and the low frequency representation shifts towards the apex. Thus, any given place in the cochlea is assumed to shift its frequency representation with age. Likely mechanisms for the shift include age-related changes in dimensions, mass, and stiffness of the basilar membrane and changes in the mechanical properties of stereocilia. This hypothesis was questioned by the finding of developmental stability of the frequency map in the chicken cochlea (review: Manley 1996). It does, however, receive partial support from experimental data on the basal and middle turn of the mammalian cochlea, but can not account for the developmental stability of mapping in the apex (see above). Another hypothesis focuses on the mediolateral gradient in cochlear development (Romand 1987), and emphasizes age-related changes in outer hair cell function. The absence of tuned responses at young age is thought to be due to immaturities at the outer hair cell level: a lack of 'active' amplification by the outer hair cells produces tuning curves of inner hair cells and auditory nerve fibers that miss the sharply tuned tips and consist of the insensitive low frequency tail region only. Development of the tip region of the tuning curve, due to maturing active amplification, progresses in a baso-apical direction and will cause an apparent shift in frequency responses that amounts to at least ½ octave.

Development of cochlear structure in horseshoe bats

Light-microscopic studies reveal that the seven half-turns of the adult horseshoe bat cochlea are established prenatally, and the diameter of the basal turn and cochlear height are adult-like although the maturation of the skull is not completed. There is no significant or systematic change in basilar membrane length between birth and adulthood. Consequently, postnatal growth of the cochlea is unlikely to contribute to physiological maturation. Measurements of thickness and width of the basilar membrane show that the specialized longitudinal gradients in basilar membrane dimensions of the horseshoe bat cochlea are already established at birth (Vater 1988). Thus, the most prominent species-specific morphological specializations are laid out prior to the biological onset of hearing. Unfortunately, there are no data on early embryonic cochlear development in bats, therefore we have no information on the principles underlying the formation of steep transitions in cochlear morphology at midbasal locations. Several studies of cochlear development in non-echolocating mammals indicate that cochlear cytodifferentiation is initiated not at the very basal end but in midbasal cochlear regions and then proceeds in both apical and basal directions. Given this leading role of midbasal regions in cochlear maturation, it may not be just coincidental that the most striking transitions in morphology of tectorial and basilar membranes, and the innervation pattern observed in the adult cochleae of horseshoe bats and mustached bats, occur at midbasal locations.

In contrast to the grossly immature structure of the organ of Corti at birth found in many small mammals with late postnatal onset of hearing (mouse around 12 days after birth, Kikuchi & Hilding 1965), the organ of Corti of the horseshoe bat is well differentiated already shortly prior to birth (Vater 1988; Vater et al. 1997). Its appearance closely corresponds to that seen in other mammals at the onset of hearing as defined with CM recordings under open bulla conditions. Consequently, the onset of hearing in horseshoe bats around postnatal days 3–5, as defined by single-cell recordings from the central auditory system using free-field stimulation, must be influenced by the status of external and middle ears. Changes in the middle ear are, however, unlikely to explain the frequency shift in tuned responses and threshold improvements of more than 80 dB sound pressure level after onset of hearing (Manley 1996).

Species-specific afferent and efferent innervation patterns of the organ of Corti in horseshoe bats are present shortly prior to birth (Vater et al. 1997). The lack of an efferent supply to the outer hair cell system through the investigated life span (prenatal to adult) excludes a modulatory role in functional development as suggested for other species.

Postnatal maturation occurs at several levels of the hydromechanical and sensorineural systems of the horseshoe bat cochlea (Henson & Rübsamen 1996; Vater et al. 1997). As in other mammals, the basal turn leads maturation, and different structural components mature with a different time course (e.g., Kraus & Aulbach-Kraus 1981; Pujol & Uziel 1988; Roth & Bruns 1992a, b; Weaver & Schweitzer 1994; review: Romand 1983; Pujol & Uziel 1988; Pujol et al. 1995). Some salient features are exemplified by a comparison of the ultrastructure of the organ of Corti in a prenatal bat and a stage at 14 days after birth (Vater et al. 1997), where the first tuned responses can be recorded from the fovea (Fig.5.13). The organ of Corti at 14 days after birth appears adult-like except that remnants of the tympanic cover layer are attached to the scala tympani side of the basilar membrane. In the late prenatal stage, cytodifferentiation is completed, the arrangements of cuticular plates and stereocilia bundles on both inner and outer hair cells are adult-like, and the tectorial membrane is fully formed. There are, however, several important differences between prenatal and later stages: 1) filaments of the basilar membrane are only poorly contrasted, the spiral vessel in the pars tecta of the basilar membrane is

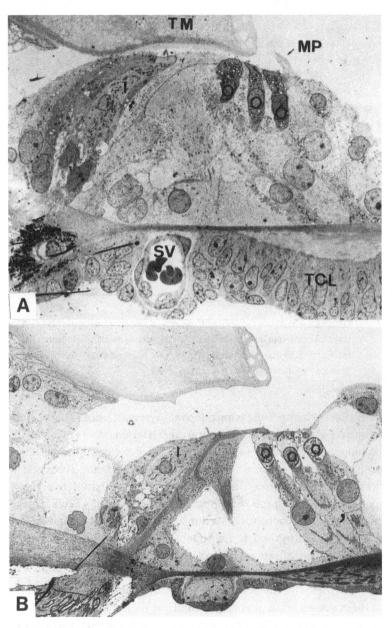

Figure 5.13. Ultrastructure of the organ of Corti in horseshoe bats prior to birth (A) and in a 2-week-old bat (B). TM – tectorial membrane, MP – marginal pillar. (After Vater *et al.* 1997.)

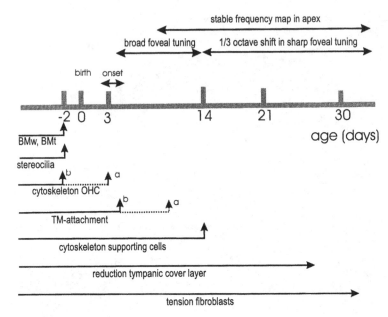

Figure 5.14. Summary of major physiological and anatomical events in cochlear maturation of horseshoe bats. BMw – width of basilar membrane, BMt – thickness of basilar membrane, OHC – outer hair cells, TM – tectorial membrane, a – apical turn, b – basal turn.

patent, and a thick cellular tympanic cover layer is attached to the vestibular surface of the basilar membrane; 2) there is an abundance of microvilli on supporting cell phalanges; 3) marginal pillars formed by microvillous protrusions of the phalanges of the outermost row of Deiter's cells attach to the free edge of the tectorial membrane and likely restrict its freedom of motion; 4) the fluid spaces of the organ of Corti are not fully formed; 5) the cytoskeleton of supporting cells is immature; 6) outer hair cells are immature as evidenced by the dark appearance of the cytoplasm, and the lack of cytoskeletal specializations of the lateral wall; 7) Hensen and Claudius cells are small.

All these structures mature postnatally. A summary of the timing of the major events in physiological and anatomical maturations of the horseshoe bat cochlea is given in Fig. 5.14 in order to evaluate and identify features that represent limiting factors in the development of hearing in the horseshoe bat. Width and thickness of the main body of the basilar membrane, and size and arrangements of receptor cell stereocilia are established prior to birth. These features do not contribute to postnatal functional maturation. The maturation of the cytoskeleton of the outer hair cells is completed throughout the cochlea at the onset of hearing. This suggests that the establishment of the structural substrate for active cochlear amplification represents a necessary prerequisite for onset of function, as in other mammals (Pujol et al. 1991; He et al. 1994), but is unlikely to contribute to the shift in frequency mapping at later ages. However, up to an age of about 14 days after birth, structurally mature outer hair cells work within the immature passive mechanical framework of supporting cells of the organ of Corti. Maturation of the cytoskeleton of Deiter's cells coincides with the first appearance of sharply tuned foveal responses (Vater et al. 1997). The shift in frequency of tuned foveal responses takes place during the time period of maturation of the tympanic cover layer of the basilar membrane, which likely acts as a mass load and limits basilar membrane displacement (Vater 1988; Vater et al. 1997), and the tension fibroblasts of the spiral ligament (Henson & Rübsamen 1996). Incorporation of actin filaments necessary for creation of radial tension on the spiral ligament–basilar membrane complex (Henson & Henson 1988) only starts in the second postnatal week and continues up to the fifth postnatal week thus coinciding with the final maturation of the cochlear frequency map.

To summarize what is known of horseshoe bats, the species-specific gradients in basilar membrane morphology as an integral part of the specialized foveal filter mechanism are established during embryonic development, independent of acoustic input. The function of the filter depends on active amplification of hydromechanical responses by outer hair cells within a patent mechanical architecture of supporting cells, and its center frequency is shifted by a change in mass load and tension of the basilar membrane. This scenario can explain the functional development for basal cochlea segments but leaves open the question of why the cochlear frequency map in the apex remains stable: maturation of supporting cells, tension fibroblasts, and tympanic cover layer occurs throughout the cochlea with no obvious apically to basally directed gradient.

Which is leading development? Cochlea or vocalization system?

The outstanding feature of the adult horseshoe bat's sonar is the precise auditory feedback control of vocalization frequency during Doppler-shift compensation. The well-matched and systematic increase in foveal frequency and echolocation signal frequency during postnatal development is unlikely to result from independent maturation of larynx and hearing system (Rübsamen & Schäfer 1990b; Pedersen 1996). This match could be created either by 1) the juveniles actively adjusting their voices to the genetically determined frequency shift in foveal tuning or 2) by an influence of the maturing vocal-motor system on the establishment of the auditory fovea (Rübsamen & Schäfer 1990b).

Two sets of experiments have been performed to test these alternatives. One set of experiments involved destruction of the cochlea in the third to fifth postnatal week, followed by a comparison of vocalizations in deafened bats with preoperative standards. Models of neuronal negative feedback control of the laryngeal motor system (Schuller et al. 1975; Metzner 1989) predict a systematic shift of vocalizations to higher frequencies after disruption of auditory input. Cochlear destruction caused a change in the CF2 frequency in all specimens but the effects were quite variable: CF2 changed between $+4$ to -14 kHz on the second postoperative day. These data demonstrate that echolocation pulses are under auditory feedback control in juveniles, but the control systems are more complex than anticipated (Rübsamen & Schäfer 1990b). The second set of experiments concerned the elimination of frequency control of the echolocation signals by cutting the superior laryngeal nerves (Schuller & Rübsamen 1981). This disruption did not alter the developmental dynamics of the tonotopic map in the auditory midbrain (Rübsamen & Schäfer 1990b). The authors conclude that maturation of the auditory fovea and shifts in foveal tuning represent an innate process. These shifts in foveal tuning alter the set point of feedback control for echolocation pulse frequency and result in a concomitant increase in vocalization frequency which amounts to 12–14 kHz.

REFERENCES

Arjmand, E., Harris, D. & Dallos, P. (1988) Developmental changes in frequency mapping of the gerbil cochlea, Comparison of two cochlear locations. Hearing Research, 32, 93–6.
Békésy, G. von (1960) Experiments in Hearing. New York: McGraw-Hill.
Bishop, A. & Henson, M. M. (1988) The efferent auditory system in Dopplershift-compensating bats. In Animal Sonar Systems, Proceedings NATO Advanced Study Institute, ed. P. E. Nachtigall & P. W. B. Moore, pp. 307–11. New York: Plenum Press.
Bruns, V. (1976a) Peripheral auditory tuning for fine frequency analysis by the CF–FM bat, Rhinolophus ferrumequinum. I. Mechanical specializations of the cochlea. Journal of Comparative Physiology, 106, 77–86.
Bruns V. (1976b) Peripheral auditory tuning for fine frequency analysis by the CF–FM bat, Rhinolophus ferrumequinum. II. Frequency mapping in the cochlea. Journal of Comparative Physiology, 106, 87–97.
Bruns, V. (1980) Basilar membrane and its anchoring system in the cochlea of the greater Horseshoe bat. Anatomy and Embryology, 161, 29–51.

Bruns, V. & Goldbach, M. (1980) Hair cells and tectorial membrane in the cochlea of the greater Horseshoe bat. *Anatomy and Embryology*, **161**, 65–83.

Bruns, V. & Schmieszek, E. (1980) Cochlear innervation in the greater Horseshoe bat, demonstration of an acoustic fovea. *Hearing Research*, **3**, 27–43.

Bruns, V., Burda, H. & Ryan, M. (1989) Ear morphology in the frog-eating bat (*Trachops cirrhosus*, Family: Phyllostomidae): apparent specializations for low-frequency hearing. *Journal of Morphology*, **199**, 103–18.

Dalland, J. I. (1965) Hearing sensitivity in bats. *Science*, **150**, 1185–6.

Dallos, P. (1992) The active cochlea. *Journal of Neuroscience*, **12**, 4575–85.

Dannhof, B. J. & Bruns, V. (1991) The organ of Corti in the bat *Hipposideros bicolor*. *Hearing Research*, **53**, 253–68.

Dannhof, B. J., Roth, B. & Bruns, V. (1991) Length of hair cells as a measure of frequency representation in the mammalian cochlea? *Naturwissenschaften*, **78**, 570–3.

Echteler, S. M., Armand, E. & Dallos, P. (1989) Developmental alterations in the frequency map of the mammalian cochlea. *Nature*, **341**, 147–9.

Echteler, S. M, Fay, R. R. & Popper, A. N. (1994) Structure of the mammalian cochlea. In *Springer Handbook of Auditory Research*, vol. 4, *Comparative Hearing, Mammals*, ed. R. R. Fay & A. N. Popper, pp.134–72. New York: Springer-Verlag.

Fay, R. R. (1988) *Hearing in Vertebrates: A Psychophysics Data Book*. Winnetka: Hill-Fay Associates.

Feng, A.-S. & Vater, M. (1985) Organization of the cochlear nucleus of rufous horseshoe bats (*Rhinolophus rouxi*), frequencies and internal connections are arranged in slabs. *Journal of Comparative Neurology*, **235**, 529–53.

Fenton, M. B. (1995) Natural history and biosonar signals. In Springer *Handbook of Auditory Research, Hearing by Bats*, ed. A. N. Popper & R. R. Fay, pp. 37–87. New York: Springer-Verlag.

Fiedler, J. (1983) Vergleichende Cochlea-Morphologie der Fledermausarten *Molossus ater*, *Taphozous nudiventris kachhensis* und *Megaderma lyra*. PhD dissertation, University of Frankfurt.

Francis, C. M. & Habersetzer, J. (1998) Interspecific and intraspecific variations in echolocation call frequency and morphology of horseshoe bats, *Rhinolophus* and *Hipposideros*. In *Bat Biology and Conservation*, ed. T. H. Kunz & P. A. Racey, pp.169–81. Washington: Smithsonian Institution Press.

Habersetzer, J. & Storch, G. (1992) Cochlea size in extant Chiroptera and middle Eocene Microchiroptera from Messel. *Naturwissenschaften*, **79**, 462–6.

Harris, D. M. & Dallos, P. (1984) Ontogenetic changes in frequency mapping of mammalian ear. *Science*, **225**, 741–3.

He, D. Z. Z., Evans, B. N. & Dallos, P. (1994) First appearance and development of electromotility in neonatal gerbil outer hair cells. *Hearing Research*, **78**, 77–90.

Henson, M. M. (1973) Unusual nerve-fiber distribution in the cochlea of the bat *Pteronotus p. parnellii* (Gray). *Journal of the Acoustical Society of America*, **53**, 1739–40.

Henson, M. M. & Henson, O. W., Jr. (1988) Tension fibroblasts and the connective tissue matrix of the spiral ligament. *Hearing Research*, **35**, 237–58.

Henson, M. M. & Henson, O. W., Jr. (1991) Specializations for sharp tuning in the mustached bat: the tectorial membrane and spiral limbus. *Hearing Research*, **56**, 122–32.

Henson, M. M. & Rübsamen, R. (1996) The postnatal development of tension fibroblasts in the spiral ligament of the horseshoe bat, *Rhinolophus rouxi*. *Auditory Neuroscience*, **2**, 3–13.

Henson, M. M., Henson, O. W., Jr. & Goldman, L. J. (1977) The perilymphatic spaces in the cochlea of the bat, *Pteronotus p. parnellii* (Gray). *Anatomical Records*, **187**, 767.

Henson, O. W., Jr. (1970) The ear and audition. In *Biology of Bats*, vol. 2, ed. W. A. Wimsatt, pp.181–256. New York: Academic Press.

Henson, O. W., Jr., Schuller, G. & Vater, M. (1985) A comparative study of the physiological properties of the inner ear in Doppler shift compensating bats (*Rhinolophus rouxi, Pteronotus parnellii*). *Journal of Comparative Physiology A*, **157**, 587–97.

Holley, M. (1996) Outer hair cell motility. In *Springer Handbook of Auditory Research*, vol. 8, *The Cochlea*, ed. P. Dallos, A. N. Popper & R. R. Fay, pp. 386–435. New York: Springer-Verlag.

Huffman, R. F. & Henson, O. W., Jr. (1993) Labile cochlear tuning in the mustached bat II. Concomitant shifts in neural tuning. *Journal of Comparative Physiology A*, **171**, 735–48.

Kikuchi, K. & Hilding, D. (1965) The development of the organ of Corti in the mouse. *Acta Otolaryngology*, **60**, 207–22.

Köppl, C., Manley, G. A. & Gleich, O. (1993) An auditory fovea in the barn owl cochlea. *Journal of Comparative Physiology A*, **171**, 695–704.

Kössl, M. (1994a) Otoacoustic emissions from the cochlea of the 'constant frequency' bats, *Pteronotus parnellii* and *Rhinolophus rouxi. Hearing Research*, **72**, 59–72.

Kössl, M. (1994b) Evidence for a mechanical filter in the cochlea of the 'constant frequency' bats *Rhinolophus rouxi* and *Pteronotus parnellii. Hearing Research*, **72**, 73–80.

Kössl, M. (1997) Sound emission from cochlear filters and foveae – does the auditory sense organ make sense? *Naturwissenschaften*, **84**, 9–16.

Kössl, M. & Russell, I. J. (1995) Basilar membrane resonance in the cochlea of the mustached bat. *Proceedings of the National Academy of Science U.S.A.*, **92**, 276–9.

Kössl, M. & Vater, M. (1985a) Evoked acoustic emissions and cochlear microphonics in the mustache bat, *Pteronotus parnellii. Hearing Research*, **19**, 157–70.

Kössl, M. & Vater, M. (1985b) The cochlear frequency map of the mustache bat, *Pteronotus parnellii. Journal of Comparative Physiolology A*, **157**, 687–97.

Kössl, M. & Vater, M. (1995) Cochlear structure and function in bats. In *Springer Handbook of Auditory Research*, vol. 5, *Hearing by Bats*, ed. A. N. Popper & R. R. Fay, pp.191–235. New York: Springer-Verlag.

Kössl, M. & Vater, M. (1996a) A tectorial membrane fovea in the cochlea of the mustached bat. *Naturwissenschaften*, **2**, 89–92.

Kössl, M. & Vater, M. (1996b) Further studies on the mechanics of the cochlear partition in the mustached bat. II. A second cochlear frequency map derived from acoustic distortion products. *Hearing Research*, **94**, 78–87.

Kössl, M., Frank, G., Faulstich, M. & Russell, I. J. (1997) Acoustic distortion products as indicator of cochlear adaptations in Jamaican mormoopid bats. In *Diversity in Auditory Mechanics*, ed. E. R. Lewis, G. R. Long, R. F. Lyon, P. M. Narins, C. R. Steele & E. Hecht-Poinar, pp.42–9. Singapore: World Scientific.

Kraus, H.-J. & Aulbach-Kraus, K. (1981) Morphological changes in the cochlea of the mouse after the onset of hearing. *Hearing Research*, **4**, 89–102.

Kuhn, B. & Vater, M. (1995) The arrangements of F-actin, tubulin and fodrin in the organ of Corti of the horseshoe bat (*Rhinolophus rouxi*) and the gerbil (*Meriones unguiculatus*). *Hearing Research*, **84**, 139–56.

Lenoir, M., Puel, J.-L. & Pujol, R. (1987) Stereocilia and tectorial membrane development in the rat cochlea. A SEM study. *Anatomy and Embryology*, **175**, 477–87.

Liberman, M. C. (1982) The cochlear frequency map for the cat, Labeling auditory-nerve fibers of known characteristic frequency. *Journal of the Acoustical Society of America*, **72**, 1441–9.

Lim, D. J. (1986) Functional structure of the organ of Corti, A review. *Hearing Research*, **22**, 117–46.

Lim, D. J. (1987) Development of the tectorial membrane. *Hearing Research*, **28**, 9–21.

Long, G. R. & Schnitzler, H.-U. (1975) Behavioural audiogram from the bat, *Rhinolophus ferrumequinum*. *Journal of Comparative Physiology A*, **100**, 211–19.

Manley, G. A. (1996) Ontogeny of frequency mapping in the peripheral auditory system of birds and mammals, A critical review. *Auditory Neuroscience*, **3**, 199–214.

Metzner, W. (1989) A possible neural basis for Doppler-shift compensation in echolocating Horseshoe bats. *Nature*, **341**, 529–32.

Müller, M. (1991*a*) Frequency representation in the rat cochlea. *Hearing Research*, **51**, 247–54.

Müller, M. (1991*b*) Developmental changes of frequency representation in the rat cochlea. *Hearing Research*, **56**, 1–7.

Müller. M., Laube, B., Burda, H. & Bruns, V. (1992) Structure and function of the cochlea in the African mole rat (*Cryptomys hottentottus*), evidence for a low frequency acoustic fovea. *Journal of Comparative Physiology A*, **171**, 469–76.

Neuweiler, G. (1990) Auditory adaptations for prey capture in echolocating bats. *Physiological Reviews*, **70**, 615–41.

Nobili, R., Mammano, F. & Ashmore, J. (1998) How well do we understand the cochlea? *Trends in Neuroscience*, **21**, 159–67.

Pedersen, S. C. (1996) Skull growth and the presence of auxiliary fontanels in rhinolophoid bats (Microchiroptera). *Zoomorphology*, **116**, 205–12.

Pollak, G. D., Henson, O. W., Jr. & Novick, A. (1972) Cochlear microphonic audiograms in the 'pure tone' bat, *Chilonycteris parnellii parnellii*. *Science*, **176**, 66–8.

Pujol, R. & Uziel, A. (1988) Auditory development, peripheral aspects. In *Handbook of Human Growth and Developmental Biology*, vol. 1, part B, ed. E. Meisami & P. S. Timiras, pp.109–30. Boca Raton: CRC Press.

Pujol, R., Lenoir, M., Ladrech, S., Tribillac, F. & Rebillard, G. (1992) Correlation between the length of outer hair cells and the frequency coding of the cochlea. In *Auditory Physiology and Perception, Advances in Biosciences*, ed. Y. Cazals, L. Demany & K. Horner, pp. 45–52. New York: Pergamon Press.

Pujol, R., Lavigne-Rebillard, M. & Lenoir, M. (1995) Development of sensory and neural structures in the mammalian cochlea. In *Springer Handbook of Auditory Research*, vol. 5, *Development of the Auditory System*, ed. E. W. Rubel, A. N. Popper & R. R. Fay. New York: Springer-Verlag.

Pujol, R., Zajic, G., Dulon, D., Raphael, Y., Altschuler, R. A. & Schacht, J. (1991) First appearance and development of motile properties in outer hair cells isolated from guinea-pig cochlea. *Hearing Research*, **57**, 129–41.

Pye, A. (1966) The structure of the cochlea in Chiroptera. A selection of Microchiroptera from Africa. *Journal of Zoology (London)*, **162**, 335–43.

Pye, J. D. (1980) Adaptiveness of echolocation signals in bats. Flexibility in behaviour and in evolution. *Trends in NeuroScience*, 232–5.

Ramprashad, F., Money, K. E., Landolt, J. P. & Laufer, J. (1978) A neuroanatomical study of the cochlea of the little brown bat (*Myotis lucifugus*). *Journal of Morphology*, **160**, 345–58.

Romand, R. (1983) Development of the cochlea. In *Development of Auditory and Vestibular Systems*, ed. R. Romand, pp.47–88. New York: Academic Press.

Romand, R. (1987) Tonotopic evolution during development. *Hearing Research*, **28**, 117–23.

Roth, B. & Bruns, V. (1992*a*) Postnatal development of the rat organ of Corti. I. General morphology, basilar membrane, tectorial membrane and border cells. *Anatomy and Embryology*, **185**, 559–69.

Roth, B. & Bruns, V.(1992*b*) Postnatal development of the rat organ of Corti. II. Hair cell receptors and their supporting elements. *Anatomy and Embryology*, **185**, 571–81.

Rubel, E. W. (1984) Ontogeny of auditory system function. *Annual Review of Physiology*, **46**, 213–29.

Rubel, E. W., Lippe, W. R. & Ryals, B. M. (1984) Development of the place principle. *Annals of Otology, Rhinology, and Laryngology*, **93**, 609–15.

Rübsamen, R. (1987) Ontogenis of the echolocation system in the rufous horseshoe bat, *Rhinolophus rouxi*. Audition and vocalization in early postnatal development. *Journal of Comparative Physiology A*, **161**, 899–913.

Rübsamen, R. (1992) Postnatal development of central auditory frequency maps. *Journal of Comparative Physiology A*, **170**, 129–43.

Rübsamen, R. & Schäfer, M. (1990a) Ontogenesis of auditory fovea representation in the inferior colliculus of the Sri Lankan rufous horseshoe bat, *Rhinolophus rouxi*. *Journal of Comparative Physiology A*, **167**, 757–69.

Rübsamen, R. & Schäfer, M. (1990b) Audiovocal interactions during development? Vocalisation in deafened young horseshoe bats vs. audition in vocalisation impaired bats. *Journal of Comparative Physiology A*, **167**, 771–84.

Rübsamen, R., Neuweiler, G. & Sripathi, K. (1988) Comparative collicular tonotopy in two bat species adapted to motion perception, *Hipposideros speoris* and *Megaderma lyra*. *Journal of Comparative Physiology A*, **163**, 271–85.

Schmidt, S., Türke, B. & Vogler, B. (1984) Behavioural audiogram from the bat, *Megaderma lyra* (Geoffroy, 1810; Microchiroptera). *Myotis*, **22**, 62–9.

Schnitzler, H.-U. (1987) Echoes of fluttering insects: information for echolocating bats. In *Recent Advances in the Study of Bats*, ed. M. B. Fenton, P. Racey & J. M. V. Rayner, pp. 226–44. Cambridge: Cambridge University Press.

Schnitzler, H.-U. & Henson, O. W., Jr. (1980) Performance of airborne animal sonar systems, I. Microchiroptera. In *Animal Sonar Systems, Proceedings NATO Advanced Study Institute*, ed. R. G. Busnel & J. F. Fish, pp. 109–81. New York: Plenum Press.

Schnitzler, H.-U. & Kalkow, E. (1998) How echolocating bats search and find food. In *Bat Biology and Conservation*, ed. T. H. Kunz & P. A. Racey, pp. 183–97. Washington: Smithsonian Institution Press.

Schuller, G. (1980) Hearing characteristics and Dopplershift compensation in South Indian CF–FM bats. *Journal of Comparative Physiology A*, **139**, 349–56.

Schuller, G. & Pollak, G. D. (1979) Disproportionate frequency representation in the inferior colliculus of Doppler-compensating greater horseshoe bats, evidence for an acoustic fovea. *Journal of Comparative Physiology A*, **132**, 47–54.

Schuller, G. & Rübsamen, R. (1981) Laryngeal nerve activity during pulse emission in the CF–FM bat, *Rhinolophus ferrumequinum*. I. Superior laryngeal nerve (External motor branch). *Journal of Comparative Physiology A*, **143**, 317–21.

Schuller, G., Beuter, K. & Rübsamen (1975) Dynamic properties of the compensation system for Doppler shifts in the bat, *Rhinolophus ferrumequinum*. *Journal of Comparative Physiology*, **97**, 113–25.

Slepecky, N. B. (1996) Structure of the mammalian cochlea. In *Springer Handbook of Auditory Research*, vol. 8, *The Cochlea*, ed. P. Dallos, A. N. Popper & R. R. Fay, pp. 44–130. New York: Springer-Verlag.

Simmons, J. A. & Stein, R. (1980) Acoustic imaging in bat sonar, echolocation signals and the evolution of echolocation. *Journal of Comparative Physiology A*, **135**, 61–84.

Suga, N. & Jen, P.-H. S. (1977) Further studies on the peripheral auditory system of CF–FM bats specialized for fine frequency analysis of Doppler shifted echoes. *Journal of Experimental Biology*, **69**, 207–32.

Steele, C. M. (1997) Three dimensional modeling of the cochlea. In *Diversity in Auditory Mechanics*, ed. E. R. Lewis, G. R. Long, R . F. Lyon, P. M. Narins, C. R. Steele & E. Hecht-Poinar, pp. 455–62. Singapore: World Scientific.

Vater, M. (1988) Light microscopic observations on cochlear development in horseshoe bats. In *Animal Sonar Systems, Proceedings NATO Advanced Study Institute*, ed. P. E. Nachtigall & P. W. B. Moore, pp. 225–40. New York: Plenum Press.

Vater, M. (1997) Evolutionary plasticity of cochlear design in echolocating bats. In *Diversity in Auditory Mechanics*, ed. E. R. Lewis, G. R. Long, R. F. Lyon, P. M. Narins, C. R. Steele & E. Hecht-Poinar, pp. 49–55. Singapore: World Scientific.

Vater, M. (1998) Adaptations of the auditory periphery of bats for echolocation. In *Bat Biology and Conservation*, ed. T. H. Kunz & P. A. Racey, pp. 231–47. Washington: Smithsonian Institution Press.

Vater, M. & Duifhuis, D. (1986) Ultra-high frequency selectivity in the Horseshoe bat: Does the bat use an acoustic interference filter? In *Auditory Frequency Selectivity, Proceedings NATO Advanced Study Institute*, ed. B. C. J. Moore & R. D. Patterson, pp. 23–31. New York: Plenum Press.

Vater, M. & Kössl, M. (1996) Further studies on the mechanics of the cochlear partition in the mustached bat. I. Ultrastructural observations on the tectorial membrane and its attachments. *Hearing Research*, **94**, 63–78.

Vater, M. & Lenoir, M. (1992) Ultrastructure of the horseshoe bat's organ of Corti. I. Scanning electron Microscopy. *Journal of Comparative Neurology*, **318**, 367–79.

Vater, M. & Rübsamen, R. (1988) Postnatal development of the cochlea in Horseshoe bats. In *Cochlear Mechanisms. Structure, Function and Models*, ed. J. P. Wilson & D. T. Kemp, pp. 217–25. New York: Plenum Press.

Vater M. & Rübsamen, R. (1992) Ontogeny of frequency maps in the peripheral auditory system of Horseshoe bats. *Proceedings Third International Congress Neuroethology*, Montreal, pp.31–2.

Vater, M. & Siefer, W. (1995) The cochlea of *Tadarida brasiliensis*, specialized functional organization in a generalized bat. *Hearing Research*, **91**, 178–95.

Vater, M., Feng, A. S. & Betz, M. (1985) An HRP-study of the frequency-place map of the Horseshoe bat cochlea, morphological correlates of the sharp tuning to a narrow frequency band. *Journal of Comparative Physiology A*, **157**, 671–86.

Vater, M., Lenoir, M. & Pujol, R. (1992) Ultrastructure of the Horseshoe bat's organ of Corti. II. Transmission electron microscopy. *Journal of Comparative Neurology*, **318**, 380–91.

Vater, M., Lenoir, M. & Pujol, R. (1997) Development of the organ of Corti in Horseshoe bats, Scanning and transmission electron microscopy. *Journal of Comparative Neurology*, **377**, 520–34.

Weaver, S. P. & Schweitzer, L. (1994) Development of gerbil outer hair cells after the onset of cochlear function, An ultrastructural study. *Hearing Research*, **72**, 44–52.

Xie, D. H., Henson, M. M., Bishop, A. L. & Henson, O. W., Jr. (1993) Efferent terminals in the cochlea of the mustached bat, quantitative data. *Hearing Research*, **66**, 81–90.

Zook, J. M. & Leake, P. A. (1989) Connections and frequency representation in the auditory brainstem of the mustache bat, *Pteronotus parnellii. Journal of Comparative Neurology*, **290**, 243–61.

Zwislocki, J. J. (1986) Analysis of cochlear mechanics. *Hearing Research*, **22**, 155–69.

6

Skull growth and the acoustical axis of the head in bats

INTRODUCTION

In this chapter, I place the extraordinary diversity in microchiropteran skull size and shape within a rather simplistic framework. That is, despite all other craniodental adaptations, the microchiropteran head must function as an efficient acoustical horn during echolocation. This truism becomes infinitely more interesting when one considers that echolocatory calls are either emitted directly from the open mouth (oral-emitters), or forced through the confines of the nasal passages (nasal-emitters). Given that oral-emission is the primitive state, the advent of nasal-emission is viewed as a complex morphological innovation that required a substantial redesign of the microchiropteran rostrum: 1) the nasal passages must be reoriented and aligned with the direction of flight, and 2) the nasal passages must exhibit dimensions that provide for the efficient transfer of sound (resonance) through the adult skull. In the following treatment, I draw examples from developmental studies and functional morphology to illustrate how evolution has solved this intriguing design problem.

Spatial competition and the packaging of the fetal head

The dynamic nature of the developing skeletal system is all too frequently overlooked in the classroom where the skull is often presented as an immutable structure into which the brain, ears, and eyes are stuffed during development. Rather, the converse is a more accurate view; cranial growth and form are 'soft tissue' phenomena affected only secondarily by osteological development. Indeed, early in development, the differential growth of the brain and pharynx governs the shape of the chondrocranium and influences the forms of the embryonic neuro- and viscerocrania (Ranly 1980; Klima 1987; Hanken & Thorogood 1993). Later, within the envelope of skin that confines the head, volumetric changes in the growing brain, brainstem, eyes, tongue, and pharynx affect bone growth via forces transmitted through the dura and periosteum to adjacent bones and sutures (Blechschmidt 1976a, b). The ensuing mechanical competition for space among the various cranial components effects a cascade of modifications throughout the growing skull in the shape, position, and orientation of distant components (Haines 1940; Silver 1962; Bosma 1976; Hanken 1983; Pedersen 1995). Evolutionarily, such patterns of differential growth among the braincase, trachea, and pharynx have been cited as driving forces behind gross morphogenetic changes in primate skulls (Baer & Nanda 1976; Blechschmidt 1976a, b; Enlow 1976; Moss 1976; Schön 1976; Schachner 1989). Such packaging constraints are accommodated by the developmental plasticity of each system (e.g., pharynx, braincase, otic capsules) in proportion to tissue composition, compliance in growth rates, the gross translation of elements or distortion of affected structures in situ (Müller 1990; Raff et al. 1990). When plastic mechanisms exceed their spatial or mechanical thresholds and fail, new morphogenetic paths may come into play that effect gross morphological changes in skull form (Herring & Lakars 1981; Hanken 1983; Herring 1985).

Bone as a plastic entity

Throughout ontogeny, the changing size, shape, and orientation of each bony element reflect the dynamic interplay between the rate at which bone tissue responds by further mineralization and/or remodeling of existing bone surfaces and the volumetric expansion of underlying capsular spaces, such as the nasal capsules or brain (Sarnat & Shanedling 1979; Koskinen-Moffett et al. 1981; Choo & Covell 1996). Whereas the shape and orientation of each element is strongly influenced by the enclosed volume, the subsequent ossification of each element is independent of capsular growth. This independence permits epigenetic remodeling of the various skull elements to track developmental variation in the enclosed soft tissues (Haines 1940; Alberch & Alberch 1981; Hanken 1983, 1984; Griffioen & Smit-Vis 1985; Smit-Vis & Griffioen 1987; Starck 1989; Ross & Ravosa 1993; Ross & Henneberg 1995). A great deal of what we know about the reciprocity between bone growth and the shape of an enclosed space has come from studies of pathological suture formation (e.g., synostoses: Koskinen-Moffett & Moffett 1989; Pedersen & Anton 1998), instances of malformation of the central nervous system (e.g., hydrocephaly, and anencephaly: Sperber 1989), and numerous studies concerning the experimental derangement of developing bone and sutures (Babler et al. 1987). I draw attention to such exaggerated examples to emphasize the epigenetic plasticity of bone and its ability to accommodate anatomical distortion, be it subtle accommodation of inter-individual variation, or gross alterations that accompany gross morphological innovation.

In addition to responding to general growth pressures, developing bone must also accommodate changes in the mechanical environment such as changes in the functional loading patterns attendant to shifts in behavior (e.g., the transition from suckling to mastication during weaning in mammals; Herring & Lakars 1981; Herring 1985). Examples of such directive interactions between muscle forces and bone development are found throughout the cranial vault. For example, the sutural complexity of the cranial vault and occiput decrease when masticatory and cervical muscles have been experimentally denervated or devascularized, and lacking appropriate muscular action, many bone features, such as the coronoid process, do not develop (Washburn 1947; Spyropoulos 1977; Von Schumacher et al. 1986, 1988; Byrd 1988; Kylamarkula 1988). Hoyte (1987) has suggested that masticatory muscles influence the shape of the cranial vault, and hence the shape and disposition of the brain itself. If true, the unique skulls of rhinolophid bats deserve further attention. Rhinolophid braincases are elongate and are distinctly 'compartmentalized,' with obvious strictures that partition the olfactory lobes from the cerebrum, and the cerebrum from the brainstem. Might this internal buttressing be a response to large, complex temporalis muscles or simply a developmental artifact of an elongate brain?

Brain growth and skull form

Interactions between the developing brain and cranial base influence a wide range of morphogenetic changes to the midface throughout development (cattle: Julian et al. 1957; rodents: bats: Sperry 1972; Moss 1976; humans: Thilander & Ingervall 1973; Moore 1983; Sperber 1989). For example, in primates, the anterior and middle cranial fossae (olfactory bulbs and cerebrum) compete with the orbits for space, thereby influencing the shape, size, position, and orientation of the interorbital septum (Enlow & McNamara 1973). Clearly, brain development has a profound impact on vertebrate craniofacial development and evolution. However, it would appear that brain volume per se does not play a strong role in the craniofacial evolution of bats (Pedersen 1993). Rather, the relative size of

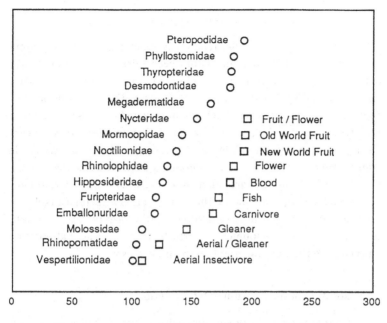

Encephalization Index

Figure 6.1 The encephalization index relates the average size of the brain in each family to the vespertilionid condition (index value of 100; after Stephan *et al.* 1981). Aerial insectivores have small brains and occupy the lower end of this index; frugivorous taxa possess larger brains and utilize more cluttered environments and dominate the upper end of the index. Brain size is closely associated with habitat use. Circles represent the average values of all known chiropteran families while squares represent the average values for foraging preference. (From Pedersen 1993.)

the adult chiropteran brain is associated with the occupation of a specific aerial niche (Eisenberg & Wilson 1978; Stephan *et al.* 1981; Jolicoeur *et al.* 1984). Aerial insectivores and foliage gleaners are found in open habitats or along forest boundaries and possess relatively small brains (Fig. 6.1) Frugivorous and carnivorous taxa have relatively larger brains and forage in more complex, cluttered habitats (Eisenberg & Wilson 1978; Stephan *et al.* 1981; Jolicoeur *et al.* 1984). Therefore, shape of the basicranium and cranial vault are affected by the gross morphology and organization of the brain rather than its size *per se* (Jolicoeur *et al.* 1984). With respect to chiropteran evolution, brain anatomy and development are clearly fertile areas for future research that have only just been touched upon (Pirlot & Bernier 1991; see Reep & Bhatnagar, this volume).

ONTOGENY OF THE CHIROPTERAN SKULL – CEPHALOMETRY

The study of the angular relationships among the various components of the skull (cephalometry) has been raised to a fine art in the
dental profession where lateral head radiographs are an indispensable
tool for planning dental treatment. Outside of the dental profession,
cephalometry has gone relatively unappreciated. This is unfortunate
because cephalometric analysis provides a rather unique method for evaluating functional morphology or for tracing evolutionary trends in
cranial evolution. Specifically, cephalometry provides a size-free description of the basic internal arrangement of the head using internal landmarks and anatomical planes that are 1) otherwise unavailable for
morphometric analysis (e.g., sella turcica, cribriform plate), and 2) relatively immune to developmental perturbations, such as malnutrition
(Pucciarelli & Dressino 1996) and sexual dimorphism.

Rotation of the rostrum in fetal mammals

One of the most obvious examples of the efficacy of cephalometry is
demonstrated in studies concerning the elevation of the mammalian
rostrum early in development. The mammalian head begins growth
tucked firmly against the chest wall from where it rotates dorsad about
the cervical axis. Simultaneously, the facial component of the skull
rotates dorsad about the braincase. The motive forces behind these rotations are complex, as are the forces that ultimately determine the location and final orientation of the adult rostrum. Certainly, translocation
of the rostrum is limited in rate and direction by the ability of adjacent
bony structures to respond (i.e., their respective abilities to get out of each
other's way). Such restrictions are clearly mediated by the adherent periosteum and sutural systems that envelop each bony element (McLain &
Vig 1983; Muhl & Gedak 1986). In primates, it has been argued that spatial
competition between the developing eyes and brain determines where
the palate comes to rest in the adult skull (Radinsky 1968; Spatz 1968;
Ross & Ravosa 1993). In other mammals, rotation has been attributed to
elongation of the snout, growth of the cranial base, and tooth eruption
(Starck 1952; Julian et al. 1957; Sperry 1972; Servoss 1973; Thilander &
Ingervall 1973; Moss 1976; Gasson & Lavergne 1977; Moore 1983;
Schachner 1989; Pedersen 1993). However, Microchiroptera interpret
their environment acoustically and follow a different set of construc-
tional rules based on the use of the facial skeleton as an acoustical horn.
As such, the angular arrangement of the various skull components pro-

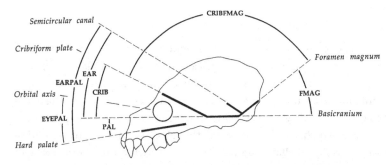

Figure 6.2. Anatomical planes and cephalometric angles are shown super-imposed on a tracing of an *Artibeus jamaicensis* skull. See text for description of each cephalometric angle and anatomical plane. (From Pedersen 1993.)

vides a unique insight into the internal form and function of the micro-chiropteran head.

Anatomical planes and cephalometric angles

Four anatomical planes are easily identified in the mammalian skull: lateral semicircular canals, palate, foramen magnum, and cribri-form plate (Fig. 6.2). Here, they are used to delineate the inertial and acoustic axes of the head and relate the general organization of the brain-case to the rest of the body. A brief explanation of each is in order.

Inner ear

The information concerning balance and orientation provided by the semicircular canals of the inner ear are of great importance to flying vertebrates, and absolutely critical to bats that often navigate through complex habitat without the aid of visual cues. Here, I use the orientation of the lateral semicircular canals to reflect the manner in which a bat holds its head during flight (de Beer 1937; Delattre & Fennart 1960).

Hard palate

The functional anatomy of the hard palate is multifaceted. It must support the dental battery and reinforce the rostrum during mastication, and it must accord sufficient space for tidal airflow across its superior surface. Because the phonal axis of the head must be aligned with the long axis of either the oral or the nasal cavities, the palate is an important landmark that lies just above, or just below this axis.

Foramen magnum

The plane of the foramen magnum represents the boundary between the occiput and the cervical axis (Baer & Nanda 1976; Schön 1976). This plane influences head posture and reflects the degree of flexibility found in the cranio-cervical axis (Fenton & Crerar 1984). The maintenance of the airway also influences the cranio-cervical angulation of the head, which, in turn, influences the relative position and orientation of the stylohyoid chain and its attendant musculature (Bjork & Skieler 1972; Gaskell 1974; Solow & Greve 1979; Herring 1985; Tallgren & Solow 1987; Behlfelt et al. 1990). For example, the stylohyoid chain must increase in length to accommodate increases in pharyngeal and esophageal diameters and the bulk of the tongue and prevertebral musculature throughout ontogeny.

Cribriform plate

The plane of the cribriform plate consists of a bony septum that lies between the nasopharynx and the brain and partitions the facial and neural components of the skull (Ranly 1980). Its relative position reflects the volumetric increase of the braincase during brain growth (Young 1959; Moss 1976), and is responsive to the structural mechanics of the midface, the volume of the olfactory lobes of the telencephalon (Frahm 1981; Jolicoeur et al. 1984; Pirlot & Bernier 1991), the development of the interorbital septum (Haines 1940), and the position of the rostrum relative to the anterior cranial base (Starck 1952).

Two cephalometric angles are useful in framing these four anatomical planes in a functional context (Pedersen 1993a, 1995). EARPAL relates the hard palate and middle ear, whereas CRIBFMAG relates the cribriform plate to the foramen magnum (Fig. 6.2). Here, and in subsequent sections, the discussion of fetal material and growth trajectories are based upon developmental (cross-sectional) series of nine species of Chiroptera (*Rousettus celebensis, Eptesicus fuscus, Lasiurus borealis, Pteronotus parnellii, Taphozous georgianus, Hipposideros armiger, H. galeritus, Rhinolophus affinus, Artibeus jamaicensis;* Pedersen 1995, 1998). All fetal material was cleared by maceration and differentially stained for bone and cartilage (Hanken & Wassersug 1981; Klymkowsky & Hanken 1991). Each fetus was staged using the index STAGE to describe the extent to which a fetal skull has reached its adult size, and to ameliorate the differences in head size and shape among species.

Cephalometry of fetal bats

Though fetuses of oral-emitting and nasal-emitting species are grossly indistinguishable early in development, the internal arrangement of their heads is well established shortly thereafter, before the skull has begun to ossify, long before the approximation of the primary suture systems, and well before the forces of mastication begin to affect skull morphogenesis. Initially, the angular constructions of oral- and nasal-emitting fetal heads are very similar and share common values of EARPAL and CRIBFMAG (80° and 120°, respectively). Species-specific skull morphology becomes increasingly more apparent, but always within the framework of either the nasal-emitting or oral-emitting cranial form. EARPAL of the four oral-emitters (*E. fuscus*, *L. borealis*, *P. parnellii* and *T. georgianus*) decreases dramatically throughout development, whereas EARPAL remains constant in the three Old World nasal-emitters (*H. armiger*, *H. galeritus* and *R. affinus*). The New World nasal-emitter and the non-emitter (*Rousettus celebensis* and *A. jamaicensis*, respectively) exhibit intermediate values of EARPAL. A similar pattern occurs in the braincase – CRIBFMAG (Fig. 6.3).

These changes show that the palate of oral-emitting fetuses rotates dorsad about the basicranium, as the occiput becomes more erect upon the floor of the braincase. Conversely, the relative positions of these anatomical planes remain fairly static throughout development in nasal-emitting taxa. The cartoon of *Eptesicus* (oral-emitter) and *Artibeus* (New World nasal-emitter) developmental series depicts this dichotomy (Fig. 6.4). As is the case with all oral-emitting taxa, the rostrum, hence hard palate, is aligned with or elevated above the basicranium (i.e., an airorhynchal skull type) and the echolocative call is forced directly out through the mouth. This basic skull form (*Bauplan*) is the plesiomorphic condition for mammals and clearly exemplified by emballonuroids and vespertilionoids. Conversely, palates of nasal-emitting bats are retained ventral to the basicranial axis, i.e., a klinorhynchal skull type (Starck 1952; Freeman 1984; Pedersen 1993*a*). In these derived forms, the echolocative call is forced out through the nostrils as exemplified by rhinolophoids and phyllostomids.

This dichotomy between oral-emitting and nasal-emitting *Baupläne* involves a change in head posture which has required a compensatory rotation of the otic capsules to keep the lateral semicircular canals aligned with the acoustical axis of the head. The orientation of the lateral semicircular canals differs dramatically between oral- (*Myotis* – 0° above the palate) and nasal-emitting taxa (*Nycteris* – 75° above the palate). As a

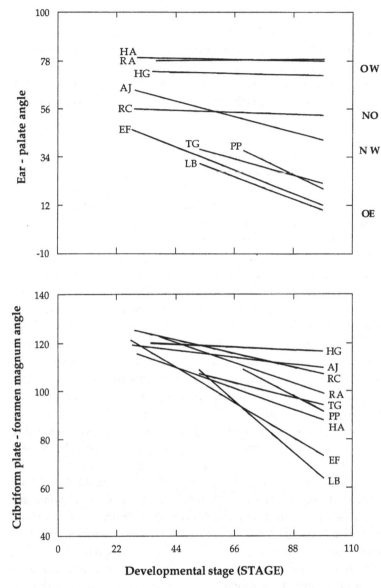

Figure 6.3. (Upper) EARPAL vs. developmental stage (STAGE) shows the clustering of all of the Old World nasal-emitting series together apart from all
oral-emitting series. All trajectories diverge from an extrapolated common
(primitive) angle between 60° and 80°. (Lower) CRIBFMAG vs. developmental
stage (STAGE) trajectories decrease from a common angle of 130° (reflecting
brain growth) and do not exhibit any emission-specific patterns. STAGE =
[{√(embryo braincase length×width) / √(embryo skull length×width)} /
{√(adult braincase length×width) / √(adult skull length×width)}]×100.

result of these otic and palatal rotations, oral-emitting fetuses 'migrate' (ontogenetically) away from the boundaries of a common 'fetal morphospace' towards various locations within the more broadly defined 'oral-emitting morphospace.' As discussed above, this migration (rostral rotation) is exhibited during the ontogeny of several non-echolocating mammals as well (e.g., rodents, rabbits, and primates: Schneiderman 1992) and reflects the primitive mammalian condition (Wimberger 1991). Of great interest, nasal-emitting fetuses remain within the 'fetal morphospace' throughout ontogeny as well as phylogeny (Fig. 6.5). This constraint, or 'morphostasis' is the derived condition in Microchiroptera and concerns only the angular arrangement of the skull and should not be confused with paedomorphic arguments.

Unlike microchiropterans, the form of non-echolocating megachiropteran skulls is not restricted by the demands of phonation. Mechanically, the relative position of the microchiropteran hard palate reflects the use of either the oral or nasal cavity as an acoustical horn. Intermediate positions of the hard palate are poorly suited biomechanically for echolocation as the call would be directed either above or below the mean flight path. Without constraints on the relative position of the hard palate, the non-echolocating megachiropteran skull appears as a morphological intermediate between the extreme oral-emitting forms and the Old World nasal-emitting forms (rhinolophids) in Figs. 6.6 and 6.7 (Pedersen 1993).

ONTOGENY OF THE CHIROPTERAN SKULL – ANATOMICAL LANDMARK DATA

The innovation and evolutionary potential of the nasal-emitting *Bauplan* have relied upon the morphogenetic plasticity of adjacent skeletal elements to accommodate changes throughout development and go on to function adequately in the adult. This balancing act is difficult given that the growth of the mammalian rostrum and pharynx are influenced by several factors: the respiratory tidal airflow (Solow & Greve

Figure 6.3 (*cont.*)

Taxonomic abbreviations are as follows: RC – *Rousettus celebensis*, EF – *Eptesicus fuscus*, LB – *Lasiurus borealis*, PP – *Pteronotus parnellii*, TG – *Taphozous georgianus*, HA – *Hipposideros armiger*, HG – *Hipposideros galeritus*, RA – *Rhinolophus affinus*, AJ – *Artibeus jamaicensis*, OW – Old World nasal-emitter, NW – New World nasal-emitter, OE – Oral-emitter, NO – Non-emitter (Megachiroptera). (From Pedersen 1995.)

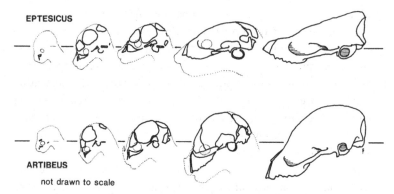

EPTESICUS

ARTIBEUS
not drawn to scale

Figure 6.4. The distinctive dorsad rotation of the orofacial complex in oral-emitters is illustrated by *Eptesicus fuscus*. The orofacial complex in nasal-emitters remains 'tucked' throughout development (*Artibeus jamaicensis*). (From Pedersen 1995.)

1979), the forces of tooth eruption and auto-occlusal mechanisms (Lakars & Herring 1980), the tissue pressures from the lips and tongue (Proffit 1978), the organization and coordination of each muscle mass in proportion to the complexity of the dentition and associated dynamics of mastication (Herring 1985), and phonation/echolocation (Roberts 1972, 1973; Hartley & Suthers 1988; Suthers *et al.* 1988). Given this complex dynamic, it is of great interest that both rhinolophoids (Old World leaf-nosed bats) and phyllostomids (New World leaf-nosed bats) exhibit the anatomical requirements for the efficient emission of ultrasound through the nostrils (Hartley & Suthers 1987, 1988, 1990; Simmons 1980; Simmons & Stein 1980; Pye 1988). Though the cephalometric data suggest a classic example of convergent evolution, it is prudent to draw upon multivariate analyses of landmark data to investigate the developmental paths by which two taxonomically distinct clades of Microchiroptera (rhinolophoids and phyllostomids) arrived at their nasal-emitting *Baupläne*.

Multivariate approach to growth trajectories

Cranial landmark data were taken from the same developmental series as described above (see Pedersen 1995 for detailed methods). These measurements (unpublished data) were grouped into distinct suites of variables according to function and/or anatomical relatedness. Each grouping was subjected to canonical analysis to identify covariance patterns among variables ($\sim$10) within each suite (jaw, ear, rostrum, cranial base, midface, temporomandibular joint). Bivariate plots of canonical

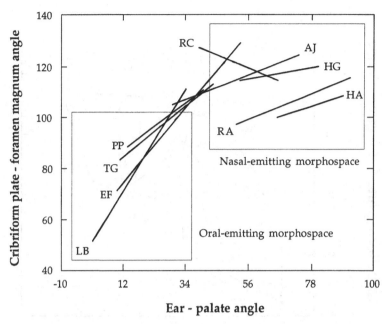

Figure 6.5. Bivariate plot of EARPAL vs. CRIBFMAG for fetuses and juveniles. Ontogenetic trajectories of oral-emitting and nasal-emitting developmental series are presented in this scatterplot. Each trajectory moves from right to left across the plot. Nasal-emitting taxa (*Hipposideros armiger, H. galeritus, Rhinolophus affinus, Artibeus jamaicensis*) are clustered together in the upper right-hand corner of the plot, whereas the trajectories of oral-emitting taxa (*Eptesicus fuscus, Lasiurus borealis, Pteronotus parnellii, Taphozous georgianus*) 'migrate' away from this nasal-emitting morphospace down towards the lower left-hand corner of the plot into an oral-emitting morphospace. (From Pedersen 1995.)

variates map each taxon's developmental path through multivariate space. Taxa that share similar covariance schemes (i.e., similar developmental patterns) co-occupy regions in multivariate space. An important capability of canonical analysis is the *post-hoc* identification of individuals. That is, known fetuses may be drawn out of the pool at random and re-identified to test the precision of the model. Suites of variables that concern the jaw, temporomandibular joint, cranial base, or ear capsule distinguish poorly among taxa. For example, fetal material was classified to the correct taxon only 52% of the time, and to the correct emission-form in only 68% of the cases using 'jaw' characters (Table 6.1). On the other hand, suites of variables concerning the rostrum or midface clearly distinguish among taxa (i.e., fetuses were correctly identified to both the

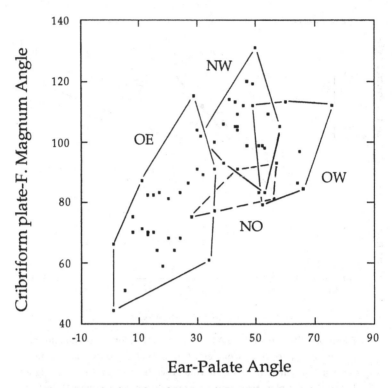

Ear-Palate Angle

Figure 6.6. Bivariate plot of EARPAL vs. CRIBFMAG for adults. Non-emitters (Megachiroptera) are intermediate between the oral-emitting and the two nasal-emitting groups. The scatter of the Old and New World nasal-emitting taxa overlaps along both axes. OW – Old World nasal-emitter, NW – New World nasal-emitter, OE – oral-emitter, NO – Megachiroptera. (From Pedersen 1998.)

correct taxon and emission form 97% of the time; Figs. 6.8, 6.9). Of interest, nasal-emitting and oral-emitting fetuses are rarely mistaken for each other, however, individual fetuses are frequently misclassified within the same emission type.

The clarity of these separations lies in the fact that the skull of Old World nasal-emitting bats is characterized by a short hard palate, large-bore choanae, and a relatively long naso-laryngo-pharynx (estimated by the distance between the posterior nasal spine and the spheno-occipital synchondrosis). This construction is related to the unique, laryngo-nasal junction between the soft palate and the cartilages of the larynx found in the Old World nasal-emitters (Matsumura 1979; Hartley & Suthers 1988). The anatomical coupling of this junction is remarkable in the completeness of its seal and presumably requires a repositioning of all musculos-

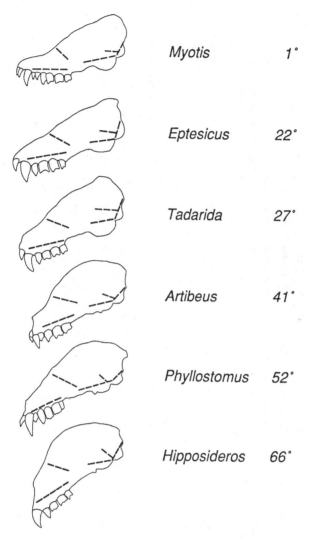

Figure 6.7. Cartoon depicting the wide range of rostral rotation (EARPAL values) within the suborder Microchiroptera. *Myotis*, *Eptesicus*, and *Tadarida* are oral-emitting taxa, while *Artibeus*, *Phyllostomus*, and *Hipposideros* are nasal-emitting taxa. (From Pedersen 1993.)

keletal elements associated with the soft palate and larynx (i.e., pterygoid width, choanal bore, pharyngeal length, and otic capsule separation) and is responsible for the clustering of the Old World nasal-emitting taxa in multivariate space. The Megachiroptera are clustered apart from the other developmental series because of their relatively large, albeit unspecialized, choanae and pterygoid complexes.

Table 6.1. *Canonical analysis post-hoc identification of individuals taken from the same developmental series using cranial landmark data. Distinct suites of variables were bundled to identify covariance patterns among variables (jaw, ear, rostrum, cranial base, midface, temporomandibular joint). For demonstration purposes, only the Jaw and Rostral groupings are presented. Tabular values are the number of individual fetuses identified for each actual–predicted (row–column) couplet. Abbreviations: EF – E. fuscus; LB – L. borealis; PP – P. parnellii; TG – T. georgianus; HG – H. galeritus; HC – H. cervinus; RA – R. affinus; AJ – A. jamaicensis; RC – R. celebensis; UNK – unidentified individuals*

		EF	LB	PP	TG	HG	HC	RA	AJ	RC	UNK	Total	Classification
Jaw structures													
Oral Emitters	E. fuscus	6				1					14	21	52% Correct by Taxon
	L. borealis	4	4			1	2					11	
	P. parnellii			13				2			2	17	
	T. georgianus		1		5	2					1	9	
Nasal Emitters	H. galeritus			2	2	2			1		1	8	68% Correct by Emission
	H. cervinus		1	3	3	1	6	1			1	16	
	R. affinus	1	2			3	3	3	1		1	14	
	A. jamaicensis	1		3			1		6	2	6	18	
No Echo	R. celebensis									4	4	9	
Total		12	8	21	10	10	12	6	8	6	30	= 123	
Rostral structures													
Oral Emitters	E. fuscus	8									13	21	97% Correct by Taxon
	L. borealis		11									11	
	P. parnellii			12							5	17	
	T. georgianus				8						1	9	

Nasal Emitters													97% Correct by Emission
Nasal Emitters	H. galeritus					7						1	8
	H. cervinus						14					2	16
	R. affinus							11				3	14
	A. jamaicensis	1								10		7	18
No Echo	R. celebensis				1						7	1	9
Total		9	11	12	9	7	14	11	10	10	7	33 =	123

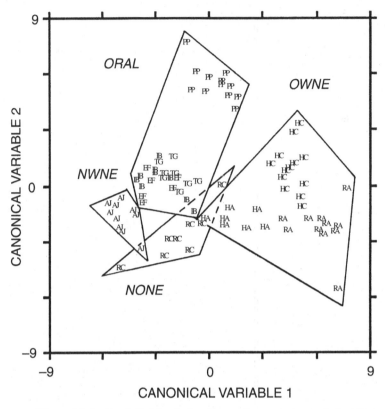

Figure 6.8. Scatterplot of the first and second canonical axes derived from the canonical analysis of the following rostral characters: canine width, palate width, external narial width, choanal diameter, temporomandibular joint width, pterygoid width, hard palate length, pharynx length, interorbital width, infraorbital foramen width, and toothrow length. Taxonomic abbreviations are as follows: OWNE– Old World nasal-emitter, NWNE– New World nasal-emitter, ORAL – oral-emitter, NONE – Megachiroptera, RC – *Rousettus celebensis*, EF – *Eptesicus fuscus*, LB – *Lasiurus borealis*, PP – *Pteronotus parnellii*, TG – *Taphozous georgianus*, HA – *Hipposideros galeritus*, HC – *H. cervinus*, RA – *Rhinolophus affinus*, AJ – *Artibeus jamaicensis*.

Regression analyses and growth trajectories

Whereas a multivariate approach is useful in identifying patterns among large numbers of variables, several individual characters were analyzed more carefully using regression analyses. In general, all features increase in size throughout development (Fig. 6.10), the notable exception being the minimal growth in the length of the hard palate in the Old

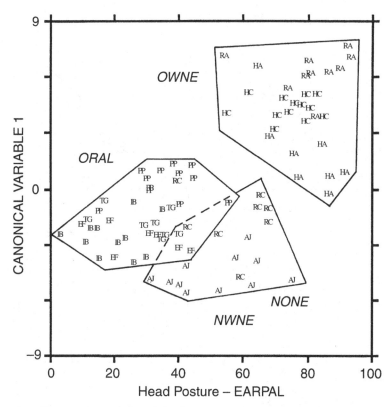

Figure 6.9. Scatterplot of the first canonical axis (previous figure) plotted against EARPAL. Note the clean separation of the OWNE and the overlay of the NWNE and NONE. Taxonomic abbreviations as for Fig. 6.8.

World nasal-emitter *Rhinolophus*. Individual structures exhibit a fair degree of variation early in development, though this variability tends to decrease after interaction and integration into other functional units in the head (Kay 1986; Zelditch & Carmichael 1989). Old World nasal-emitting taxa share a cohesive developmental trajectory. Oral-emitters are equally cohesive amongst themselves, but the trajectories of the New World nasal-emitter (*Artibeus*) and the megachiropteran (*Rousettus*) exhibit an unpredicted similarity with each other. This resemblance is most likely related to the fact that both *Rousettus* and *Artibeus* exhibit well-developed visual and olfactory acuities.

The simplistic dichotomy between oral- and nasal-emitting skulls must therefore be revised to include the observation that there are two distinct developmental paths that lead to the nasal-emitting construct. Subsequently, there are, in fact, three fundamental evolutionary programs

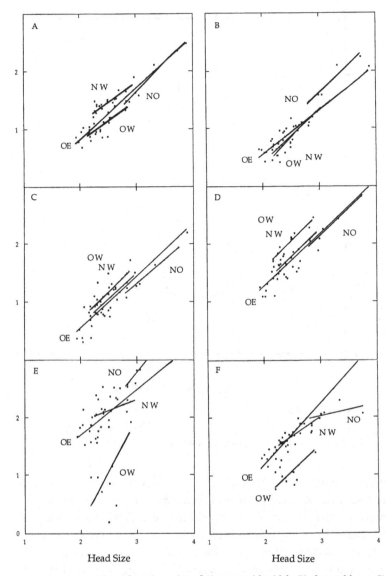

Figure 6.10. Growth trajectories of A) pterygoid width, B) choanal bore, C) narial width, D) pharyngeal length, E) palatal length, and F) interorbital width plotted against log head size (the cube root of the greatest length of skull × zygomatic breadth × midfacial depth). Taxonomic abbreviations are as follows: OW – Old World nasal-emitter, NW – New World nasal-emitter, OE – oral-emitter, NO – non-emitter (Megachiroptera).

for the microchiropteran skull: 1) relatively unmodified skulls (oral-emitters), 2) highly derived forms built around expansive nasal cavities (Old World nasal-emitters; see below), and 3) forms in which olfactory and visual (rather than acoustic) factors dominate the form-function of the facial component of the skull (i.e., New World nasal-emitting taxa).

OSSIFICATION OF THE CRANIUM

Given that developing cranial bone is a plastic entity, responsive to *in utero* and neonatal function (Lanyon 1980; Herring 1993), materiel availability (Müller & Streicher 1989; Presley 1989), brain development (Smith 1997), and spatial constraints (Haines 1940; Sarnat & Shanedling 1979), has the gross redesign of the microchiropteran skull altered the number of ossification centers or their sequence of appearance in any systematic pattern?

Fortunately, skeletal development is easily observed by clearing and differentially staining in whole-mount embryos. The presence of Alizarin red S stain in a skeletal element is evidence of the first macroscopic appearance of calcification. Yet this macroscopic calcification is a proliferation of pre-existing microscopic bone foci that may have appeared several developmental stages earlier (Hanken & Hall 1988). This time-lag between bone differentiation and bone proliferation means that the ossification sequences presented in this study are relative rather than absolute in nature. To register the degree to which each fetal skull had developed osteologically, I constructed an index to describe the proportion of cranial ossification foci present in each fetus relative to the number of foci (bones) expected in the adult skull. Observed bone foci include the dentary, maxilla, premaxilla, squamosal, parietal, frontal, nasal, jugal, palatine, pterygoid, basioccipital, prearticular, stylohyal, tympanic, basisphenoid, interparietal, vomer, supraoccipital, exoccipital, alisphenoid, prootic, orbitosphenoid, malleus, opisthotic, epiotic, lacrimal, incus, and stapes. The jugal is absent in *Artibeus* and the lacrimal was absent in the three Old World nasal-emitters, therefore, the ossification index was recalculated as a percent of 27 rather than 28 bone foci in these four taxa.

The general sequence of cranial ossification exhibited by the nine taxa follow the common mammalian pattern (de Beer 1937; Smith 1997). The basicranial elements ossify in the correct posterior-anterior sequence typical of mammals while the auditory bullae and the ossicular chain are almost always the last series of bones to appear (de Beer 1937). In each taxa, every bony element had appeared before the skull reached 66% of its expected adult size (Fig. 6.11).

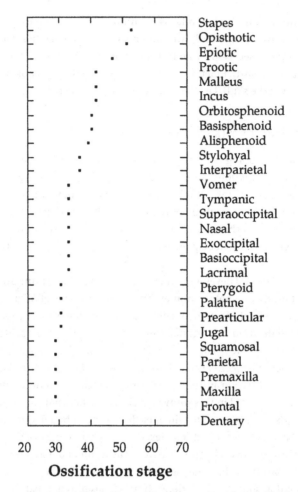

Stapes
Opisthotic
Epiotic
Prootic
Malleus
Incus
Orbitosphenoid
Basisphenoid
Alisphenoid
Stylohyal
Interparietal
Vomer
Tympanic
Supraoccipital
Nasal
Exoccipital
Basioccipital
Lacrimal
Pterygoid
Palatine
Prearticular
Jugal
Squamosal
Parietal
Premaxilla
Maxilla
Frontal
Dentary

20   30   40   50   60   70

Ossification stage

Figure 6.11. Plot indicating first appearance of an ossification center of each bone vs. developmental stage. All ossification foci (combined taxa) are present in the fetal skull by the time it has achieved two-thirds of its expected adult size. (After Pedersen 1995.)

Exceptions of note are related to the unique pharynx of rhinolophoid bats (Simmons & Stein 1980; Hartley & Suthers 1987, 1988, 1990; Pye 1988; Pedersen 1993a, 1995, 1996). I will return to this in a later section, but the rhinolophoid rostrum has been extensively modified to function as a 'resonator' for the echolocative call (Roberts 1972, 1973; Hartley & Suthers 1988; Suthers et al.1988; Pedersen 1996). The great expansion of the nasal passages has led to a local derangement of tissues including: 1) the presence of a large fontanel between the nasal, maxillary, and frontal

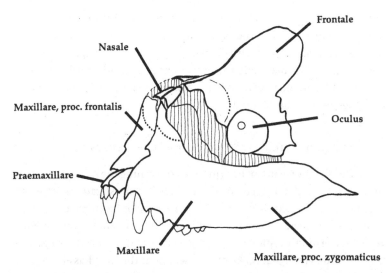

Figure 6.12. Line drawing taken from photograph of the rostral fontanel in a fetal *Hipposideros galeritus*, AMNH #152700; greatest length of skull = 12.38 mm, preserved body mass = 1.30 g, forearm length = 12.00 mm. Dotted lines represent the extent of the dorsal nasal chambers. Shaded regions represent the membranous median wall of the orbit and the rostral fontanel. (From Pedersen 1996.)

bones that persists until well after birth (Fig. 6.12), 2) the apparent loss of the lacrimal bone (as observed with the present technique), and 3) a reduction in the size of the nasal bones. Such elemental losses, translations, and/or distortions are not uncommon to the midface in either developmental or evolutionary terms (Haines 1940; Presley & Steel 1976; Presley 1981; Hanken 1983, 1984; Pedersen 1996). However, the unique coupling and suspension of the larynx and soft palate in rhinolophids manifests itself in characteristic shift in the ossification sequence of the stylohyoid chain, i.e., the ossification of all stylohyal elements is delayed until *after* the pterygoid processes are ossified. These features do not appear to be shared with the phyllostomids which exhibit the more generic sequence in which the chain ossifies before the pterygoid processes (de Beer 1937).

As an aside, the deciduous (milk) teeth of *Hipposideros caffer* are either resorbed or shed before parturition and the pups are born with a full battery of adult teeth ready to cut through the gums. This interesting feature is found in many chiropteran lineages (see Phillips, this volume). Here, however, it is important to note that the overall appearance of the perinatal skull of *Hipposideros* (and *Rhinolophus*) are exceedingly bizarre to say the least. Lateral to the nasal passages, the maxillary corpus is dominated by dental crypts to the extent that it has the appearance of a thin-walled, elongate bag, stuffed with ball-bearings (Fig. 6.12). The appearance of the mandible is similar. Presently, work is being done to describe this unique relationship between maxillofacial growth and odontogenesis in rhinolophids.

THE ACOUSTICAL AXIS AND SKULL MECHANICS

Despite the wealth of information concerning the mechanics of the mammalian skull, there have been relatively few investigations of the biomechanics of bat skulls (Marshall & Butler 1966; Freeman 1979, 1984, 1988, 1992; Czarnecki & Kallen 1980). Apart from some experimental work concerning mastication (*Myotis*: Kallen & Gans 1972; *Pteropus*: de Gueldre & de Vree 1984, 1988), most research in this area has either gone unpublished (Mohl 1971; Sperry 1972), is incidental to other studies (Matsumura 1979), or can be sampled in the present volume (see chapters by Pedersen, Phillips, and Vater, this volume). Bat crania exhibit a stunning range of morphological diversity that reflects their diverse dietary specializations. However, the orofacial component of the skull is exceedingly well integrated *regardless* of its relative position on the braincase (Starck 1952; Czarnecki & Kallen 1980; Freeman 1984; Pedersen 1993a). Despite this, does the dichotomy in skull form affect general skull mechanics in any predictable manner?

Mechanics of the midface

During mastication, the rostrum and zygomatic arches brace the palate against the braincase posteriorly, which then transfers these forces to the occiput and cervical spine. Strongly klinorhynchal skulls are thought to transfer forces from the palate to the basicranium ventrally with little involvement of the interorbital midface (e.g., *Daubentonia*: Cartmill 1977; Ross 1996). The distribution of bony mass in the strongly ventro-flexed skulls of many rhinolophids suggests a skull poorly reinforced to resist torsional forces; the rhinolophid zygoma is relatively weak, while the rostrum is extremely narrow dorsally. The bulk of the bony midface is distributed lateral to the choanae (pterygoid complex) where it may convey some degree of reinforcement to the palate akin to what is seen in other extremely klinorhynchal skulls (above). Conversely, bat skulls that are strongly aryrhynchal (dorsi-flexed rostra) are constructed such that forces are passed directly from the rostrum to the braincase via the frontal bones (i.e., *Mormoops*) through a rather broad, robust interorbital midface. Little work has been done along these lines, but one might predict that bat species with extremely klinorhynchal or airorhynchal skulls will not exhibit robust masticatory forces. Rather, durophages (species that eat hard-bodied food items) should exhibit more moderate skull angulations (i.e., EARPAL) within their oral- or nasal-emitting construct (*Cheiromeles* and *Vampyrum* respectively; see also Freeman 1984).

Nasal septum and the Old World nasal-emitters

The inner dimensions of the nasal passages and the composition and mass of the nasal septum (vomer, ethmoid, maxilla, vomeronasal cartilages, turbinates, etc.) are most often absent from discussions concerning bending and torsional forces within the rostrum. Extreme forms, such as Old World leaf-nosed bats (Rhinolophidae), exhibit unusual septa that are confounded by the adjacent dorsal nasal chambers and underlain by an abbreviated hard palate. Mechanically, this organization of the skull is not optimized for robust masticatory function. Among the many extraordinary features of this system is a kinetic premaxillary segment that is unique among mammals. Drs Bhatnagar and Wible have produced some intriguing evidence concerning a reinforced vomeronasal complex within the rhinolophoid nasal septum, possible sequelae of the kinetic premaxilla (personal communication). Furthermore, given the spatial restrictions of the midface (Haines 1940; Hanken 1984; Hoyte 1987), it seems unlikely that a nasal-emitting skull could exhibit both large olfactory fossae and resonating chambers within the interorbital septum. In the balance, phyllostomids may have retained olfaction at the cost of loudness of the call, while rhinolophoids may have emphasized loudness of the call at the expense of olfaction.

RHINOLOPHOID SKULL DESIGN AND ACOUSTICS

Here, I will shift emphasis to focus upon the highly modified skulls of the Old World nasal-emitters. The rostra of rhinolophoid bats are dominated by large outpocketings of the nasal cavity, the dorsal nasal chambers. These chambers are thought to act as resonators to amplify the echolocative call (Roberts 1972, 1973; Hartley & Suthers 1988; Suthers *et al.* 1988). Phyllostomids do not possess these resonant chambers and are therefore incapable of such call amplification. This feature led Griffin (1958) to label these New World nasal-emitters as 'whispering bats.' Despite these tremendous differences in the external appearance of their skulls, rhinolophoid (Infraorder: Yinochiroptera; Koopman 1984) and phyllostomid bats (Infraorder: Yangochiroptera) share the same cranial infrastructure in which the head is constructed around a ventrally deflected palate (Pedersen 1993a, 1995). However, both types of nasal-emitting bats are faced with the same problem of projecting sound through the restrictive nasal passages.

Acoustical considerations

The sound emitted by the larynx consists of a fundamental frequency (f1) that may be accompanied by several overtones, or harmonics. Vocalizations are modified in the vocal tract by the differential filtering and amplification of various frequency combinations. This filtering is effected by abrupt changes in the diameter of the vocal tract and discontinuities in the pharyngeal wall (i.e., diverticular bursae and orifices). Adult Old World nasal-emitting bats that utilize constant-frequency (CF) calls typically emphasize the second harmonic (f2), eliminate the fundamental (f1), and de-emphasize the remaining overtones. In these animals, the acoustics of the vocal tract have been modeled with some success (Roberts 1972; Hartley & Suthers 1988; Suthers *et al.* 1988). Models of the vocal tract in non-adults are complicated by the constantly changing proportions of the growing larynx, pharynx, and rostrum. Nevertheless, the anatomical dimensions of fetal and juvenile vocal tracts (*R. affinus, H. armiger* and *H. galeritus*) were fitted to mathematical equations concerning acoustics theory (Pedersen 1996). In general, there is a mismatch between laryngeally produced sound and the acoustic properties of the developing rostrum in all three species until well after parturition (Fig. 6.13).

Fundamental frequency

As the rhinolophid juveniles shift from their multi-harmonic, orally-emitted, isolation calls to their first nasally emitted calls, several harmonics (including the fundamental) are lost from the call structure. The space between the adult glottis and the laryngo-nasal junction is one quarter of a wavelength (f1) in length (*Rhinolophus*), and shaped like an expansion chamber resonator. Such chambers specifically absorb quarter wavelengths making this the most likely location at which the fundamental is suppressed in the adult supraglottal vocal tract. The nasopharynx never reaches a length capable of supporting a resonant f1 at any age. As such, the apparent loss of f1 from echolocation calls may be a by-product of simply closing the mouth.

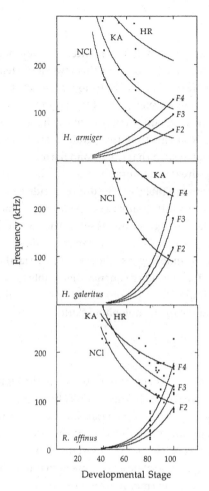

Figure 6.13. Hypothetical length-frequency curves. Based on the dimensions of the fetal nasal passages, hypothetical tuning/filter curves were calculated for the second (simple resonance equations), third (Helmholtz resonance within the dorsal nasal chambers), and fourth harmonics (*KA* filtering; see Kinsler & Frey 1962). Literature values for the second (Table 6.2), third, and fourth harmonics are included in successive plots for three rhinolophoid developmental series. Note: 1) hypothetical (calculated from measurements of fetal dimensions) and actual (literature) values for the second harmonic (*F* 2) do not intersect until the pups have reached near-adult size, 2) dimensions of the dorsal nasal chambers in *R. affinus* are sufficient to suppress the third harmonic by way of Helmholtz resonance, and 3) *KA* filtering is capable of removing the fourth harmonic (*F*4) from adult calls in each of the three species.

Second harmonic

If sound is to pass with minimal impedance (i.e., resonance) through the nasal cavity, there is a fixed relationship between the length of the cavity and the frequency (e.g., the second harmonic resonates when the length of the chamber is one wavelength). As the larynx grows, its sounds are not matched by the resonant properties of the nasal cavities until well after birth. It is only at this time that adult frequencies and adult-sized resonant filters match such that the second harmonic comes to dominate the CF call structure (Matsumura 1979; Konstantinov 1989). The results of this predictive model (Table 6.2, Fig. 6.13) coincide with the timing of the first appearance of adult f2s noted by several other studies. It would appear that a juvenile must 'grow into' its second harmonic, and the dominance of the second harmonic in CF calls may not lie in some unique feature of a second harmonic *per se* (Roberts 1973; Hartley & Suthers 1987), but rather that it is the only frequency that is able to match the anatomical constraints imposed on echolocation by the nasal passages throughout the evolution of the rhinolophoid skull.

Third harmonic

Adult rhinolophoid bats emit their echolocative calls through an extensively modified nasopharynx. One obvious modification is the gross distortion of the nasal passages to form 'dorsal nasal chambers.' The presence of an f3 in rhinolophid juveniles and its frequent absence in adults (e.g., *R. hildebrandti*) indicates an additional filter must develop elsewhere in the nasal passages, and it is the dorsal nasal chambers that have been hypothesized to remove the third harmonic from the vocalization by acting as Helmholtz resonators (Suthers *et al.* 1988). Hartley & Suthers (1988) rejected Helmholtz suppression of f3 in *R. hildebrandti*. However, dimensions of the dorsal nasal chambers in 3-week-old *R. affinus* pups were clearly capable of Helmholtz suppression of the third harmonic. Helmholtz suppression of f3 does not appear to be a consideration for *H. armiger* nor *H. caffer* as the dorsal nasal chambers in these two species never achieve the necessary dimensions (Fig. 6.13; Pedersen 1996).

Anatomical correlates of the dorsal nasal chambers

Of great interest are the effects of the dilation of these chambers on craniofacial development and cranial evolution in rhinolophid bats. Developmental sequelae of the expansion of these chambers include:

Table 6.2. *Second harmonic frequencies of some Old World nasal-emitting bats*

	Neonate $f2$	Adult $f2$	Literature source
Asellia tridens (Geoffroy, 1812)	–	111, 120	Obrist et al. 1993; Pye 1972; Roberts 1972
Hipposideros caffer (Sundervall, 1846)	–	138, 150	Obrist et al. 1993; Pye 1972; Roberts 1972
Hipposideros clivosus	–	70	Obrist et al. 1993
Hipposideros commersoni (Geoffroy, 1813)	–	56, 69	Pye 1972; Roberts 1972
Hipposideros diadema	40	58, 63	Brown & Grinnel 1980; Roberts 1972
Hipposideros lankadiva	–	70	Obrist et al. 1993
Rhinolophus alcyone (Temminck, 1852)	.	90	Roberts 1972
Rhinolophus eloquens	–	80	Obrist et al. 1993
Rhinolophus euryale (Blasius, 1853)	–	104	Pye 1972
Rhinolophus ferrumequinum (Schreber, 1775)	40	82, 86	Konstantinov 1973, 1987; Pye 1972; Roberts 1972, 1973
Rhinolophus f. nippon	48	72	Brown & Grinnel 1980; Matsumura 1979, 1981
Rhinolophus f. ferrumequinum	25	83	Brown & Grinnel 1980
Rhinolophus fumigatus	–	45	Roberts 1972
Rhinolophus hildebrandti (Peters, 1878)	–	46	Hartley & Suthers 1988; Suthers et al. 1988
Rhinolophus hipposideros (Bechstein 1801)	50	110, 114	Konstantinov 1987; Pye 1972; Roberts 1972
Rhinolophus landeri (Marti 1837)	-	121	Roberts 1972
Rhinolophus luctus	-	42, 45	Roberts 1972, 1973
Rhinolophus rouxii	20	72, 78	Obrist et al. 1993; Rübsamen 1987, 1988
Triaenops afer	-	79	Roberts 1972

Note: All values are taken directly from the literature or approximated from figures. Neonate f 2s are typically 'isolation calls' emitted from the mouth.

1) gross dorsolateral distortion of the maxillary corpus, 2) the apparent loss of the lacrimal bone as a separate ossification center (see above), 3) the transformation of the infraorbital foramen into an infraorbital canal, 4) the reduction in mass of the floor of the nasal cavity (e.g., the premaxillary segment is kinetic, and the maxillary contribution to the hard palate is very small), and 5) the appearance of supernumerary fontanels centered over each of the paired dorsal nasal chambers (see above, Fig. 6.12). These fontanels expand with volumetric changes in the underlying chambers and remain open until 2–3 weeks postpartum. Their influence on the construction of the midface is a clear example of how capsular growth is independent of the growth of adjacent or overlying bone(s) early in development (Presley & Steel 1976; Presley 1981; Hanken 1983; Pedersen 1991, 1993a, b). Similar observations have been reported in frogs (Hanken 1984), artiodactyls, lagomorphs, rodents (Hoyte 1987), marsupials, carnivores, and primates (Haines 1940).

Additional interpretations of the function of the dorsal nasal chambers include the possibility that they may act as spacing elements within the maxillary corpus and affect the relative position of the posterior nasal septum and choanae. In this manner, the chambers indirectly determine the dimensions of the resonant chamber and the length of the standing wave of the second harmonic. In addition, the characteristic bossing of the maxillae above and anterior to the dorsal nasal chambers may also help support the adjacent noseleaf.

Physiological integration

The physiological integration of vocalization with audition is a gradual process that is primarily dependent on the maturation of the head and its function as an acoustical horn and receiver. Neither the rostrum nor the pinnae (Obrist *et al.* 1993) exhibit dimensions capable of adult acoustics (certainly not the dominant spectral component of the adult call − f 2) until well after birth. Ironically, the dimensions of the cochlea and the basal membrane are established at birth before physiological responses from the brain can be measured (Konstantinov 1973; Brown & Grinnel 1980; Vater 1988). Eventually, a neural-auditory filter focused at the second harmonic appears, but only at those frequencies used by the young animals. Gradually, a progressive tuning of the auditory filter and the maturation of the larynx shift upwards into the higher frequencies used by adults (Konstantinov 1973; Rübsamen 1987, 1988). Although Rübsamen (1987, 1988) put forth the excellent argument that maturation of neurological processing and cochlear function are the dominant processes during the ontogeny of echolocation in rhinolophids (i.e., *Rhinolophus rouxi*), it is clear that neurological development is secondary to the maturation and 'tuning' of the head. Indeed, it is difficult to conceive of a control mechanism that regulates cranial growth such that the skull grows to match cochlear dynamics. Rather, a more parsimonious explanation would suggest that as the rostrum grows, neurophysiology must adapt to those sounds produced by the larynx and 'permitted' by the supraglottal vocal tract (but see Vater, this volume).

EVOLUTIONARY SYNTHESIS

Distinct developmental paths

Echolocation has been a primary factor in microchiropteran evolution, but the developmental and evolutionary mechanisms by which nasal-emission evolved are speculative. The divergence between oral- and nasal-emitting forms of the adult skull is most likely an exaptation (Gould & Vrba 1982), resulting from selective forces acting upon the form and function of the pharynx during echolocation rather than the result of selection on cranial shape or head posture *per se*. Developmentally, the microchiropteran skull has been canalized into three distinct evolutionary paths: 1) oral-emitting forms (primitive state: Jepsen 1966; Smith 1976; Van Valen 1979; Habersetzer & Storch 1987, 1992), 2) simple nasal-emitting forms (New World nasal-emitters), and 3) derived nasal-emitting forms (resonator-equipped Old World nasal-emitters). The latter two provide a clear example of convergent evolution in that both nasal-emitting forms share three distinct (albeit functionally related) features: 1) behavior: the nasal-emission of the echolocative call, 2) soft-tissue: the fleshy, often elaborate, flap of skin projecting around the nostrils (noseleaf) that helps focus the call as it is emitted through the nostrils (Möhres 1966*a*, *b*; Simmons & Stein 1980; Hartley & Suthers 1987, 1988, 1990; Pye 1988; Arita 1990), and 3) osteology: the reorganization of the skull about the nasal passages (Fig. 6.14; Pedersen 1993*a*). As such, the fine details of microchiropteran skull morphology could be perceived as simple epigenetic accommodations to the fundamental use of the head as an acoustical horn. Similar examples of taxonomic radiations coincident with developmental pattern shifts are not uncommon (Needham 1933; Gould 1977; Alberch *et al.* 1979; Alberch & Alberch 1981; Müller 1990).

Adaptive landscapes

To visualize morphological evolution, taxa have been pictured as migrating through an adaptive landscape whose topology represents

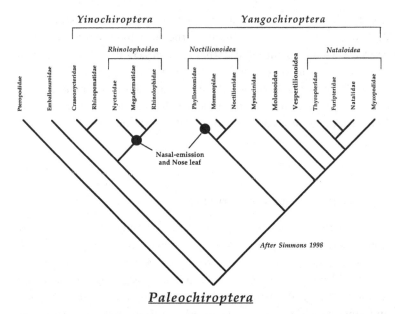

Paleochiroptera

Figure 6.14. Convergent evolution of nasal-emitting forms in the Microchiroptera. Extant Microchiroptera are separated into two infraorders based upon the mobility of the premaxillae: the Yinochiroptera (premaxillae free from the maxillae) and the Yangochiroptera (premaxillae fused to the maxillae; Koopman 1984; see Simmons, this volume). Nasal-emitting groups have evolved three unique features independently: nasal-emission of the echolocative call, rearrangement of the skull about the nasal cavity, and possession of a noseleaf. (After Simmons 1998.)

different morphological constructs (Simpson 1944, 1953). Adaptive morphological constructs appear at the tops of peaks spotted throughout an evolutionary landscape. Accordingly, I reworked the bivariate representation of oral-emitting and nasal-emitting adult skull shapes (Fig. 6.6) as an adaptive landscape (Fig. 6.15) that clearly depicts two adaptive peaks representing the dichotomy between the nasal and oral phonal axes of the head.

Developmentally, taxa migrate through a morphogenetic landscape as well (Figs. 6.5, 6.15). Here, selection *in utero* is based upon mechanical feasibility and compatible histogenesis within the ontogenetic milieu (Katz *et al.* 1981; Alberch 1982; Katz 1982; Müller 1990) such that structural accommodation throughout development may be more important than Darwinian selection after parturition (Schmalhausen 1949; Kuhn 1987; Bonner 1988; Maier 1989; Presley 1989). While the spaces between peaks in the adult landscape are often considered 'non-adaptive' (Simpson 1944), inter-peak regions in the developmental landscape may

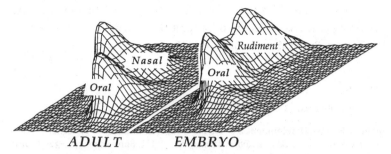

ADULT   EMBRYO

Figure 6.15. Adaptive landscapes. These landscapes are three-dimensional reconstructions of Figs. 6.5 and 6.6 designed to illustrate the two basic forms of the chiropteran skull (oral- and nasal-emitting peaks). The shallow saddle between peaks in the developmental landscape is occupied by oral-emitting fetuses as they migrate from the rudimentary morphospace toward the oral-emitting morphospace. (From Pedersen 1995.)

be traversed or transiently occupied by individuals sufficiently integrated to survive gestation (Cheverud 1982, 1989). Shifts from one peak to the other require the rerouting of an ancestral developmental path (oral axis) onto a new morphogenetic trajectory (nasal axis) – a 'key innovation' (Frazzetta 1975; Cracraft 1990; Müller 1990; Raff *et al.* 1990). This 'new' *Bauplan* is set in place before internal selective forces affect morphogenesis (mechanical integration of musculoskeletal elements surrounding the pharynx), and certainly before the first neonatal vocalizations.

Because this dichotomy is clearly associated with the co-opting of the nasopharynx as an acoustical horn in nasal-emitting species, intermediate states would be quickly weeded out early in the developmental program because of the precise anatomical and physiological requirements for the efficient emission of ultrasound by adult bats (Simmons & Stein 1980; Pye 1988; Suthers *et al.* 1988). The strictness of this selection suggests that the shift from oral-emission to nasal-emission must have occurred quickly both in developmental and evolutionary terms (Lewin 1986; Price *et al.* 1993). Continuing studies of this conjunction of developmental, behavioral, and morphological data provides a tremendous springboard from which to evaluate the genetic, paleontological, and biogeographical underpinnings of chiropteran evolution.

ACKNOWLEDGMENTS

I wish to thank Drs S. Herring, J. Hanken, R. Adams, and K. Rafferty for comments on the manuscript. I wish to acknowledge the curators and staff of the zoological collections at the following institutions: University

of Nebraska State Museum, American Museum of Natural History,
University of Kansas Museum of Natural History, and the University of
Washington Burke Museum of Natural History.

REFERENCES

Alberch, P. (1982) Developmental constraints in evolutionary processes. In *Evolution and Development: Dahlem Konferenzen, 1982*, ed. J. T. Bonner, pp. 313–32. Berlin: Springer-Verlag.

Alberch, P. & Alberch, J. (1981) Heterochronic mechanisms of morphological diversity and evolutionary change in the neotropical salamander, *Bolitoglossa occidentalis* (Amphibia: Plethodontidae). *Journal of Morphology*, **167**, 249–64.

Alberch, P., Gould, S. J., Oster, G. F. & Wake, D. B. (1979) Size and shape in ontogeny and phylogeny. *Paleobiology*, **5**, 296–317.

Arita, H. (1990) Noseleaf morphology and ecological correlates in phyllostomid bats. *Journal of Mammalogy*, **71**, 36–47.

Babler, W. J., Persing, J. A., Nagorsky, M. J. & Lane, J. A. (1987) Restricted growth at the frontonasal suture: Alterations in craniofacial growth in rabbits. *American Journal of Anatomy*, **17**, 90–8.

Baer, M. J. & Nanda, S. K. (1976) A commentary on the growth and form of the cranial base. In *Development of the Basicranium*, Department of Health, Education and Welfare Publication No. (NIH) 76–989, ed. J. F. Bosma, pp. 515–36. Washington: Government Printing Office.

Behlfelt, K., Linder-Aronsen, S. & Neander, P. (1990) Posture of the head, the hyoid bone, and the tongue in children with and without enlarged tonsils. *European Journal of Orthodontics*, **12**, 458–67.

Bjork, A. & Skieler, V. (1972) Facial development and tooth eruption; an implant study at the age of puberty. *American Journal of Orthodontics*, **62**, 339–83.

Blechschmidt, E. (1976a) Principles of biodynamic differentiation. In *Development of the Basicranium*, Department of Health, Education and Welfare Publication No. (NIH) 76–989, ed. J. F. Bosma, pp. 54–76. Washington: Government Printing Office.

Blechschmidt, M. (1976b) Biokinetics of the developing basicranium. In *Development of the Basicranium*, Department of Health, Education and Welfare Publication No. (NIH) 76–989, ed. J. F. Bosma, pp. 44–53. Washington: Government Printing Office.

Bonner, J. (1988). *The Evolution of Complexity*. Princeton: Princeton University Press.

Bosma, J. F. (1976) Introduction to the symposium. In *Development of the Basicranium*, Department of Health, Education and Welfare Publication No. (NIH) 76–989, ed. J. F. Bosma, pp. 3–28. Washington: Government Printing Office.

Brown, P. & Grinnell, A. (1980) Echolocation ontogeny in bats. In *Animal Sonar Systems, Proceedings NATO Advanced Study Institute*, ed. R. Busnel & J. F. Fish, pp. 355–80. New York: Plenum Press.

Byrd, K. (1988) Craniofacial sequelae of lesions to facial and trigeminal motor nuclei in growing rats. *American Journal of Physical Anthropology*, **76**, 87–103.

Cartmill, M. (1977) *Daubentonia, Dactylopsila*, woodpeckers and klinorynchy. In *Prosimian Anatomy, Biochemistry and Evolution*, ed. R. Martin, A. Doyle & A. Walker, pp. 655–70. London: Duckworth.

Cheverud, J. (1982) Phenotypic, genetic, and environmental morphological integration in the cranium. *Evolution*, **36**, 499–516.

Cheverud, J. (1989) The evolution of morphological integration. *Trends in Vertebrate Morphology*, **35**, 196–7.

Choo, J. & Covell, D. A. (1996) Effects of inhibiting periosteal migration of the growth of the guinea pig mandible. In *Biological Mechanisms of Tooth Movement and Craniofacial Adaptaton*, ed. Z. Davidovitch & L. Norton, pp. 529–36. Cambridge: Harvard University Press.

Cracraft, J. (1990) The origin of evolutionary novelties: patterns and process at different hierarchical levels. In *Evolutionary Innovations*, ed. M. Nitecki, pp. 21–47. Chicago: University of Chicago Press.

Crocker, M. (1987) Acoustic noise. In *Acoustics Source Book*, ed. S. P. Parker, pp. 150–8. New York: McGraw-Hill.

Czarnecki, R. T. & Kallen, F. C. (1980) Craniofacial, occlusal, and masticatory anatomy in bats. *Anatomical Record*, **198**, 87–105.

de Beer, G. R. (1937) *Development of the Vertebrate Skull* (republished 1995) Chicago: University of Chicago Press.

de Gueldre, G. & de Vree, F. (1984) Movement of the mandibles and tongue during mastication and swallowing in *Pteropus giganteus* (Megachiroptera): a cineradiographical study. *Journal of Morphology*, **179**, 95–114.

de Gueldre, G. & de Vree, F. (1988) Quantitative elecromyography of the masticatory muscles of *Pteropus giganteus* (Megachiroptera). *Journal of Morphology*, **196**, 73–106.

Delattre, A. & Fennart, R. (1960) *L'Hominisation du crâne*. Paris: Editions du Centre National de la Recherche Scientifique.

Eisenberg, J. F. & Wilson, D. E. (1978) Relative brain size and feeding strategies in the Chiroptera. *Evolution*, **32**, 740–51.

Enlow, D. H. (1976) The prenatal and postnatal growth of the human basicranium. In *Development of the Basicranium*, Department of Health, Education and Welfare Publication No. (NIH) 76-989, ed. J. F. Bosma, pp. 192–203. Washington: Government Printing Office.

Enlow, D. H. & McNamara, J. A. (1973) The neurocranial basis for facial form and pattern. *Angle Orthodontist*, **43**, 256–70.

Fenton, M. & Crerar, L. (1984) Cervical vertebrae in relation to roosting posture in bats. *Journal of Mammalogy*, **65**, 395–403.

Frahm, H. (1981) Volumetric comparison of the accessory olfactory bulb in bats. *Acta Anatomica*, **109**, 173–83.

Frazzetta, T. (1975) *Complex Adaptations in Evolving Populations*. Sunderland: Sinauer & Associates.

Freeman, P. W. (1979) Specialized insectivory: beetle-eating and moth-eating molossid bats. *Journal of Mammalogy*, **60**, 467–79.

Freeman, P. W. (1984) Functional analysis of large animalivorous bats (Microchiroptera). *Biological Journal of the Linnean Society*, **21**, 387–408.

Freeman, P. W. (1988) Frugivorous and animalivorous bats (Microchiroptera): dental and cranial adaptations. *Biological Journal of the Linnean Society*, **33**, 249–72.

Freeman, P. W. (1992) Canine teeth of bats (Microchiroptera): size, shape and role in crack propagation. *Biological Journal of the Linnean Society*, **45**, 97–115.

Gaskell, C. (1974) The radiographic anatomy of the pharynx and larynx of the dog. *Journal of the Small Animal Practitioner*, **14**, 89–100.

Gasson, N. & Lavergne, J. (1977) Maxillary rotation during human growth: annual variation and correlations with mandibular rotation. *Acta Odontologica Scandinavica*, **35**, 13–21.

Gould, S. J. (1977) *Ontogeny and Phylogeny*. Cambridge: Belknap Press.

Gould, S. J. & Vrba, E. S. (1982) Exaptation – a missing term in the science of form. *Paleobiology*, **8**, 4–15.

Griffin, D. (1958) *Listening in the Dark*. New Haven: Yale University Press.

Griffioen, F. M. M. & Smit-Vis, J. H. (1985) The skull: mould or cast? *Acta Morphologica Neerlando-Scandinavica*, **23**, 325–35.

Habersetzer, J. & Storch, G. (1987) Ecology and echolocation of the Eocene Messel bats. In *European Bat Research 1987*, ed. V. Hanák, I. Horácek & J. Gaisler, pp. 213–33. Prague: Charles University Press.

Habersetzer, J. & Storch, G. (1992) Cochlea size in extant Chiroptera and middle Eocene microchiropterans from Messel. *Naturwissenschaften*, **79**, 462–6.

Haines, W. (1940) The interorbital septum in mammals. *Journal of the Linnaean Society of London*, **41**, 585–607.

Hanken, J. (1983) Miniaturization and its effects on cranial morphology in plethodontid salamanders, genus *Thorius* (Amphibia, Plethodontidae): II. The fate of the brain and sense organs and their role in skull morphogenesis and evolution. *Journal of Morphology*, **177**, 255–68.

Hanken, J. (1984) Miniaturization and its effects on cranial morphology in plethodontid salamanders, genus *Thorius* (Amphibia: Plethodontidae). I. Osteological variation. *Biological Journal of the Linnean Society*, **23**, 55–75.

Hanken, J. & Hall, B. K. (1988) Skull development during anuran metamorphosis: I. Early development of the first three bones to form – the exoccipital, the parasphenoid, and the frontoparietal. *Journal of Morphology*, **195**, 247–56.

Hanken, J. & Thorogood, P. (1993) Evolution and development of the vertebrate skull: the role of pattern formation. *Trends in Ecology and Evolution*, **8**, 9–15.

Hanken, J. & Wassersug, R. (1981) The visible skeleton. *Functional Photography*, **4**, 22–6.

Hartley, D. & Suthers, R. (1987) The sound emission pattern and the acoustical role of the noseleaf in the echolocating bat, *Carollia perspicillata*. *Journal of the Acoustic Society of America*, **82**, 1892–1900.

Hartley, D. & Suthers, R. (1988) The acoustics of the vocal tract in the horseshoe bat, *Rhinolophus hildebrandti*. *Journal of the Acoustic Society of America*, **84**, 1201–13.

Hartley, D. & Suthers, R. (1990) Sonar pulse radiation and filtering in the mustached bat, *Pteronotus parnellii rubiginosus*. *Journal of the Acoustic Society of America*, **87**, 2756–72.

Herring, S. (1985) The ontogeny of mammalian mastication. *American Zoologist*, **25**, 339–49.

Herring, S. (1993) Epigenetic and functional influences on skull growth. In *The Skull*, vol. 1, *Development*, ed. J. Hanken & B. Hall, pp. 153–206. Chicago: University of Chicago Press.

Herring, S. W. & Lakars, T. C. (1981) Craniofacial development in the absence of muscle contraction. *Journal of Craniofacial Genetics and Developmental Biology*, **1**, 341–57.

Hoyte, D. A. N. (1987) Muscles and cranial form. In *Mammalia Depicta: Morphogenesis of the Mammalian Skull*, ed. H. Kuhn & U. Zeller, pp.123–44. Hamburg: Verlag Paul Parely.

Jepsen, G. (1966) Early Eocene bat from Wyoming. *Science*, **154**, 1333–9.

Jolicoeur, J., Pirlot, P., Baron, G. & Stephan, H. (1984) Brain structure and correlation patterns in Insectivora, Chiroptera, and Primates. *Systematic Zoology*, **33**, 14–33.

Julian, L., Tyler, W., Hage, T. & Gregory, P. (1957) Premature closure of the spheno-occipital synchondrosis in the Horned Hereford Dwarf of the 'short headed' variety. *American Journal of Anatomy*, **100**, 269–87.

Kallen, F. & Gans, C. (1972) Mastication in the little brown bat, *Myotis lucifugus*. *Journal of Morphology*, **136**, 385–420.

Katz, M. (1982) Ontogenetic mechanisms: the middle ground of evolution. In *Evolution and Development: Dahlem Konferenzen, 1982*, ed. J. T. Bonner, pp. 207–12. Berlin: Springer-Verlag.

Katz, M., Lasek, R. & Kaiser-Abramof, J. (1981) Ontophyletics of the nervous system: eyeless mutants illustrate how ontogenetic buffer mechanisms channel evolution. *Proceedings of the National Academy of Science, U.S.A.*, **7**, 397–401.

Kay, E. (1986) The phenotypic interdependence of the musculoskeletal characters of the mandibular arch in mice. *Journal of Embryology and Experimental Morphology*, **98**, 123–36.

Kinsler, L. & Frey, A. (1962) *Fundamentals of Acoustics*. New York: John Wiley.

Klima, M. (1987) Morphogenesis of the nasal structures of the skull in toothed whales (Odontoceti). In *Mammalia Depicta: Morphogenesis of the Mammalian Skull*, ed. H. Kuhn & U. Zeller, pp. 105–21. Hamburg: Verlag Paul Parely.

Klymkowsky, M. & Hanken, J. (1991) Whole-mount staining of *Xenopus* and other vertebrates. *Methods in Cell Biology*, **36**, 419–41.

Konstantinov, A. I. (1973) Development of echolocation in bats in postnatal ontogenesis. *Periodicum Biologicum*, **75**, 13–19.

Konstantinov, A. I. (1989) The ontogeny of echolocation functions in horseshoe bats. In *European Bat Research 1987*, ed. V. Hanák, I. Horácek & J. Gaisler, pp. 271–80. Prague: Charles University Press.

Koopman, K. (1984) A synopsis of the families of bats – Part VII. *Bat Research News*, **25**, 25–7.

Koskinen-Moffett, L. & Moffett, B. (1989) Sutures and intrauterine deformation. In *Scientific Foundations and Surgical Treatment for Craniosynostosis*, ed. J. Persing, pp. 96–106. Baltimore: Williams & Wilkins.

Koskinen-Moffett, L., McMinn, R., Isotupa, K. & Moffett, B. (1981) Migration of the craniofacial periosteum in rabbits. *Proceedings of the Finnish Dental Society*, **77**, 83–8.

Kuhn, H. (1987) Introduction. In *Mammalia Depicta: Morphogenesis of the Mammalian Skull*, ed. H. Kuhn & U. Zeller, pp. 9–16. Hamburg: Verlag Paul Parely.

Kylamarkula, S. (1988) Growth changes in the skull and upper cervical skeleton after partial detachment for neck muscles. An experimental study in the rat. *Journal of Anatomy*, **159**, 197–205.

Lakars, T. & Herring, S. (1980) Ontogeny and oral function in hamsters (*Mesocricetus auratus*). *Journal of Morphology*, **165**, 237–54.

Lanyon, L. E. (1980) The influence of function on the development of bone curvature. An experimental study on the rat tibia. *Journal of Zoology (London)*, **192**, 457–66.

Lewin, R. (1986) Punctuated equilibrium is now old hat. *Science*, **231**, 672–3.

Maier, W. (1989) Ala temporalis and alisphenoid in therian mammals. *Trends in Vertebrate Morphology*, **35**, 396–400.

Marshall, P. M. & Butler, P. M. (1966) Molar cusp development in the bat, *Hipposideros beatus*, with reference to the ontogenetic basis of occlusion. *Archives of Oral Biology*, **11**, 949–65.

Matsumura, S. (1979) Mother–infant communication in a horseshoe bat (*Rhinolophus ferrumequinum nippon*): development of vocalization. *Journal of Mammalogy*, **60**, 76–84.

McLain, J. B. & Vig, P. S. (1983) Transverse periosteal sectioning and femur growth in the rat. *Anatomical Record*, **207**, 339–48.

Mohl, N. (1971) Craniofacial relationships and adaptations in bats. PhD dissertation, State University of New York, Buffalo.

Möhres, F. P. (1966a) Ultrasonic orientation in megadermatid bats. In *Animal Sonar Systems – Biology and Bionics, Proceedings NATO Advanced Study Institute*, ed. R. Busnel, pp. 115–28. Jouy-en-Josas: Laboratoire de physiologie acoustique, Institut national de la recherche agronomique.

Möhres, F. P. (1966b) General characters of acoustic orientation sounds and performance of sonar in the order of Chiroptera. In *Animal Sonar Systems – Biology and*

Bionics, Proceedings NATO Advanced Study Institute, ed. R. Busnel, pp. 401–7. Jouy-en- Josas: Laboratoire de physiologie acoustique, Institut national de la recherche agronomique.

Moore, K. L. (1983) *Before We Are Born*, 2nd edn. Philadelphia: W. B. Saunders.

Moss, M. (1976) Experimental alteration of basisynchondrosal cartilage growth in rat and mouse. In *Development of the Basicranium*, Department of Health, Education and Welfare Publication No. (NIH) 76–989, ed. J. F. Bosma, pp. 541–69. Washington: Government Printing Office.

Muhl, Z. F. & Gedak, G. K. (1986) The influence of periosteum on tendon and ligament migrations. *Journal of Anatomy*, **145**, 161–71.

Müller, G. B. (1990) Developmental mechanisms: a side-effect hypothesis. In *Evolutionary Innovations*, ed. M. Nitecki, pp. 99–130. Chicago: University of Chicago Press.

Müller, G. B. & Streicher, J. (1989) Ontogeny of the syndesmosis tibiofibularis and the evolution of the bird hindlimb: a caenogenetic feature triggers phenotypic novelty. *Anatomy & Embryology*, **179**, 327–39.

Needham, J. (1933) On the dissociability of the fundamental process in ontogenesis. *Biological Review*, **8**, 180–223.

Novacek, M. J. (1991) Aspects of the morphology of the cochlea in microchiropteran bats: an investigation of character transformation. *Bulletin of the American Museum of Natural History*, **206**, 84–100.

Obrist, M., Fenton, M. B., Eger, J. & Schlegel, P. (1993) What ears do for bats: a comparative study of pinna sound pressure transformation in Chiroptera. *Journal of Experimental Biology*, **180**, 119–52.

Pedersen, S. (1991) Dental morphology of the cannibal morph in the tiger salamander, *Ambystoma tigrinium*. *Amphibia–Reptilia*, **12**, 1–14.

Pedersen, S. (1993*a*) Cephalometric correlates of echolocation in the Chiroptera. *Journal of Morphology*, **218**, 85–98.

Pedersen, S. (1993*b*) Skull growth in cannibalistic Tiger salamanders, *Ambystoma tigrinum*. *Southwestern Naturalist*, **38**, 316–24.

Pedersen, S. (1995) Cephalometric correlates of echolocation in the Chiroptera II: fetal development. *Journal of Morphology*, **225**, 107–23.

Pedersen, S. (1996) Skull growth and the presence of an auxiliary fontanels in rhinolophid bats (Microchiroptera). *Zoomorphology*, **116**, 205–12.

Pedersen, S. (1998) Morphometric analysis of the chiropteran skull with regard to mode of echolocation. *Journal of Mammalogy*, **79**, 91–103.

Pedersen, S. & Anton S. (1998) Bicoronal synostoseis in a child from Historic Omaha Cemetary 25DK10. *American Journal of Physical Anthropology*, **105**, 369–76.

Pirlot, P. & Bernier, R. (1991) Brain growth and differentiation in two fetal bats: qualitative and quantitative aspects. *American Journal of Anatomy*, **190**, 167–81.

Presley, R. (1981) Alisphenoid equivalents in placentals, marsupials, monotremes and fossils. *Nature*, **294**, 668–70.

Presley, R. (1989) Ala temporalis: function or phylogenetic memory. *Trends in Vertebrate Morphology*, **35**, 392–5.

Presley, R. & Steel, F. (1976) On the homology of the alisphenoid. *Journal of Anatomy*, **121**, 441–59.

Price, T., Turelli, M. & Slatkin, M. (1993) Peak shifts produced by correlated response to selection. *Evolution*, **47**, 280–90.

Proffit, W. (1978) Equilibrium theory revisited: factors influencing position of the teeth. *Angle Orthodontist*, **48**, 175–86.

Pucciarelli, H. M. & Dressino, V. (1996) Orthocephalization in the postweaning Squirrel monkey. *American Journal of Physical Anthropology*, **101**, 173–81.

Pye, J. (1988) Noseleaves and bat pulses. In *Animal Sonar – Processes and Performance*,

Proceedings NATO Advanced Study Institute, ed. P. E. Nachtigall & P. W. B. Moore, pp. 791–6. New York: Plenum Press.

Radinsky, L. B. (1968) A new approach to mammalian cranial analysis, illustrated by examples of prosimian primates. Journal of Morphology, 124, 167–80.

Raff, R., Parr, B., Parks, A. & Wray, G. (1990) Heterochrony and other mechanisms of radical evolutionary change in early development. In Evolutionary Innovations, ed. M. Nitecki, pp. 71–98. Chicago: University of Chicago Press.

Ranly, D. (1980) A Synopsis of Craniofacial Growth. Norwalk: Appleton–Century–Crofts.

Roberts, L. (1972) Variable resonance in constant frequency bats. Journal of Zoology (London), 166, 337–48.

Roberts, L. (1973) Cavity resonances in the production of orientation cries. Periodicum Biologicum, 75, 27–32.

Ross, C. (1996) Adaptive explanation for the origins of the Anthropoidea (Primates). American Journal of Primatology, 40, 205–30.

Ross, C. & Henneberg, M. (1995) Basicranial flexion, relative brain size, and facial kyphosis in Homo sapiens and some fossil hominids. American Journal of Physical Anthropology, 98, 575–93.

Ross, C. & Ravosa, M. (1993) Basicranial flexion, relative brain size, and facial kyphosis in nonhuman primates. American Journal of Physical Anthropology, 91, 305–24.

Rübsamen, R. (1987) Ontogenesis of the echolocation system in the rufous horseshoe bat, Rhinolophus rouxi (audition and vocalization in early postnatal development). Journal of Comparative Physiology A, 161, 899–913.

Rübsamen, R. (1988) Ontogeny of the echolocation system in rhinolophoid CF–FM bats: audition and vocalization in early postnatal development. In Animal Sonar – Processes and Performance, Proceedings NATO Advanced Study Institute, ed. P. E. Nachtigall & P. W. B. Moore, pp. 335–40. New York: Plenum Press.

Sarnat, B. G. & Shanedling, P. D. (1979) Increased orbital volume after periodic intrabulbar injections of silicone in growing rabbits. American Journal of Anatomy, 140, 523–32.

Schachner, O. (1989) Raising of the head and allometry: morphometric study on the embryonic development of the mouse skeleton. Trends in Vertebrate Morphology, 35, 291–8.

Schmalhausen, I. (1949) Factors of Evolution: The Theory of Stabilizing Selection (republished 1986). Chicago: University of Chicago Press.

Schneiderman, E. (1992) Facial Growth in the Rhesus Monkey. Princeton: Princeton University Press.

Schön, M. (1976) Adaptive modification in the basicranium of Howling monkeys (Alouatta). In Development of the Basicranium, Department of Health, Education and Welfare Publication No. (NIH) 76-989, ed. J. F. Bosma, pp. 664–75. Washington: Government Printing Office.

Schumacher, G. H. Von, Fanghanel, J., Koster, D. & Mierzwa, J. (1988) Craniofacial growth under the influence of blood supply: II. Scoliosis of the skull. Anatomischer Anzeiger, Jena, 165, 303–9.

Schumacher, G. H. Von, Schoof, S., Fanghanel, J., Mildschlag, G. & Kannmann, F. (1986) Skull deformities following unilateral mandibular dysbalance: I. General review of secondary changes. Anatomischer Anzeiger, Jena, 161, 105–11.

Servoss, J. M. (1973) An in vivo and in vitro autoradiographic investigation of growth in synchondrosal cartilage. American Journal of Anatomy, 136, 479–86.

Silver, P. A. S. (1962) In ovo experiments concerning the eye, the orbit and certain juxta-orbital structures in the chick embryo. Journal of Embryology and Experimental Morphology, 10, 423–50.

Simmons, J. & Stein, R. (1980) Acoustic imaging in bat sonar: echolocation signals and the evolution of echolocation. *Journal of Comparative Physiology A*, **135**, 61–84.

Simmons, J. (1980) Phylogenetic adaptations and the evolution of echolocation in bats. In *Proceedings of the Fifth International Bat Research Conference A*, ed. D. E. Wilson & A. L. Gardner, pp. 267–78. Lubbock: Texas Tech Press.

Simpson, G. G. (1944) *Tempo and Mode in Evolution* (republished 1984). New York: Columbia University Press.

Simpson, G. G. (1953) *The Major Features of Evolution*. New York: Columbia University Press.

Smit-Vis, J. H. & Griffioen, F. M. M. (1987) Growth control of neurocranial height of the rat skull. *Anatomischer Anzeiger, Jena*, **163**, 401–6.

Smith, J. (1976) Chiropteran evolution. *Special Publications of the Museum, Texas Tech University*, **10**, 49–70.

Smith, K. K. (1997) Comparative patterns of craniofacial development in eutherian and metatherian mammals. *Evolution*, **51**, 1663–78.

Sokal, R. R. & Rohlf, F. J. (1981) *Biometry: The Principles and Practice of Statistics in Biological Research*, 2nd edn. New York: W. H. Freeman.

Solow, B. & Greve, E. (1979) Craniocervical angulation and nasal respiratory resistance. In *Naso-Respiratory Function and Craniofacial Growth*, Craniofacial growth series, Monograph No. 9, ed. J. A. McNamara, pp. 87–119. Ann Arbor: University of Michigan.

Spatz, W. von (1968) Die Bedeutung der Augen für die sagittale Gestaltung des Schädels von *Tarsius* (Prosimiae, Tarsiiformes). *Folia Primatologica*, **99**, 22–40.

Sperber, G. H. (1989) *Craniofacial Embryology*, 4th edn. London: Wright Press.

Sperry, T. (1972) Development of the specialized craniofacial complex in the bats of the genera *Mormoops* and *Chilonycteris* (Family Mormoopidae). PhD dissertation, University of Illinois Medical Center, Chicago.

Spyropoulos, M. N. (1977) The morphogenetic relationship of the temporal muscle to the coronoid process in human embryos and fetuses. *American Journal of Anatomy*, **150**, 395–410.

Starck, D. (1952) Form und Formbildung der Schädelbasis bei Chiropteran. Ergänzungsheft zum 99 Band; Ver. Anatomische Ges., 50th Versammlung Marburg. *Anatomischer Anzeiger*, **99**, 114–21.

Starck, D. (1989) Considerations on the nature of skeletal elements in the vertebrate skull, especially in mammals. *Trends in Vertebrate Morphology*, **35**, 375–85.

Stephan, H., Nelson, J. & Frahm, H. (1981) Brain size in Chiroptera. *Zeitschrift für zoologische Systematik und Evolutionsforschung*, **19**, 195–222.

Suthers, R., Hartley, D. & Wenstrup, J. (1988) The acoustical role of tracheal chambers and nasal cavities in the production of sonar pulses by the horseshoe bat, *Rhinolophus hildebrandti*. *Journal of Comparative Physiology A*, **162**, 799–813.

Tallgren, A. & Solow, B. (1987) Hyoid bone position, facial morphology and head posture in adults. *European Journal of Orthodontics*, **9**, 1–8.

Thilander, B. & Ingervall, B. (1973) The human spheno-occipital synchondrosis II. A histological and microradiographic study of its growth. *Acta Odontologica Scandinavica*, **31**, 323–36.

Thomason, J. J. & Russell, A. P. (1986) Mechanical factors in the evolution of the mammalian secondary palate: a theoretical analysis. *Journal of Morphology*, **189**, 119–213.

Thorogood, P. (1988) The developmental specification of the vertebrate skull. *Development* (supplement), **103**, 141–53.

Van Valen, L. (1979) The evolution of bats. *Evolutionary Theory*, **4**, 103–21.

Vater, M. (1988) Light microscopic observations on cochlear development in Horseshoe bats (*Rhinolophus rouxi*). In *Animal Sonar – Processes and Performance*,

Proceedings NATO Advanced Study Institute, ed. P. E. Nachtigall & P. W. B. Moore, pp. 341–5. New York: Plenum Press.

Washburn, S. L. (1947) The relation of the temporal muscle to the form of the skull. *Anatomical Record*, **99**, 239–48.

Wimberger, P. (1991) Plasticity of jaw and skull morphology in the neotropical cichlids *Geophagus brasiliensis*, and *G. steindachneri*. *Evolution*, **45**, 1545–63.

Young, R. (1959) The influence of cranial contents on postnatal growth of the skull in the rat. *American Journal of Anatomy*, **105**, 383–415.

Zelditch, M. & Carmichael, A. A. (1989) Ontogenetic variation in patterns of developmental and functional integration in skulls of *Sigmodon fulviventer*. *Evolution*, **43**, 814–24.

JOHN R. WIBLE & DIANNE L. DAVIS

7

Ontogeny of the chiropteran basicranium, with reference to the Indian false vampire bat, *Megaderma lyra*

INTRODUCTION

The mammalian basicranium, including the hard- and soft-tissue structures of the auditory region, has long provided information deemed to be of systematic value (Kampen 1905; Klaauw 1931; Novacek 1977, 1993; MacPhee 1981). In fact, basicranial characters have played crucial roles historically in assessing relationships within certain orders, such as Carnivora (Flower 1869; Hunt 1974) and Primates (Gregory 1920; MacPhee & Cartmill 1986). Although the adult basicranium has been the object of most systematic studies, there are dangers in this approach. For example, sutural fusion can mask the number of osseous elements in the adult auditory bulla (Klaauw 1930). Also, cartilaginous elements, which often are components of the basicranium (Klaauw 1923, 1931), are generally lacking from macerated adult skulls. Because of such pitfalls, ontogeny has played a critical role in the study of the basicranium, by providing an additional level of analysis for evaluating the homologies of both hard- and soft-tissue structures in the adult (MacPhee 1977, 1979, 1981; Presley 1979; Wible 1986). Additionally, for phylogenetic systematics, ontogeny has provided characters in the form of transformational and/or temporal sequences (Velhagen 1997) and has been used to polarize characters (Nelson 1978) and to order multistate characters (Mabee 1993).

Chiroptera is one mammalian order within which the basicranium has not been a major contributor of systematic information, in spite of some detailed comparative treatises on selected aspects of this anatomical region (e.g., Grosser 1901; Bondy 1907; Henson 1961). There may be practical reasons why the chiropteran basicranium has been generally overlooked: many osseous auditory structures in bats are delicate and loosely held in place, and, therefore, are damaged or lost in macerated

214

adult skulls. One recent contribution stands out: Novacek (1980) investigated the distribution of 18 auditory characters across the living families of bats, exclusive of Craseonycteridae and Myzopodidae, and produced a cladogram of interfamilial relationships. His analysis was not undertaken to provide a new classification for Chiroptera and, in fact, was highly discordant with previous phylogenetic hypotheses. Yet, it represented the first systematic attempt to use basicranial data to assess relationships within Chiroptera. In fact, nine of Novacek's (1980) 18 auditory characters have been used by Simmons (1998) in her recent reappraisal of chiropteran interfamilial relationships based on 192 mainly morphological characters (see also Simmons & Geisler 1998).

Whereas basicranial characters thus far have had limited impact on relationships within Chiroptera, they have played a role recently in the controversy about the monophyletic versus diphyletic origin of bats. Wible & Novacek (1988) observed several unique or highly unusual basicranial features in the few mega- and microchiropterans for which serially sectioned specimens were available (see also Wible 1992; Wible & Martin 1993). These features concerned the components of the tympanic floor (auditory bulla) and roof, and the course of two branches of the stapedial arterial system. Pettigrew et al. (1989) and King (1991) disagreed with the distributions of some of these characters and concluded they do not represent chiropteran synapomorphies. However, addition of these basicranial characters to a data matrix of other morphological features has strongly supported bat monophyly (Simmons 1993).

Basicranial characters have also played a role recently in the controversy about the echolocating ability of the earliest known bats. Novacek (1985, 1987, 1991) suggested that Eocene bats were specialized echolocators, because their cochlear size (cochlear width to skull length) falls in the range of living microchiropterans. However, Habersetzer & Storch (1992) disagreed with Novacek's measure of body size, because skull length also reflects dietary habits and dentition type. Instead, they employed basicranial width as a measure of body size. Their results differed from Novacek's, with the cochlear size of Eocene bats falling in the zone of overlap between mega- and microchiropterans or just within vespertilionids. Habersetzer & Storch (1992) concluded that the echolocation performance in Eocene bats was less advanced than in living microchiropterans (see also Simmons & Geisler 1998).

In light of the foregoing, it is apparent that the chiropteran basicranium represents a valuable source of information of systematic value, but one that has only begun to be investigated. Little is still known about variability in most basicranial features at both higher and lower systematic levels within Chiroptera. It is the goal of this review to summarize the current state of knowledge regarding the ontogeny of the bat basicranium. Included as the centerpiece of this chapter is new information from the Indian false vampire bat *Megaderma lyra* (Davis 1998), a previously poorly known form for which prenatal specimens were available to us. *Megaderma lyra* is one of five Recent species of the Old World microchiropteran family Megadermatidae (Koopman 1993). It is our hope that others will join in the investigation of the basicranium across the vast diversity of Chiroptera, with the ultimate goal of providing morphological characters for assessing bat interrelationships.

For the present study, pregnant *Megaderma lyra* were collected in Nagpur, India by Dr K. B. Karim of the Institute of Science. Two prenatal specimens were sent to Dr K. P. Bhatnagar of the Department of Anatomical Sciences and Neurobiology of the University of Louisville, Louisville, Kentucky (ULASN). In preparation for paraffin sectioning, the heads of the prenatal specimens were treated with formic acid decalcifier (Bhatnagar & Kallen 1974). These are ULASN M1 (crown–rump length = 21.5 mm; head length = 14.5 mm) and ULASN M4 (crown–rump length 27.5 mm; head length = 18 mm). Serial coronal sections (10 μm) were made and stained by the Gomori trichrome method (Gomori 1950). Illustrations in Figs. 7.2, 7.3, and 7.6 were redrawn from computer-generated reconstructions made with three-dimensional reconstruction software from Jandel Scientific, Sausalito, California (PC3D, version 5.0). Every third section (ULASN M1) and every sixth section (ULASN M4) were scanned into Adobe Photoshop (version 3.0) and printed. After aligning the sections with visual best-fit methods (Gaunt & Gaunt 1978), the outlines of relevant structures were traced using a Jandel digitizing tablet. The PC3D program running on an IBM PS2 stacked the outlines to produce a three-dimensional representation that could be viewed in any specified orientation.

Features of the adult basicranium were observed in two macerated skulls of *Megaderma lyra* borrowed from the Department of Mammalogy of the American Museum of Natural History in New York (AMNH). These were: AMNH 39590 (labeled *Lyroderma sinensis*, a synonym of *M. lyra*, Koopman 1993) and 208823 (labeled *M. lyra lyra*). The right auditory bulla in the latter specimen is missing in part, exposing some of the features within the middle-ear space. In the former specimen, the skull base is fairly clean of connective-tissue fibers, revealing the structure of the auditory bulla. Comparative materials were drawn from the literature (Table 7.1). Taxonomic usages follow Koopman (1993).

Table 7.1. *Previous studies of the cartilages and bones of the basicranium in sectioned prenatal bats*

Taxa	Crown–rump length, mm	Head length, mm	Reference
Pteropodidae			
Pteropus vampyrus	32	–	Klaauw (1922)
Pteropus giganteus	14.15	8.7	Wible & Martin (1993)
"	26.95	18.1	Wible & Martin (1993)
Pteropus sp.	93.5	–	Wible (1992); Wible & Martin (1993)
Rousettus amplexicaudatus	16	–	Klaauw (1922)
"	–	9.78	Klaauw (1922)
"	–	12.96	Klaauw (1922)
"	–	16.24	Klaauw (1922)
"	–	22.5	Klaauw (1922)
"	–	29.2	Klaauw (1922)
Rousettus egyptiacus	27	15	Fehse (1990)
"	31	19	Fehse (1990)
"	32	21	Wible (1984)
"	150[a]	–	Jurgens (1963)
"	175[a]	–	Jurgens (1963)
Rousettus leschenaulti	15	–	Starck (1943)
"	18.5	–	Martin (1991)
"	34.1	–	Martin (1991)
Rhinolophidae			
Rhinolophus rouxii	15	–	Sitt (1943)
Phyllostomidae			
Artibeus jamaicensis	21	14	Hartmann (1990)
Vespertilionidae			
Miniopterus schreibersii	17	–	Fawcett (1919)
Myotis capaccinii	10	–	Frick (1954)
Myotis myotis	13	8	Frick (1954); Wible (1984)
"	18	13	Frick (1954); Wible (1984)
"	23	14	Frick (1954); Wible (1984)
"	26	15	Frick (1954); Wible (1984)
"	30	–	Frick (1954); Wible (1984)
Scotophilus kuhlii	23	18	Koch (1950)

Note: [a] Snout to tip of tail length in mm.

DESCRIPTIONS OF BASICRANIAL STRUCTURES

For the purposes of this review, we focus on those aspects of the basicranium that have received the most attention in mammals in general:
the components of the middle-ear floor and roof, and the basicranial
arteries. By way of introduction, the basicranium of a generalized fetal
placental is shown in Fig. 7.1. The most conspicuous feature is the promontorium of the petrosal, the portion of the ossified auditory capsule
that houses the cochlea (Fig. 7.1A). Lateral to it is the roof of the auditory
ossicles and cavum tympani, the air-filled diverticulum of the pharynx
that completely fills the adult middle-ear cavity. The roof is composed of
bone (ossified tegmen tympani and squamosal) and of a connective-tissue
membrane (sphenoobturator membrane; Gaupp 1908) occupying the gap
here called the piriform fenestra (Fig. 7.1A, C). Ventrolateral to the promontorium is the floor of the middle ear. Its major occupant is the ectotympanic bone, a ring-shaped intramembranous element that supports
the tympanic membrane and is usually incomplete laterally (Fig. 7.1B, C).
Also in the floor is a connective-tissue membrane (fibrous membrane of
the tympanic cavity; MacPhee 1981) stretched between the promontorium and ectotympanic (Fig. 7.1C). Not shown here, but piercing the
fibrous membrane and running across the promontorium is the internal
carotid artery (Presley 1979; Wible 1986), which enters the cranial cavity
to become the major supplier of blood to the brain. The internal carotid's
principal extracranial branch is the stapedial artery, which penetrates
the stapes and then runs forward beneath the middle-ear roof (Wible
1987). The stapedial artery provides two branches that accompany components of the trigeminal nerve (Tandler 1902; Wible 1987): a ramus superior, which runs endocranially sending branches to the meninges and into
the orbit with the ophthalmic nerve; and a ramus inferior, which sends

Figure 7.1. Basicrania of a generalized late-fetal placental. A) Left side in
ventral view, showing the components of the skull base. B) Left side in ventral
view, with the following additional elements: ectotympanic, Reichert's cartilage, malleus, Meckel's cartilage, gonial, and cartilage of the auditory tube.
C) Frontal section at the level of the dashed line in B, showing the connective-tissue membranes contributing to the presumptive tympanic floor and
roof. Open circles represent cartilage; the remaining stippled elements are
osseous. Two arrangements of the caudal tympanic process are illustrated:
in A, it encloses the stapedius fossa within the tympanic cavity; in B, it is
incomplete and the origin of the stapedius muscle extends posterior to the
tympanic cavity (the part of the gray stipple behind the dashed line). (After
Wible & Martin 1993.)

A

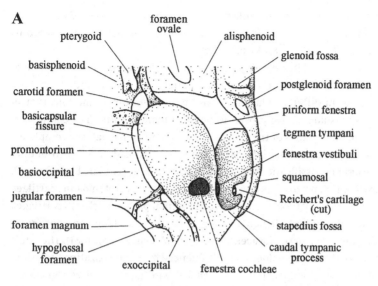

pterygoid

foramen ovale

alisphenoid

basisphenoid

glenoid fossa

carotid foramen

postglenoid foramen

basicapsular fissure

piriform fenestra

tegmen tympani

promontorium

fenestra vestibuli

basioccipital

jugular foramen

squamosal

Reichert's cartilage (cut)

foramen magnum

stapedius fossa

hypoglossal foramen

exoccipital

fenestra cochleae

caudal tympanic process

B

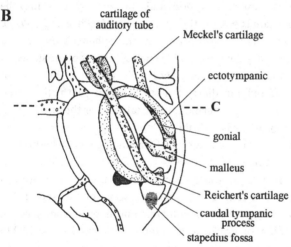

cartilage of auditory tube

Meckel's cartilage

ectotympanic

C

gonial

malleus

Reichert's cartilage

caudal tympanic process

stapedius fossa

C

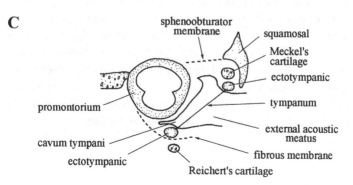

sphenoobturator membrane

squamosal

Meckel's cartilage

ectotympanic

promontorium

tympanum

cavum tympani

external acoustic meatus

ectotympanic

fibrous membrane

Reichert's cartilage

branches with the maxillary and mandibular nerves. Also variably present among fetal placentals is a ramus posterior to the stapedius muscle (MacPhee 1981; Wible 1987).

During later developmental stages, the components of the fetal middle-ear roof, floor, and arterial pattern are transformed to varying degrees in different placentals, and it is these transformations that in many instances have proven to be of systematic value. The connective-tissue membranes of the roof and floor are filled partially or completely by different combinations of the following: processes from surrounding basicranial ossifications (i.e., alisphenoid, basioccipital, basisphenoid, exoccipital, petrosal, pterygoid, and squamosal), ectotympanic, cartilage of the auditory tube, tympanohyal (proximal part of Reichert's cartilage), and independent elements called entotympanics (Kampen 1905, 1915; Klaauw 1922, 1931; Novacek 1977, 1993; MacPhee 1979, 1981). The adult state of the roof and floor is also influenced by pneumatization (inflation) of the cavum tympani (MacPhee 1981). Regarding the arteries, some or all components of the internal carotid and stapedial systems may be enclosed in bony canals within the middle ear (MacPhee 1981; Wible 1986, 1987). Additionally, ossifications in the sphenoobturator and fibrous membranes may define foramina enclosing parts of the stapedial and internal carotid arteries respectively. Finally, some or all branches of the internal carotid and stapedial systems may involute, with other arteries, such as the vertebral or external carotid, taking over the distribution areas (Tandler 1899; Bugge 1974; Wible 1987).

Implicit in the usage of anatomical terminology is a hypothesis of homology, which in reality requires testing via phylogenetic analysis as homologous structures are those characterizing monophyletic groups (Patterson 1982). Such testing is beyond the scope of this report. We caution that our usage of the term entotympanic, in particular, does not necessarily imply homology. The term entotympanic has been applied to a broad category of independent tympanic-floor elements occurring in diverse placental mammals; these grow from a cartilaginous precursor, with one possible exception in the three-toed sloth *Bradypus tridactylus* (Schneider 1955). Two major sorts of entotympanics are generally recognized (rostral and caudal) based on their positional relationships to the developing tympanic floor (Klaauw 1922, 1931; MacPhee 1979, 1981). Our usage of the terms rostral and caudal entotympanic follows this in merely referring to their positional relationships to one another.

MEGADERMA LYRA

Reconstructions of the developing basicranium were made for both prenatal specimens. In the smaller reconstructed specimen, ULASN M1, the major ossification centers are present in the chondrocranium (Fig. 7.2A), including exoccipital, basioccipital, and sphenoid (the basi- and alisphenoid being merged). The auditory capsule is largely unossified, with a single petrosal ossification center in the middle third. In addition, the auditory capsule is connected to the central stem of the chondrocranium posteriorly by a broad posterior basicapsular commissure and anteriorly by an even broader element. The latter element may best be referred to as a conjoint anterior basicapsular and alicochlear commissure, because it encloses the carotid foramen. The cavum tympani shows little inflation and occupies less than half of the potential middle-ear space. Although not a subject of this report, the most conspicuous feature on the endocranial surface of the auditory capsule is the deep subarcuate fossa, which houses the paraflocculus of the cerebellum. In contrast to the condition reported in some other microchiropterans (e.g., *Miniopterus schreibersii*, Fawcett 1919; *Myotis myotis*, Frick 1954), the subarcuate fossa in ULASN M1 (and M4) does not penetrate to the external surface of the auditory capsule.

In the larger reconstructed specimen, ULASN M4, the auditory capsule is largely ossified (Fig. 7.3A); cartilage persists in the mastoid area and at the crista parotica, the crest to which Reichert's cartilage is attached. The auditory capsule's connections to the central stem of the chondrocranium are considerably narrower, and, therefore, the basicapsular fissure is considerably elongated. In fact, the cartilaginous bar anterior to the basicapsular fissure is here best referred to as an alicochlear commissure, because of its position (postero)lateral to the carotid foramen. Lastly, the cavum tympani is not much inflated over the condition in ULASN M1 (Fig. 7.4).

Tympanic floor

ULASN M1

The major constituents of the tympanic floor are the fibrous membrane and the ectotympanic (Fig. 7.2B). The fibrous membrane has several outer attachments: anteriorly, the cartilage of the auditory tube and the broad anterior basicapsular–alicochlear commissure; medially, the ventralmost extent of the promontorium; and posteriorly, the caudal tympanic process and Reichert's cartilage. From these attachments, the

A

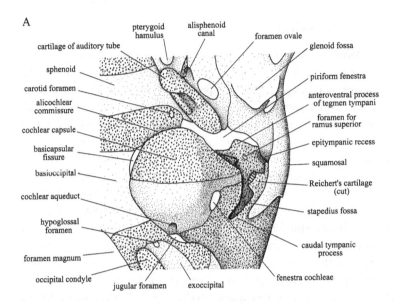

- pterygoid hamulus
- alisphenoid canal
- foramen ovale
- cartilage of auditory tube
- glenoid fossa
- sphenoid
- piriform fenestra
- carotid foramen
- anteroventral process of tegmen tympani
- alicochlear commissure
- foramen for ramus superior
- cochlear capsule
- epitympanic recess
- basicapsular fissure
- squamosal
- basioccipital
- Reichert's cartilage (cut)
- cochlear aqueduct
- stapedius fossa
- hypoglossal foramen
- caudal tympanic process
- foramen magnum
- occipital condyle
- jugular foramen
- exoccipital
- fenestra cochleae

B

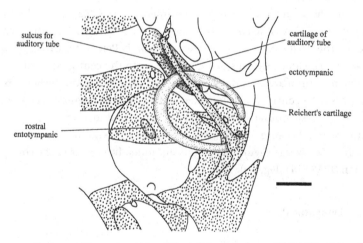

- sulcus for auditory tube
- cartilage of auditory tube
- ectotympanic
- Reichert's cartilage
- rostral entotympanic

Figure 7.2. Reconstruction of the left basicranium of fetal *Megaderma lyra* (ULASN M1) in ventral view. A) Cartilages (open circles) and endochondral and intramembranous ossifications of the skull base, plus the cartilage of the auditory tube. Closing the piriform fenestra is the sphenoobturator membrane. B) As above, but with the tympanic-floor elements added. Closing the tympanic floor around the outer contour of the ectotympanic is the fibrous membrane. The fenestra vestibuli is hidden by the bulging promontorium, medial to the stapedius fossa. Scale bar equals 1.0 mm.

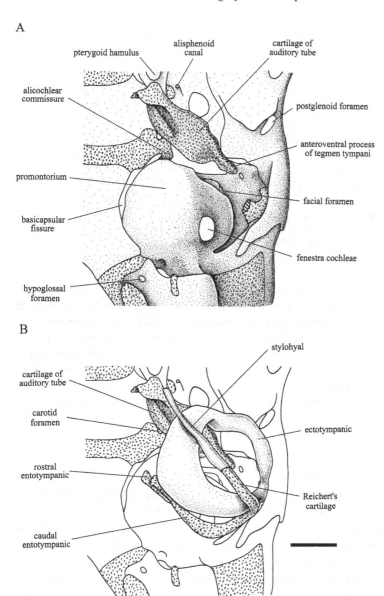

Figure 7.3. Reconstruction of the left basicranium of fetal *Megaderma lyra*
(ULASN M4) in ventral view. A) Cartilages (open circles) and endochondral
and intramembranous ossifications of the skull base, plus the cartilage of
the auditory tube. Closing the piriform fenestra is the sphenoobturator
membrane. B) As above, but with the tympanic-floor elements added.
Closing the tympanic floor around the outer contour of the ectotympanic is
the fibrous membrane. The cartilage of the auditory is fused with the anteroventral process of the tegmen tympani, which is still cartilaginous. As in
Fig. 7.2, the fenestra vestibuli is hidden by the bulging promontorium,
medial to the stapedius fossa. Scale bar equals 1.0 mm.

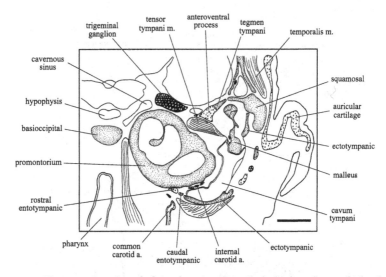

Figure 7.4. Drawing of a frontal section through the basicranium at the level of the epitympanic recess of a fetal *Megaderma lyra* (ULASN M4: slide 354–3). Open circles represent cartilage; stipple indicates osseous elements. In the tympanic floor, rostral and caudal entotympanics and expanded ectotympanic are present. In the tympanic roof, the osseous part of the tegmen tympani is continuous with the cartilaginous anteroventral process. Accompanying the internal carotid artery beneath the promontorium of the petrosal are internal carotid (sympathetic) nerves. The cavum tympani is only partially inflated and does not fill the middle-ear space. Scale bar equals 1.0 mm. Abbreviations: a – artery, m – muscle.

fibrous membrane runs ventrally to and merges with the periosteum of the ectotympanic. The ectotympanic is simple, ring-shaped, and incomplete laterally. It does not contact any elements of the dermatocranium or chondrocranium (except Reichert's cartilage), and lies at a fairly low angle to the skull base (approximately 52° measured on slide 97, where the anterior tendon of the tensor tympani muscle attaches to the malleus). Even at this stage in development, the ectotympanic appears to exhibit the same proportional size relationship to the auditory capsule as it does in the adult.

   In the medial part of the tympanic floor is a small, rod-shaped cartilage (approximately 200 μm in length), which following Klaauw (1922, 1931) we designate a rostral entotympanic (Fig. 7.2B). This element is encircled by the fibrous membrane and throughout its length abuts the ventral surface of the promontorium, well anterior to the internal carotid artery's entrance into the middle ear. Posterolateral to the rostral entotympanic, where the fibrous membrane contacts the ectotympanic, there is a high concentration of mesenchymal cells that represent the procartilaginous mass for a second floor element, a caudal entotympanic. This mass reaches laterally nearly to the point where Reichert's cartilage runs posteroventral to the ectotympanic.

ULASN M4

The fibrous membrane has the same arrangement as in the earlier stage, but more of the tympanic floor is filled in with bone and cartilage (Fig. 7.3B). The ectotympanic is a complete ring, and its posterior crus, in particular, is expanded both medially and laterally, though more so in the former direction (Fig. 7.4). Its anterior crus abuts the squamosal posteromedial to the postglenoid foramen. The ectotympanic lies at a higher angle to the skull base than in ULASN M1 (approximately 73° measured at slide 354, where the anterior tendon of the tensor tympani attaches to the malleus). The rostral entotympanic remains essentially rod-shaped (Fig. 7.3B), but it has increased in size (more than 600 μm in length). Its posterior half abuts the promontorium (Fig. 7.4), whereas anteriorly it is separated somewhat from the promontorium's anterior pole. The other major element in the tympanic floor is the caudal entotympanic, which like the rostral one is encircled by the fibrous membrane and is largely rod-shaped (Fig. 7.3B). In the medial tympanic floor, the caudal entotympanic abuts the ectotympanic and lies between that element and the rostral entotympanic (Fig. 7.4); it does not extend as far anteriorly as the rostral entotympanic. In the posterior tympanic floor, the caudal entotympanic is separated from both the ectotympanic and the surface of the auditory capsule by narrow gaps that are completed by the fibrous membrane. In the posterolateral corner of the tympanic floor (Fig. 7.3B), the caudal entotympanic is continuous with Reichert's cartilage immediately anterior to the exit of the facial nerve, and there is no noticeable break in morphology between the two.

AMNH 39590 and 208823

The adult basicranium (Fig. 7.5) has a nearly complete, osseous auditory bulla that covers the lateral half of the promontorium as well as the epitympanic recess (the space over the mallear–incudal articulation). Careful scrutiny of the osseous bulla reveals that the three elements present in ULASN M4 are identifiable in the adult.

The principal bullar component is the ectotympanic, which

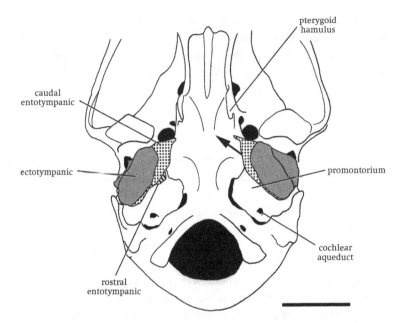

Figure 7.5. Basicranium of adult *Megaderma lyra* (AMNH 39590) in ventral view. Patterns are used to delimit the components of the nearly complete osseous auditory bulla. The rostral entotympanic (parallel diagonal lines) is visible on the specimen's right side only; it likely fell out on the opposite side. The caudal entotympanic (dots) forms a floor beneath the exit of the auditory tube from the tympanic cavity (arrow on specimen's left side) and fills the space between the promontorium of the petrosal and the ectotympanic (gray). The ectotympanic also contributes to a floor beneath the proximal part of the external acoustic meatus. Scale bar equals 1.0 mm.

remains a complete ring. Most of the expansion in the ectotympanic is in the anterior, medial, and posterior tympanic floor, but there is even some expansion of the ectotympanic laterally to contribute a floor beneath the proximalmost external acoustic meatus. Anterolaterally and laterally, the ectotympanic contacts the posterior surface of the postglenoid process of the squamosal, behind the postglenoid foramen. Anteromedially, medially, and posteriorly, the ectotympanic contacts the second largest bullar element, which in turn contacts the skull base (see below). The tympanum is visible on the right side of AMNH 208823, which is missing the medial part of the auditory bulla. At the level where the dried tensor tympani muscle attaches to the malleus, we estimate the tympanum (and so the ectotympanic) to be at 73° to the skull base.

The second largest bullar element, the caudal entotympanic of

ULASN M4, is fairly narrow and sandwiched between the ectotympanic and skull base in the anteromedial, medial, and posteromedial tympanic floor. Although this element is certainly osseous, it is more opaque (denser) than the neighboring, translucent ectotympanic. It has several contacts with the skull base: anteromedially, the sphenoid, posterior to the foramen ovale; and medially, the sphenoid, posterior to the carotid foramen, the third bullar element (see below), and the promontorium of the petrosal. It does not contact the skull base posteriorly, but the fibrous membrane (visible in AMNH 39590) connects it to the caudal tympanic process, enclosing the stapedius fossa within the middle ear. A direct connection is lacking between the caudal entotympanic and Reichert's cartilage, which is represented in the adult by a tiny, ossified prong of tympanohyal at the posterolateral corner of the bulla (visible on the right side of AMNH 39590). At the anteromedial corner of the tympanic floor is an elongated aperture that transmits the auditory tube from the middle ear (visible in AMNH 39590 in which more soft tissue has been removed). The anterior and ventral walls of this aperture are formed by the caudal entotympanic with the remaining borders formed by the sphenoid. This aperture is separated from the pterygoid hamulus by a considerable distance. To transform the caudal entotympanic of ULASN M4 into the adult structure, the following steps are required (not necessarily in the order given): 1) expansion anteriorly, ventral to the auditory tube's exit from the middle ear, 2) resorption of the connection with Reichert's cartilage, and 3) ossification.

The third bullar element, the rostral entotympanic of ULASN M1 and M4, is a small, rod-shaped ossification in the anteromedial tympanic floor, visible on the right side of AMNH 39590 and the left side of AMNH 208823. It is wedged between and contacts the caudal entotympanic and the promontorium; it approximates but does not contact the sphenoid, being held to that bone by the fibrous membrane. Without prior knowledge of the existence of a rostral entotympanic in the prenatal specimens, we likely would have overlooked this small structure in the adult (and in fact we did for quite some time). It is partly obscured by the thick mass of connective tissue fibers in the anteromedial part of the fibrous membrane. To transform the rostral entotympanic of ULASN M1 and M4 into the adult structure, the only step that is required is ossification. It is uncertain whether the presence of the rostral entotympanic on only one side of the adult skulls studied here is indicative of post-mortem loss or intra-individual variation.

On the right side of AMNH 208823, there is another mass of connective tissue fibers that connects the anteromedial surface of the caudal entotympanic with the pterygoid hamulus and defines the floor of a second, more medial aperture for the auditory tube. We believe that this membrane indicates the location of the cartilage of the auditory tube, which as is often the case with macerated specimens is not preserved. Supporting this interpretation is the intimate relationship between the cartilage of the auditory tube and pterygoid hamulus in both ULASN M1 and M4.

Tympanic roof

ULASN M1

The roof over the auditory ossicles is formed largely by the tegmen tympani, the cartilaginous portion of the auditory capsule lateral to the promontorium (Fig. 7.2A). The squamosal contributes the lateral wall of the fossa incudis, which houses the crus breve of the incus. Anterior to that, the squamosal has a horizontal epitympanic wing that forms most of the roof and lateral wall of the epitympanic recess, the space over the mallear–incudal articulation. At the level of the foramen for the ramus superior in the tympanic roof, the tegmen tympani and the epitympanic wing of the squamosal are essentially in the same plane. However, anterior to that aperture, the tegmen tympani sends a narrow process antero-ventromedially, which we term the anteroventral process following Martin (1991). This process is triangular in cross-section, forms a medial wall to the epitympanic recess, and contacts the tensor tympani muscle. The cartilage of this process lacks lacunae and is not as mature as that of the tegmen tympani proper or other chondrocranial elements.

In front of the tegmen tympani is a gap in the skull base, the piriform fenestra. This gap is filled by the sphenoobturator membrane, which is a diffuse mass of collagen fibers and mesenchymal cells. Anteriorly, two separate planes develop in the sphenoobturator membrane: a dorsal one, which is coplanar with the epitympanic wing of the squamosal, and a ventral one, which is coplanar with the anteroventral process of the tegmen tympani. The anterior margin of the piriform fenestra is formed by the horizontal epitympanic wing of the sphenoid, which appears in the dorsal plane of the sphenoobturator membrane, and the cartilage of the auditory tube, which appears in the ventral plane.

The cartilage of the auditory tube is an obliquely oriented, roughly oblong-shaped element. The structure of the cartilage resembles that of the anteroventral process of the tegmen tympani, with organized, rounded cells lacking lacunae. Posteriorly, the cartilage of the auditory tube is fairly flat, with its medial edge providing attachment area for the tensor tympani muscle and its lateral edge abutting the gonial, the intramembranous element that forms part of the anterior process of the malleus in the adult (see Fig. 7.1B). Anteriorly, the cartilage of the auditory tube is more complex. Its lateral surface provides attachment for the tensor veli palatini muscle. Ventrally, there are longitudinal crests on the medial and lateral margins, which demarcate a deep groove for the auditory tube. Near the anterior limit of the cartilage of the auditory tube, the lateral crest is inflected medially, forming a floor beneath the auditory tube. Arising from the medial surface of this inflected crest is the levator veli palatini muscle. Anterior to the cartilage of the auditory tube and connected to it by a mass of fibers and mesenchymal cells is the pterygoid hamulus.

ULASN M4

Although the tegmen tympani remains the major constituent of the roof over the auditory ossicles, it is here ossified (Fig. 7.3A), the exception being the anteroventral process (see below). The tegmen tympani forms the medial wall and the medial half of the roof of the fossa incudis and the epitympanic recess, and the squamosal forms the lateral half of the roof and lateral wall of both spaces (Fig. 7.4). As in the earlier stage, the anteroventral process of the tegmen tympani is triangular in cross-section, forms the medial wall of the epitympanic recess, and provides attachment for the tensor tympani muscle on its posteromedial surface (Fig. 7.4).

Dorsomedial to the anteroventral process of the tegmen tympani, there is a small, horizontal lamina of bone that extends laterally from the ossified cochlear capsule into the sphenoobturator membrane. It lies in the same plane as the tegmen tympani and the epitympanic wing of the squamosal and sphenoid. We identify it as an epitympanic wing of the petrosal. Given that this process has no cartilaginous precursor in ULASN M1, it most likely formed by appositional bone growth.

In ULASN M1, the anteroventral process of the tegmen tympani and the cartilage of the auditory tube are separated by a sizable gap (Fig. 7.2A), although they are connected by fibers of the sphenoobturator membrane. In ULASN M4, these two cartilaginous structures are fused (Fig. 7.3B), with no obvious morphological transition zone between them. Anteriorly, the narrow, triangular anteroventral process gives way to a broad cartilage of the auditory tube, which is crescent-shaped in its posterior half. The anterior half of the cartilage of the auditory tube closely resembles that described above in ULASN M1. The contacts of the cartilage of the auditory tube are with the pterygoid hamulus anteriorly and with the ectotympanic (Fig. 7.3B) and gonial laterally. In addition to providing the sole origin for the tensor and levator veli palatini and attachment area for tensor tympani, the cartilage of the auditory tube has the superior pharyngeal constrictor attached to its anteroventral margin. The bulk of this muscle, however, arises from the pterygoid hamulus.

AMNH 208823

Part of the right auditory bulla is lacking, thus exposing the contents of the middle ear. Lateral to the promontorium is the dried tensor tympani muscle, with its two tendons of insertion on the malleus visible (Wassif 1950). This muscle obscures the features of the tympanic roof that have been described above from the sectioned prenatal specimens. This highlights a problem frequently encountered in studying features of the tympanic roof in macerated adult skulls.

Arterial system

ULASN M4

As is typical of mammals (Tandler 1899), the major blood supply to the head in *Megaderma lyra* is through the common carotid and vertebral arteries. The latter is in fact the principal supplier to the brain, as the internal carotid artery is considerably smaller. The external carotid artery supplies extracranial tissues, with the exception of the occiput, jaws, and orbit, which are fed through branches of the internal carotid artery. The major basicranial branches of the external and internal carotid arteries in ULASN M4 are shown in Figs. 7.6A and B, respectively, and are described below. Differences in the arterial system exhibited in the smaller ULASN M1 are also noted.

Ascending through the neck, anterior to the thyroid gland, the common carotid artery supplies a superior thyroid branch medially to the thyroid and larynx. Below the ear region, at the posteromedial aspect of the ectotympanic, the common carotid artery bifurcates into an internal and a slightly larger external carotid artery, with a small ascending pharyngeal artery arising from the bifurcation and running anteromedially along the dorsal wall of the pharynx.

The external carotid artery runs anterolaterally beneath the ectotympanic and sends a lingual artery anteromedially, which en route to the

A

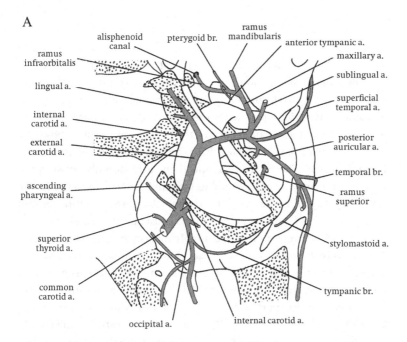

B

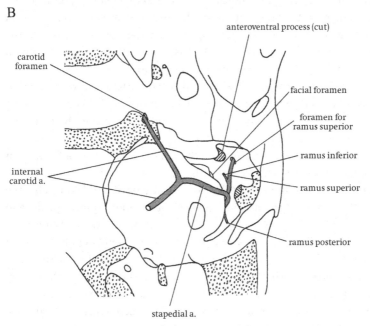

Figure 7.6. Reconstruction of the left basicranium of fetal *Megaderma lyra* (ULASN M4) in ventral view. A) As in Fig. 7.3B, but with the major basicranial arteries added. B) As in Fig. 7.3A, but with the major intratympanic arteries added. The origin of the ramus posterior from the stapedial artery is hidden by the bulging promontorium. Abbreviations: a – artery, br – branch.

tongue supplies a muscular branch to the larynx. The external carotid
then turns laterally, ventral to the stylohyal, and gives off a trunk that
supplies superficial temporal and posterior auricular arteries. The latter
supplies branches to the parotid gland, the temporalis muscle, and a stylomastoid artery that approaches the exit of the facial nerve from the
middle ear. The next branch beyond the superficial temporal/posterior
auricular trunk is a sublingual, which continues forward ventromedial
to the mandible. The external carotid then turns dorsally between the
ectotympanic and the postglenoid (retromandibular) vein as the maxillary artery. Medial to the jaw joint, the maxillary artery bifurcates into
subequal branches, a ramus mandibularis and a ramus infraorbitalis. The
ramus mandibularis runs forward, supplying deep temporal and inferior
alveolar branches. The ramus infraorbitalis turns laterally dorsal to
Meckel's cartilage and medial to the mandibular nerve, supplying a
branch to the pterygoid muscles, which in turn sends a tiny anterior tympanic artery (lacking in ULASN M1). The anterior tympanic runs posteriorly a short distance ventral to the foramen ovale, between the chorda
tympani nerve laterally and the otic ganglion medially. Upon reaching
the skull base, the ramus infraorbitalis turns forward in a deep groove in
the sphenoid and enters the alisphenoid canal, which opens into the
cranial cavity. The ramus infraorbitalis ultimately enters the orbit with
the maxillary nerve via the sphenorbital fissure.

The first branch of the internal carotid artery is a well-developed
occipital artery, which en route to the occiput supplies branches to the
digastric and prevertebral muscles, and a tympanic branch that
approaches the stylomastoid foramen. Beyond the origin of the occipital
artery, the internal carotid enters the middle ear through a foramen in
the medial aspect of the tympanic floor. The boundaries of this posterior
carotid foramen are anteriorly, the rostral entotympanic; dorsally, the
promontorium of the petrosal; ventrally, the caudal entotympanic; and
posteriorly, the fibrous membrane of the tympanic cavity. Within the
middle ear, the internal carotid artery travels anterolaterally ventral to
the promontorium (Fig. 7.4) and bifurcates anterior to the fenestra cochleae (round window). The smaller branch at the bifurcation is the contin-
uation of the internal carotid. It runs anteromedially along the
ventrolateral aspect of the promontorium and then ventral to the alicochlear commissure. It enters the cranial cavity via the carotid foramen
located in the posterior edge of the sphenoid. In its course within the
middle ear, the internal carotid artery is accompanied by a sympathetic
nerve (Fig. 7.4) derived from the superior cervical ganglion below the
jugular foramen. Although this internal carotid nerve enters the cranial

cavity, it does not do so with the artery via the carotid foramen, but via the anteromedial corner of the piriform fenestra.

The larger branch arising from the internal carotid anterior to the fenestra cochleae is the stapedial artery. It runs laterally, slightly posteriorly, and then dorsally through the obturator foramen in the stapes. Beyond the stapes, a small ramus posterior is sent off towards the back of the middle ear, ventral to the facial nerve into the stapedius muscle; this branch is lacking in ULASN M1. The stapedial artery then turns forward ventral to the ossified tegmen tympani, lateral to the primary facial foramen. It supplies a small artery, which we identify as a ramus inferior, that travels forward within the middle ear just ventral to where the tegmen tympani abuts the promontorium. This vessel is very short in ULASN M1, but in M4 it continues forward beneath the piriform fenestra with the lesser petrosal nerve and sends tiny branches into the tensor tympani muscle and into the cranial cavity via the piriform fenestra towards the trigeminal ganglion.

The main stem of the stapedial artery, the ramus superior, enters the cranial cavity through a foramen in the tegmen tympani. The main occupant of this foramen, however, is a vein that drains from the capsuloparietal emissary vein (which quits the cranial cavity via the postglenoid foramen) into the middle ear, and then back into the cranial cavity via the piriform fenestra, ultimately into the cavernous sinus medial to the trigeminal ganglion. Within the cranial cavity, the ramus superior bifurcates into unequal anterior and posterior branches. The smaller anterior branch, the ramus temporalis, runs dorsolaterally, passing through the parietal bone into the temporalis muscle. The larger posterior branch first travels back medial to the capsuloparietal emissary vein and then makes a broad U-turn, which directs it forward along the inner surface of the parietal, well up the side wall of the braincase. Ultimately, it reaches the orbit as the ramus supraorbitalis, accompanying branches of the ophthalmic nerve.

AMNH 39590 and 208823

As described above for ULASN M4, the basicranial arterial pattern does not leave many markings on the prenatal skull (i.e., grooves, canals, and foramina). The following are preserved in the adult skulls: 1) The carotid foramen is visible on the right side of AMNH 208823, in which part of the bulla is missing. The foramen is not pristinely preserved, but it appears to be entirely within the sphenoid with contributions from no other elements, the condition in ULASN M1 and M4. 2) A narrow posterior carotid foramen is visible in the posteromedial tympanic floor of AMNH 39590, between the caudal entotympanic and promontorium. This transmits the internal carotid artery and nerve into the middle ear. 3) A small foramen is visible in the side wall of the braincase in AMNH 39590, above the external acoustic meatus. This transmits the ramus temporalis into the temporal fossa. The bone(s) forming this foramen is indeterminate because of sutural fusion. 4) A groove (canal?) on the inner surface of the side wall of the braincase (parietal?) is visible through the translucent bone in AMNH 39590 and 208823. This channel is a broad, inverted U. The posterior leg is posterodorsal to the external acoustic meatus, and the anterior leg opens in the orbit at a foramen (between the parietal and ethmoid?). It transmits the ramus superior into the orbit as the ramus supraorbitalis. 5) An alisphenoid canal opens on the skull base in AMNH 39590 and 208823, anterior to the foramen ovale and lateral to the pterygoid plate. It transmits the ramus infraorbitalis into the cranial cavity en route to the orbit.

One feature specifically not visible in either adult skull is the foramen for the ramus superior in the tegmen tympani. It is most likely present but obscured by other structures (e.g., the dried tensor tympani muscle).

COMPARISONS OF BASICRANIAL STRUCTURES

Our study of basicranial ontogeny in *Megaderma lyra* is designed to provide information for interpreting the adult morphology and for assessing the interrelationships of bats. Regarding the former, we document that ontogeny is critical for identifying the components of the adult tympanic floor and roof. Without prior knowledge of the existence of two entotympanics, it is likely that the small rostral element, partly obscured within the fibrous membrane, would be overlooked. The complex details of tympanic roof anatomy are essentially undetectable in macerated adult skulls because of their delicacy and intimate relationship with soft tissues. Yet, our study of ontogeny is incomplete. Without access to serially sectioned postnatal specimens, for example, we do not know the fate of the entire cartilage of the auditory tube and tegmen tympani in the adult.

Regarding the interrelationships of bats, the logical next step in such an analysis would be to add the information gleaned from the above descriptions on *M. lyra* to a data base of other bats. Unfortunately, few bats have been studied in detail to the degree that we present here for *M. lyra*. In lieu of a proper phylogenetic analysis, we present here selected features and their known distributions within Chiroptera, with the goal that future studies can add to this data base.

Tympanic floor

Other microchiropterans

Although there are numerous anecdotal accounts of the condition of the tympanic floor in prenatal microchiropterans (references in Table 7.1), the ontogeny of the floor is known in detail in only the vespertilionid *Myotis myotis* (Frick 1954; Wible 1984; Wible & Martin 1993). As in *Megaderma*, there is a nearly complete osseous auditory bulla in the adult, whose principal component is the ectotympanic. However, the ectotympanic in *Myotis* differs in that it is not a complete ring (Frick 1954), and the osseous bulla obscures more of the ventral surface of the promontorium. As in *Megaderma*, two entotympanics form in *Myotis*, but there are striking differences between these elements in the two taxa. In *Megaderma*, the small, rod-shaped rostral element is the first entotympanic to appear (Fig. 7.2B), and it remains an insignificant component of the bulla throughout ontogeny (Figs. 7.3B, 7.5). On the other hand, in *Myotis*, the rostral element appears later than the caudal one, and by the adult it is substantial, occupying the space between the sphenoid and petrosal beneath the carotid foramen and forming a complete ring around the internal carotid artery (Wible 1984). The caudal element in each taxon is superficially similar in position, but that in *Megaderma* differs in that it is continuous with Reichert's cartilage (Fig. 7.3B) and it expands farther forward, ventral to the auditory tube in the adult (Fig. 7.5).

Rostral and caudal entotympanics have also been reported for prenatal specimens of the vespertilionid *Pipistrellus pipistrellus* and the molossid *Tadarida brasiliensis* (Wible & Novacek 1988), but details enabling comparison with the elements in *Megaderma* and *Myotis* have not been published. Fawcett (1919) reported an entotympanic equivalent in position to a caudal one for a prenatal specimen of the vespertilionid *Miniopterus schreibersii*. However, the presence of a single entotympanic may not be indicative of the adult condition, because the specimen in question may have been too young to possess a rostral element.

Megachiropterans

The most comprehensive accounts on the ontogeny of the tympanic floor in megachiropterans (Table 7.1) are those on various species of

Pteropus (Klaauw 1922; Wible 1984, 1992; Wible & Martin 1993) and *Rousettus* (Klaauw 1922; Jurgens 1963; Wible 1984; Martin 1991). In general, the ontogeny in these forms is reminiscent of that in *Myotis myotis*, with an ectotympanic incomplete laterally and a well-developed caudal entotympanic appearing before a smaller rostral one. The rostral entotympanic forms in close association with the anterior exit of the internal carotid artery from the middle ear, though not surrounding that vessel as in *Myotis*. A rostral entotympanic partially or fully forming a canal for the internal carotid artery beneath the carotid foramen has been proposed as a synapomorphy of bats (Wible & Novacek 1988; Wible 1992). There is considerable variability in the relationships of the mega-chiropteran entotympanics to neighboring structures. The caudal ento-tympanic has been reported to fuse to the following: 1) posterolaterally to Reichert's cartilage (*Pteropus* sp., Kampen 1915; *Rousettus amplexicaudatus*, Klaauw 1922; *R. egyptiacus*, Jurgens 1963, Wible 1984; *R. leschenaulti*, Martin 1991); 2) anteriorly with the rostral entotympanic (*R. amplexicaudatus*, Klaauw 1922; *R. egyptiacus*, Jurgens 1963; *R. leschenaulti*, Martin 1991); and 3) posteriorly with the cochlear capsule behind the fenestra cochleae (*R. amplexicaudatus*, Klaauw 1922; *R. leschenaulti*, Martin 1991). The rostral entotympanic has been reported to fuse with cartilaginous components of the tympanic roof (see below).

Along with the ectotympanic, the two entotympanics contribute to the formation of a nearly complete auditory bulla in adult *Pteropus* and *Rousettus*. Whereas the ectotympanic is far and away the largest bullar element in adult *Megaderma lyra* and *Myotis myotis*, that bone is not signifi-cantly expanded over the late fetal condition in adult *Pteropus* and *Rousettus*, and the entotympanics are much more substantial structures. In fact, the caudal entotympanic is reminiscent of that in adult *Megaderma lyra* in that it reaches forward ventral to the auditory tube. The rostral entotympanic apparently ossifies in the adult, but the caudal one may remain cartilaginous or it may ossify to varying degrees with one or more centers of ossification (Kampen 1905; Novacek 1980; Wible 1984).

Tympanic roof

Other microchiropterans

As with the ontogeny of the tympanic floor, that of the roof is known in detail in only the vespertilionid *Myotis myotis* (Frick 1954; Wible 1984; Wible & Martin 1993). As in *Megaderma*, the principal elements in the tympanic roof in *Myotis* are the tegmen tympani, including its anteroventral process, and the cartilage of the auditory tube, but there is a subtle difference in the formation of the latter structure in the two taxa. In *Megaderma*, the cartilage of the auditory tube forms from one center of chondrification, which subsequently fuses to the cartilaginous antero-ventral process of the tegmen tympani (Figs. 7.2A, 7.3A). On the other hand, in *Myotis*, two centers of chondrification, anteromedial and postero-lateral, contribute to the formation of the cartilage of the auditory tube, which ultimately fuses to the cartilaginous anteroventral process of the tegmen tympani. The fate of the cartilage of the auditory tube is known for adult *Myotis*, as a specimen has been serially sectioned (see Frick 1954; Wible 1984). That element remains cartilaginous in the adult, whereas the anteroventral process of the tegmen tympani to which it is fused ossifies as part of the petrosal.

An anteroventral process of the tegmen tympani (the tuberculum tympani of some authors) that forms the medial wall of the epitympanic recess has been noted in all prenatal microchiropterans for which reports exist, including representatives of Rhinolophidae, Vespertilionidae, Molossidae, and Phyllostomidae (Fawcett 1919; Koch 1950; Sitt 1943; Wible & Novacek 1988; Hartmann 1990; Wible & Martin 1993). In addition, fusion of the anteroventral process and cartilage of the auditory tube has been reported for the vespertilionid *Pipistrellus pipistrellus*, the molossid *Tadarida brasiliensis*, and the phyllostomid *Carollia perspicillata* (Wible & Novacek 1988; Wible & Martin 1993). It is uncertain whether the studies in which fusion is not mentioned are instances of stages prior to fusion (such as ULASN M1), of true non-fusion, or of lack of observation. Authors describing the chondrocranium frequently neglect to consider the cartilage of the auditory tube.

Megachiropterans

The most comprehensive accounts of the ontogeny of the tympanic roof in megachiropterans are those on various species of *Pteropus* (Klaauw 1922; Wible 1984, 1992; Wible & Martin 1993) and *Rousettus* (Klaauw 1922; Starck 1943; Jurgens, 1963; Wible 1984; Fehse 1990; Martin 1991). As in microchiropterans, an anteroventral process of the tegmen tympani forming the medial wall of the epitympanic recess has been reported for all prenatal megachiropterans studied to date, and it has been proposed as a chiropteran synapomorphy (Wible & Novacek 1988; Wible 1992). Two other cartilages appear in the prenatal sphenoobturator membrane in the same plane as the anteroventral process of the tegmen tympani: a small cartilage of the auditory tube is located anteriorly, where the auditory tube joins the nasopharynx, and in the gap between the cartilage of the auditory tube and tegmen tympani is a contribution to the tympanic roof from the rostral entotympanic. A third small piece of cartilage ventral to the tensor tympani muscle has been described in *Pteropus vampyrus* (Klaauw 1922) and *Pteropus* sp. (Wible & Martin 1993). Of these various cartilages, the rostral entotympanic is known to fuse with the cartilage of the auditory tube in *Rousettus amplexicaudatus* (Klaauw 1922) and *P. medius* (Wible 1984), the only taxa for which the ontogeny of these elements has been sufficiently traced. The ultimate fate of these cartilages in the adult is not well understood.

Arterial system

Other microchiropterans

The major branches of the cranial arterial system have been reported for the following microchiropterans: embryos and adults of the rhinolophid *Rhinolophus hipposideros* and the vespertilionids *Vespertilio murinus* (Grosser 1901) and *Myotis myotis* (Wible 1984) studied from sections; adults of the vespertilionid *Nyctalus noctula* (= *Vesperugo noctula*), *Pipistrellus pipistrellus* (= *Vesperugo pipistrellus*), *Myotis mystacinus* (= *Vespertilio mystacinus*), and *Plecotus auritus* studied from sections (Grosser 1901); adult of the vespertilionid *Myotis lucifugus* studied by dissection (Kallen 1977); and neonates and adults of the phyllostomid *Artibeus lituratus* studied by dissection (Buchanan & Arata 1969). All of the above forms exhibit a cranial arterial pattern that in its general features is similar to that reported here for *Megaderma lyra*. In fact, all of these forms share with *M. lyra* the following features that are considered to be derived compared with the primitive placental pattern (Tandler 1899; Bugge 1974; Wible 1984, 1987): 1) great reduction (to absence) in the internal carotid's contribution to the cerebral vasculature along with increase in the size of the vertebral artery; 2) internal carotid origin for the occipital artery; 3) absence of the arteria diploëtica magna; and 4) great reduction (to absence) of the ramus inferior of the stapedial artery. Also, all microchiropterans studied have a functional stapedial artery, which distinguishes them from megachiropterans (see below), but not from the pattern thought to be primitive for placentals. In addition to the taxa named above, a patent stapedial artery at the obturator foramen has been reported for the following species based on serial sections: by Henson (1961) for the rhinolophid *Rhinolophus ferrumequinum*, and by Samano Bishop (1965) for the emballonurid *Balantiopteryx plicata*; the phyllostomids *Macrotus waterhousii, Anoura geoffroyi, Glossophaga soricina, Leptonycteris nivalis, Carollia perspicillata, Artibeus jamaicensis,* and *Desmodus rotundus*; the natalid *Natalus stramineus*; the molossids *Molossus* sp. and *Tadarida brasiliensis*; the mormoopids *Mormoops megalophylla, Pteronotus davyi,* and *P. parnellii*; the vespertilionid *Lasiurus borealis*; and the noctiolionid *Noctilio leporinus*.

The major differences among the microchiropterans whose cranial arteries have been studied to date concern components of the stapedial system. The ramus posterior has been reported for only *Myotis myotis* (Wible 1984, 1987) and *Megaderma lyra*, though this small vessel has generally been overlooked in the older literature. The proximal stump of the ramus inferior has been reported in only the same two taxa. However, this vessel exhibits a different relationship to the tympanic roof in the two: it arises from the stapedial artery ventral to the tegmen tympani in *M. lyra*, but dorsal to the tegmen tympani's anteroventral process in *Myotis myotis* (Wible 1984). The course of the ramus infraorbitalis to the orbit differs dramatically: it runs medial to the mandibular nerve and into the cranial cavity via the foramen ovale in vespertilionids (Grosser 1901; Kallen 1977; Wible 1984); medial to the mandibular nerve and into the cranial cavity via the alisphenoid canal in *Megaderma lyra*; lateral to the mandibular nerve and ventral to the skull base in some *Rhinolophus hipposideros* (Grosser 1901); medial to the mandibular nerve and ventral to the skull base in other *R. hipposideros* (Grosser 1901); and with parallel channels in *Artibeus lituratus*, one intracranial entering via the alisphenoid canal and the other extracranial lateral to the mandibular nerve (Buchanan & Arata 1969). An extracranial course for the ramus infraorbitalis is also present in the phyllostomid *Desmodus rotundus*, though its positional relationship to the mandibular nerve was not noted (Kallen 1977).

One feature of the arterial system distinguishes *M. lyra* from all microchiropterans studied to date. It is the only species with a foramen in the tegmen tympani for a branch of the stapedial system. In the other forms the artery enters the cranial cavity via the piriform fenestra, medial to the anteroventral process of the tegmen tympani.

Megachiropterans

The major branches of the cranial arterial system have been reported for the following megachiropterans: prenatal specimens of *Pteropus* sp. (Wible 1992), *Rousettus leschenaulti* (Martin 1991), *P. medius,* and *R. egyptiacus* (Wible 1984) studied from sections; and adults of *P. vampyrus* (=*P. edulis*) (Tandler 1899) and *R. egyptiacus* (=*Cynonycteris aegyptiaca*)

(Grosser 1901) studied by dissection. The major difference between the prenatal specimens and adults is the total absence of the stapedial artery in the latter. As in all other placentals studied to date (Tandler 1902; Wible 1984, 1987), the stapedial artery forms in early ontogenetic stages of megachiropterans. However, the stapedial artery involutes in later stages, and its end branches are annexed to the external carotid system via the maxillary artery.

Megachiropterans share the following features that are considered to be derived compared with the primitive placental pattern (Tandler 1899; Bugge 1974; Wible 1984, 1986, 1987): 1) an extracranial course for the internal carotid along the anteromedial surface of the promontorium, either in a transpromontorial or perbullar position (Wible 1986); 2) absence of the stapedial artery in the adult; 3) a course for the ramus inferior and origin of the ramus superior dorsal to the tympanic roof; and 4) an intracranial course for the ramus infraorbitalis, either via the foramen ovale or a separate alisphenoid canal. *M. lyra* shares one of these derived features with megachiropterans: an intracranial course for the ramus infraorbitalis. A major difference between megachiropterans on the one hand and *M. lyra* (and other microchiropterans?) on the other concerns the location of the carotid foramen; it is between the petrosal and sphenoid in the former and within the sphenoid in the latter.

Miscellaneous

Our study and comparisons of *Megaderma lyra* provide relevant information and/or amendments for several miscellaneous basicranial features previously used in or suggested for phylogenetic analyses. These include the following: 1) Ectotympanic inclination – semi-inclined (0) or nearly vertical (1). In their phylogenetic analyses, Novacek (1980) and Simmons (1998; see also Simmons & Geisler 1998) employed the inclination of the adult ectotympanic, that bone being nearly vertical in most microchiropterans, but only semi-inclined in megachiropterans and most other mammals. Before this character can be used in such analyses, however, more data must be collected to ascertain whether this should be a binary or multistate character and what angles demarcate the states. For example, it is unclear whether the ectotympanic in *M. lyra* at 73° should be scored as semi-inclined or nearly vertical. 2) Cochlear aqueduct – large (0) or small or absent (1). Novacek (1980) and Simmons (1998; see also Simmons & Geisler 1998) used a character concerning the size of the cochlear aqueduct, which transmits the perilymphatic duct into the inner ear, with megadermatids scored as small or absent. Rather than small or absent, we believe that the cochlear aqueduct is quite large in *M. lyra* (AMNH 39590; Fig. 7.5), at least on the same scale as the phyllostomid *Carollia perspicillata* (AMNH 184709), with the Phyllostomidae being scored as large by the above authors. 3) Fossa for stapedius muscle – indistinct (0), shallow and broad (1), or deep, constricted in area, often a crescent-shaped fissure (2). Simmons (1998) and Simmons & Geisler (1998) divided the stapedius fossa into a multistate character, with the condition in megadermatids as shallow and broad. In the sectioned prenatal specimens of *M. lyra*, the stapedius fossa is narrow and deep, definitely not shallow and broad (Figs. 7.2A, 7.3A). Additionally, the condition of the stapedius fossa in the adult skulls appears reminiscent of that in the phyllostomid *C. perspicillata* (AMNH 184709), with the Phyllostomidae being scored as deep by Simmons (1998) and Simmons & Geisler (1998). 4) Caudal tympanic process of petrosal – does (0) or does not surround the stapedius fossa (1). Wible & Martin (1993) noted the distribution of these two states (see Figs. 7.1A, B) among various archontans (e.g., bats, primates, tree shrews, and colugos). They scored both Megachiroptera and Microchiroptera as having a caudal tympanic process that does not surround the stapedius fossa. However, both the prenatal specimens and adults of *M. lyra* clearly have a low caudal tympanic process that forms a wall behind the stapedius fossa (Figs. 7.2A, 7.3A). This is also true for the phyllostomid *C. perspicillata* (AMNH 184709). Consequently, this character needs to be re-evaluated within microchiropterans before it can be used in phylogenetic analysis. 5) Epitympanic wing of the petrosal – present (0) or absent (1). Wible & Martin (1993) noted the absence of an epitympanic wing of the petrosal in both Megachiroptera and Microchiroptera. Although the incidence of this process could be neither confirmed nor denied in the adult skulls of *M. lyra*, a small epitympanic wing was clearly present in ULASN M4.

CONCLUSIONS

Because the ontogeny of the basicranium is known in detail in so few bats, sweeping generalizations are premature to say the least. Nevertheless, our study of *Megaderma lyra* has reinforced some previously held notions about the distribution of basicranial features in bats and has raised questions about others. Evaluating the systematic impact of these features has not been attempted here, but requires broadscale phylogenetic analysis well beyond our current scope.

Megaderma lyra, on the one hand, has the following two basicranial features that have been hypothesized to distinguish bats from other placentals (Wible & Novacek 1988; Wible 1992): 1) an anteroventral process of the tegmen tympani that forms the medial wall of the epitympanic recess, and 2) a ramus infraorbitalis of the stapedial artery that runs through the cranial cavity en route to the orbit. On the other hand, *M. lyra* lacks two other features hypothesized as bat synapomorphies (Wible & Novacek, 1988; Wible, 1992): 1) the ramus inferior runs ventral rather than dorsal to the tegmen tympani, and 2) the rostral entotympanic does not form a groove or canal for the internal carotid artery.

Differences exist that, given the current state of comparative knowledge, appear to distinguish mega- and microchiropterans. In the tympanic floor, the relative contributions of ectotympanic and entotympanics to the auditory bulla differ, with the entotympanics more substantial in megachiropterans and the ectotympanic more substantial in microchiropterans. In the tympanic roof, the anteroventral process of the tegmen tympani is fused to the cartilage of the auditory tube in microchiropterans but not in megachiropterans. In the arterial system, the stapedial artery is present in microchiropterans but lacking in megachiropterans. It is unclear how these differences might relate to the dramatic differences in auditory function between mega- and microchiropterans, the former being non-echolocators, with the exception of some *Rousettus* (Medway 1978; Herbert 1985), and the latter echolocators.

Differences also exist between the two microchiropterans for which basicranial ontogeny is known in detail: the vespertilionid *Myotis myotis* (Frick 1954; Wible 1984; Wible & Martin 1993) and the megadermatid *Megaderma lyra* (Davis 1998; this report). In the tympanic floor, the ectotympanic is a complete ring and the rostral entotympanic forms later than the caudal in *Myotis*, whereas the ectotympanic is a complete ring and the rostral entotympanic forms before the caudal in *Megaderma*. In the tympanic roof, the cartilage of the auditory tube forms from a single chondrification center in *Megaderma*, but from two centers in *Myotis*. In the arterial system, the ramus inferior runs ventral to the tympanic roof and the ramus infraorbitalis enters the cranial cavity via the alisphenoid canal in *Megaderma*; in contrast, the ramus inferior is dorsal to the roof and the ramus infraorbitalis occupies the foramen ovale in *Myotis*. In light of the paucity of detailed comparative data, it is unclear whether these differences relate somehow to the sorts of differences that Pedersen (1993) reported between the two types of echolocators, *Myotis* being an oral-emitter and *Megaderma* a nasal-emitter.

Our study has also highlighted the importance of ontogenetic studies for understanding the details of adult basicranial anatomy, in particular, for identifying the components of the tympanic floor and roof. Without access to sectioned specimens, our understanding of both the tympanic floor and roof in adult *Megaderma lyra* would have been severely compromised; the incidence of both rostral and caudal entotympanics and of the anteroventral process of the tegmen tympani most likely would have been disregarded.

Finally, we believe that our study has also emphasized the potential that the basicranium contains for future phylogenetic studies within bats. Our observations of this anatomical domain have uncovered numerous features of the ear region and its vasculature that vary at different taxonomic levels. We hope to address this variability in the future through phylogenetic analysis.

ACKNOWLEDGMENTS

We are grateful to the following individuals and institutions for access to the specimens described above: Dr K. P. Bhatnagar, Department of Anatomical Sciences and Neurobiology, University of Louisville; Dr K. B. Karim, Department of Zoology, Institute of Science, Nagpur; and Dr N. B. Simmons, Department of Mammalogy, American Museum of Natural History. We thank Nikki Dennard and Lucinda Schultz for technical assistance. Finally, the editors must be acknowledged for their invitation to join in this volume and their patience.

REFERENCES

Bhatnagar, K. B. & Kallen, F. C. (1974) Morphology of the nasal cavities and associated structures in *Artibeus lituratus* and *Myotis lucifugus*. *American Journal of Anatomy*, **139**, 167–90.

Bondy, G. (1907) Beiträge zur vergleichenden Anatomie des Gehörorgans der Säuger. (Tympanicum, Membrana Shrapnelli und Chordaverlauf). *Anatomische Hefte*, **35**, 293–408.

Buchanan, G. D. & Arata, A. A. (1969) Cranial vasculature of a neotropical fruit-eating bat, *Artibeus lituratus*. *Anatomischer Anzeiger*, **124**, 314–25.

Bugge, J. (1974) The cephalic arterial system in insectivores, primates, rodents and lagomorphs, with special reference to the systematic classification. *Acta Anatomica*, **87**, suppl. 62, 1–160.

Davis, D. L. (1998) The ontogeny of the basicranium of the Indian false vampire bat, *Megaderma lyra* (Megadermatidae, Chiroptera). MSc. thesis, University of Louisville.

Fawcett, E. (1919) The primordial cranium of *Miniopterus schreibersi* at the 17 millimetre total length stage. *Journal of Anatomy*, **53**, 315–50.

Fehse, O. (1990) Zur Craniogenese bei *Rousettus aegyptiacus* E. Geoffroy, 1810 (Megachiroptera, Mammalia). Diplomarbeit Fachbereich Biologie, Universität Erlangen-Nürnberg.

Flower, W. H. (1869) On the value of the characters of the base of the cranium in the classification of the Order Carnivora, and on the systematic position of

Bassaris and other disputed forms. *Proceedings, Zoological Society of London*, 1869, 4–37.

Frick, H. (1954) *Die Entwicklung und Morphologie des Chondrokraniums von* Myotis *Kaup*. Stuttgart: Georg Thieme Verlag.

Gaunt, W. A. & Gaunt, P. N. (1978) *Three Dimensional Reconstruction in Biology*. Baltimore: University Park Press.

Gaupp, E. (1908) Zur Entwicklungsgeschichte und vergleichenden Morphologie des Schädels von *Echidna aculeata* var. *typica*. *Semons zoologische Forschungsreisen in Australien*, **6**, 539–788.

Gomori, G. (1950) A rapid one-step trichrome stain. *American Journal of Clinical Pathology*, **20**, 661–4.

Gregory, W. K. (1920) On the structure and relations of *Notharctus*, an American Eocene primate. *Memoirs, American Museum of Natural History*, **5**, 51–243.

Grosser, O. (1901) Zur Anatomie und Entwicklungsgeschichte des Gefässsystemes der Chiropteran. *Anatomische Hefte*, **17**, 203–424.

Habersetzer, J. & Storch, G. (1992) Cochlea size in extant Chiroptera and middle Eocene microchiropterans from Messel. *Naturwissenschaften*, **79**, 462–6.

Hartmann, L. (1990) Das Primordialcranium von *Artibeus jamaicensis* Leach 1821 (21 mm SSL) (Phyllostomidae; Microchiroptera). Diplomarbeit Fachbereich Biologie, Universität Erlangen-Nürnberg.

Henson, O. W., Jr. (1961) Some morphological and functional aspects of certain structures of the middle ear in bats and insectivores. *University of Kansas Science Bulletin*, **42**, 151–255.

Herbert, H. (1985) Echoortungsverhalten des Flughundes *Rousettus aegyptiacus* (Megachiroptera). *Zeitschrift für Säugetierkunde*, **50**, 141–52.

Hunt, R. M., Jr. (1974) The auditory bulla in Carnivora: an anatomical basis for reappraisal of carnivore evolution. *Journal of Morphology*, **143**, 21–76.

Jurgens, J. D. (1963) Contributions to the descriptive and comparative anatomy of the cranium of the Cape fruit-eating bat. *Annals of the University of Stellenbosch*, **38**, 3–37.

Kallen, F. C. (1977) The cardiovascular system of bats: structure and function. In *Biology of Bats*, vol. 3, ed. W. A. Wimsatt, pp. 289–483. New York: Academic Press.

Kampen, P. N. van (1905) Die tympanalgegend des Säugetierschädels. *Gegenbaurs morphologisches Jahrbuch*, **34**, 321–722.

Kampen, P. N. van (1915) De Phylogenie van het Entotympanicum. *Tijdschrift nederlandsche dierkundige Vereenigung*, **14**, xxiv.

King, A. J. (1991) Re-examination of the basicranial anatomy of the Megachiroptera. *Acta Anatomica*, **140**, 313–18.

Klaauw, C. J. van der (1922) Über die Entwicklung des Entotympanicums. *Tijdschrift nederlandsche dierkundige Vereenigung*, **18**, 135–74.

Klaauw, C. J. van der (1923) Die Skelettstückchen in der Sehne des Musculus stapedius und nahe dem Ursprung der Chorda tympani. *Zeitschrift für Anatomie und Entwicklungsgeschichte*, **69**, 32–83.

Klaauw, C. J. van der (1930) On mammalian auditory bulla showing an indistinctly complex structure in the adult. *Journal of Mammalogy*, **11**, 55–60.

Klaauw, C. J. van der (1931) On the auditory bulla in some fossil mammals. *Bulletin, American Museum Natural History*, **62**, 1–352.

Koch, L. (1950) Zur Morphologie des Craniums von *Scotophilus temmincki*. Inaugural Dissertation Medizin, Universität Frankfurt am Main.

Koopman, K. F. (1993) Order Chiroptera. In *Mammal Species of the World*, ed. D. E. Wilson & D. M. Reeder, pp. 137–241. Washington: Smithsonian Institution Press.

Mabee, P. M. (1993) Phylogenetic interpretation of ontogenetic change: sorting out the actual and artefactual in an empirical case study of centrarchid fishes. *Zoological Journal of the Linnean Society*, **107**, 175–291.

MacPhee, R. D. E. (1977) Ontogeny of the ectotympanic–petrosal plate relationship in strepsirhine prosimians. *Folia Primatologica*, **27**, 245–83.

MacPhee, R. D. E. (1979) Entotympanics, ontogeny and primates. *Folia Primatologica*, **31**, 23–47.

MacPhee, R. D. E. (1981) Auditory regions of primates and eutherian insectivores: morphology, ontogeny and character analysis. *Contributions to Primatology*, **18**, 1–282.

MacPhee, R. D. E. & Cartmill, M. (1986) Basicranial structures and primate systematics. In *Comparative Primate Biology*, vol. 1, *Systematics, Evolution, and Anatomy*, ed. D. R. Swindler & J. Erwin, pp. 219–75. New York: Alan R. Liss.

Martin, J. R. (1991) The development of the auditory region of the rousette fruit bat *Rousettus leschenaulti* (Mammalia, Megachiroptera). MSc. thesis, University of Louisville.

Medway, Lord (1978) *The Wild Mammals of Malaya (Peninsular Malaysia) and Singapore*. Kuala Lumpur: Oxford University Press.

Nelson, G. J. (1978) Ontogeny, phylogeny, paleontology, and the biogenetic law. *Systematic Zoology*, **27**, 324–45.

Novacek, M. J. (1977) Aspects of the problem of variation, origin and evolution of the eutherian auditory bulla. *Mammal Review*, **7**, 131–49.

Novacek, M. J. (1980) Phylogenetic analysis of the chiropteran auditory region. In *Proceedings of the Fifth International Bat Research Conference*, ed. D. E. Wilson & A. L. Gardner, pp. 347–65. Lubbock: Texas Tech Press.

Novacek, M. J. (1985) Evidence for echolocation in the oldest known bats. *Nature*, **315**, 140–1.

Novacek, M. J. (1987) Auditory features and affinities of the Eocene bats *Icaronycteris* and *Paleochiropteryx* (Microchiroptera, incertae sedis). *American Museum Novitates*, **2877**, 1–18.

Novacek, M. J. (1991) Aspects of the morphology of the cochlea in microchiropteran bats: an investigation of character transformation. *Bulletin, American Museum Natural History*, **206**, 84–100.

Novacek, M. J. (1993) Patterns of skull diversity in the mammalian skull. In *The Skull*, vol. 2, *Patterns of Structural and Systematic Diversity*, ed. J. Hanken & B. K. Hall, pp. 438–545. Chicago: University of Chicago Press.

Patterson, C. (1982) Morphological characters and homology. In *Problems of Phylogenetic Reconstruction*, ed. K. A. Joysey & A. E. Friday, pp. 21–74. New York: Academic Press.

Pedersen, S. C. (1993) Cephalometric correlates of echolocation in the Chiroptera. *Journal of Morphology*, **218**, 85–98.

Pettigrew, J. D., Jamieson, B. G. M., Robson, S. K., Hall, L. S., McAnally, K. I. & Cooper, H. M. (1989) Phylogenetic relations between microbats, megabats, and primates (Mammalia: Chiroptera and Primates). *Philosophical Transactions, Royal Society of London*, B**325**, 489–559.

Presley, R. (1979) The primitive course of the internal carotid artery in mammals. *Acta Anatomica*, **103**, 238–44.

Samano Bishop, A. (1965) Persistencia de la arteria estapedica en Quiropteros (murcielagos). *Anales del Instituto de Biología Mexico*, **36**, 303–17.

Schneider, R. (1955) Zur Entwicklung des Chondrocraniums der Gattung *Bradypus*. *Gegenbaurs morphologisches Jahrbuch*, **95**, 209–301.

Simmons, N. B. (1993) The importance of methods: archontan phylogeny and cladistic analysis of morphological data. In *Primates and their Relatives in Phylogenetic Perspective*, ed. R. D. E. MacPhee, pp. 1–61. New York: Plenum Press.

Simmons, N. B. (1998) A reappraisal of interfamilial relationships of bats. In *Bat Biology and Conservation*, ed. T. H. Kunz & P. A. Racey, pp. 3–26. Washington: Smithsonian Institution Press.

Simmons, N. B. & Geisler, J. H. (1998) Phylogenetic relationships of *Icaronycteris*, *Archaeonycteris*, *Hassianycteris*, and *Paleochiropteryx* to extant bat lineages, with comments on the evolution of echolocation and foraging strategies in Microchiroptera. *Bulletin, American Museum of Natural History*, **235**, 1–182.

Sitt, W. (1943) Zur Morphologie des Primordialcraniums und des Osteocraniums eines Embryos von *Rhinolophus rouxii* von 15 mm Scheitel-Steiß-Länge. *Gegenbaurs morphologisches Jahrbuch*, **88**, 268–342.

Starck, D. (1943) Beitrag zur Kenntnis der Morphologie und Entwicklungsgeschichte des Chiropterancraniums. Das Chondrocranium von *Pteropus seminudus*. *Zeitschrift für Anatomie und Entwicklungsgeschichte*, **112**, 588–633.

Tandler, J. (1899) Zur vergleichenden Anatomie der Kopfarterien bei den Mammalia. *Denkschriften der kaiserlichen Akademie der Wissenschaften Wien mathematisch-naturwissenschaften Klasse*, **67**, 677–784.

Tandler, J. (1902) Zur Entwicklungsgeschichte der Kopfarterien bei den Mammalia. *Gegenbaurs morphologisches Jahrbuch*, **30**, 275–373.

Velhagen, W. A., Jr. (1997) Analyzing developmental sequences using sequence units. *Systematic Biology*, **46**, 204–10.

Wassif, K. (1950) The tensor tympani muscle of bats. *Annals and Magazine of Natural History*, **3**, 811–12.

Wible, J. R. (1984) The ontogeny and phylogeny of the mammalian cranial arterial pattern. PhD dissertation, Duke University.

Wible, J. R. (1986) Transformations in the extracranial course of the internal carotid artery in mammalian phylogeny. *Journal of Vertebrate Paleontology*, **6**, 313–25.

Wible, J. R. (1987) The eutherian stapedial artery: character analysis and implications for superordinal relationships. *Zoological Journal of the Linnaean Society*, **91**, 107–35.

Wible, J. R. (1992) Further examination of the basicranial anatomy of the Megachiroptera: a reply to A. J. King. *Acta Anatomica*, **143**, 309–16.

Wible, J. R. & Martin, J. R. (1993) Ontogeny of the tympanic floor and roof in archontans. In *Primates and their Relatives in Phylogenetic Perspective*, ed. R. D. E. MacPhee, pp. 111–48. New York: Plenum Press.

Wible, J. R. & Novacek, M. J. (1988) Cranial evidence for the monophyletic origin of bats. *American Museum Novitates*, **2911**, 1–19.

8

A theoretical consideration of dental morphology, ontogeny, and evolution in bats

INTRODUCTION

Bat dentitions have been examined intensely by many scientists, beginning with the comparative anatomists and natural historians of the 19th century. Aside from understandable curiosity about the teeth of these seemingly unusual animals, much of the original substantive work can best be explained in terms of classical taxonomy. As in most other mammals, the formulae and morphology of teeth in bats provide many characters useful in classification. Interspecies diversity in dental morphology and numbers of teeth were recognized early and bat dentition thus was examined with great care. Qualitative (and even some quantitative) data were carefully recorded and often emphasized in the early papers. The early classic monographs by Dobson (1878), Andersen (1912), and Miller (1907) offered excellent illustrations of dental morphology in a great variety of species. The quality of these pen-and-ink or wash drawings is such that time will not diminish their utility for the student of chiropteran anatomy, evolution, and systematics. Likewise, some of the conclusions and taxonomic hypotheses drawn by these early workers have stood the test of time.

Dentition and dental characters are still important to modern bat systematists. Dental character states (presence or absence of individual teeth or morphological features of crowns) figure prominently in some of the most recent attempts to posit genealogical hypotheses (e.g., Kirsch et al. 1995; Springer et al. 1995; Simmons & Geisler 1998). Sometimes, taxonomic or systematic data sets have been used to make inferences about evolutionary trends in diet. For example, Kirsch et al. (1995) mapped a combination of megachiropteran dental morphology and field data on diet onto phylogenetic trees and used the outcome to argue whether frugivory or nectarivory evolved first or whether one or the other evolved independently.

A book about bat natural history by Hill & Smith (1986) provides an example of how simple dietary categories (i.e., 'insectivory,' 'frugivory') can be associated with generalized descriptions of dental morphology. Superficially, at least, it almost appears that bat dental morphology is indicative of diet. At the same time, scientific studies of the relationships among dental morphology and dietary categories in bats have produced somewhat mixed results. For instance, when Dumont (1995) tested the hypothesis that enamel thickness correlated positively with hardness of diet, she found that such was not the case in either bats or primates (although she did find a correlation between thickness and acidity of the diet). Analyzing species of molossid bats, Freeman (1979) correlated shearing cusps with feeding on hard (chitinous) beetles, as opposed to softer-bodied prey. Looking at a different set of insectivorous species, Strait (1993) correlated shearing cusps with soft-bodied prey. Simmons & Geisler (1998: 52) recently concluded that it 'seems premature to draw any functional conclusions from transformations in talonid structure in bats.'

It would be convenient if single organ systems evolved in independence from others. It also would be convenient if bats were truly obligate feeders or if all insects posed the same challenges (rather than having different toxins or different amounts of chitin in their exoskeletons). It would be easier if all tropical fruits and flowers were identical, or if particular species of bats ate only one type of fruit or one type of nectar. In fact, however, dietary categories such as 'insectivorous' or 'frugivorous' are so broad that it is difficult to relate them to dental morphology beyond a superficial level. Thus, by far the most promising morphological analyses are those that take a functional approach and, especially, those that treat diets in terms of their physical, structural, character rather than in terms of their biological character (e.g., Freeman 1988, 1992, 1998). Interestingly, similar comments can be made with regard to the morphology of mammalian salivary glands in relationship to diet. As with dental morphology one might logically hypothesize a correlation between salivary gland morphology and dietary categories, but once again the relationship is superficial at best (Tandler & Phillips 1998).

The problem is further complicated by the fact that diets of bats (with a few notable exceptions) are imperfectly known *except* in terms of broad categories (e.g., Gardner 1977). Moreover, there is ample evidence that some species feed on a variety of nutritionally different items or shift diets seasonally or at various times of day, or seek supplemental nutrient resources (e.g., Fleming 1988; Handley *et al.* 1991*b*; Kirsch *et al.* 1995; Whitaker *et al.* 1996; Tan *et al.* 1998). At first glance, such evidence runs counter to ideas about ecological niche, interspecific competition, and resource partitioning. At the same time, there is considerable detailed evidence that chiropteran salivary glands and digestive tracts are the evolutionary playing fields that reflect competition, or competitive edge, for access to, and use of, specific nutrient resources (Mennone *et al.* 1986; Nagato *et al.* 1998; Phillips *et al.* 1998). The physiological complexities of the interspecies war over nutrient resources is well illustrated by a comprehensive study of the Neotropical fruit bat, *Artibeus jamaicensis* (Handley *et al.* 1991a; Studier & Wilson 1991). Among bats, the physiological and metabolic challenges of some diets is further illustrated by evolution of novel types of cells (Tandler *et al.* 1997a, b; Nagato *et al.* 1998) and cellular specializations associated with ionic balance (Tandler *et al.* 1990; Phillips *et al.* 1993).

One might ask: if many bat species are somewhat 'generalist' in their dietary habits, or change diets seasonally, or if salivary glands and digestive tracts are the organ systems that best respond to dietary subtleties, then what should we make of their dental morphologies? Surely teeth are not merely passive passengers in chiropteran adaptive radiation. Or are they?

The main objective of the present essay is to call attention to alternative ways to think about bat dentition (or dentition of any species of mammal). Rather than think of teeth as independent entities, or as morphological features fine-tuned to diet, I will offer an alternative, integrated, way of understanding dentition. This approach is hardly novel. Textbooks of oral biology (even editions that are 20 years old, e.g., Shaw *et al.* 1978) illustrate that dentition in human beings is best understood in broad biological context. In order to understand the significance of dental morphology in bats and, especially, in order to develop an evolutionary model with testable predictions, it is necessary to think of teeth in terms of the entire organism. This means that one must at least consider dental ontogeny, the oral environment, digestive tract, biochemical qualities of diet, metabolic rates, and life history. In short, dentition makes the most sense when approached in broad biological context and the least sense when treated independently (Phillips *et al.* 1977).

To achieve the objective of the essay, the remainder is organized into several sections. The first reviews the concept of placing dentition into a broader biological context. The second section explores dental ontogeny in bats, with an emphasis on genetic control and testable predictions about the role of development and control in the evolution of teeth. The final section applies aspects of dental interaction and ontogeny to a discussion of dental evolution in bats.

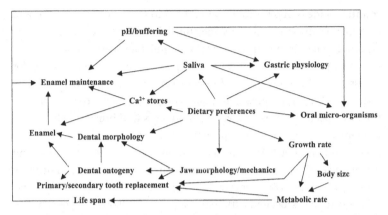

Figure 8.1. This network diagram places dentition in the context of oral biology. 'Teeth' thus are seen as just one component of an entire system and clearly are not independent entities in an evolutionary sense.

DENTITION IN A BIOLOGICAL CONTEXT

Interactions that must be considered in order to discuss bat dentition in a biological context beyond diet *per se* can be seen in Fig. 8.1. Dentition is subdivided into dental morphology, ontogeny, tooth replacement processes, enamel, and enamel maintenance. It would be possible to make further subdivisions, or to organize components in terms of ontogenetic components – e.g., ectodermal versus mesenchymal tissues or levels of genetic control insofar as they are currently understood.

In its simplest form, one might say that bat dentition is affected by at least seven basic (but interrelated) issues: 1) physical structure of items manipulated by the teeth; 2) abrasiveness of diet; 3) reproductive life span of the individual (time-in-service for the teeth); 4) availability of dietary Ca (and, to a lesser extent, K); 5) animal size (space available in oral cavity) and jaw mechanics; 6) growth rate and maxillofacial developmental dynamics; and 7) chemical and microbial characteristics of the oral environment. The first consideration, physical structure of the items in a diet, clearly is important to tooth morphology and jaw mechanics and, in turn, the anatomy of the skull and bony palate (Freeman 1998). Many dental features make functional sense when food items are considered in terms of hardness, softness, brittleness, or combinations of internal and external physical features (e.g., Freeman 1988). A logical extension of this approach is to replace oversimplified descriptors such as 'insectivorous' with functionally meaningful descriptors such as durophagous (eating hard brittle food) or jusophagous (juicy foods) or elasticophagous (chewy foods) (Freeman 1988). This approach has considerable merit, especially in view of the fact that it directs us to think of dental evolution in a way that it can be integrated to other physical and mechanical aspects of head anatomy. As an aside, it also should be noted that 'digestibility' of diet has been postulated as a factor affecting dilambdodonty in bat (and primate) teeth (Kay & Sheine 1979; Freeman 1998). Ordinarily one might think of digestibility in terms of having the appropriate enzymes to break down ingested macromolecules, but in the present case it refers to the idea that crushed, broken, or finely divided dietary items are *easier* to digest than are whole items. Taken in this way, digestibility places physical rather than biological demands on dental morphology.

As a related example, we can consider the combination of abrasive diet and life span. In an animal with a long life span (> 10 years), abrasive diet, and non-ever-growing teeth (the case with all bats), the dentition must be substantial (in relative size), wear resistant, and repairable (through post-eruption surface mineralization). Moreover, one would expect the oral environment to be conducive to maintenance of enamel surfaces (e.g., buffered fluids, neutral pH, limited opportunities for colonization by acidic bacteria, limited dental plaque formation). To support such teeth over time, the maxillary and dentary bones and, especially, the alveolar bone and associated periodontal ligaments would have to accommodate to physical stresses of dental function. The dynamic nature of these processes is readily seen when one compares cementum and dentin in bats of various ages (Phillips *et al.* 1982). Support of skeletal elements and continuous mineralization of the enamel surfaces of the teeth both require the availability of dietary calcium or Ca^{2+} ions.

A potential evolutionary interaction among salivary gland gene regulation of proline-rich proteins, Ca^{2+} ions, diet, life span, and dentition was discussed recently (Phillips 1996). It appears that the presence of acidic proline-rich proteins (APRPs) in saliva is one evolutionary mechanism for maintaining enamel mineral content, and hardness (Phillips & Tandler 1996). APRPs inhibit calcium phosphate precipitation by maintaining saliva in a supersaturated state relative to the enamel hydroxyapatite (Oppenheim *et al.* 1987) and hypothetically these salivary proteins limit gradual demineralization over the life-time of the individual (Bennick 1987). Given all of this, one might expect that all mammals should produce saliva rich in APRPs. In fact, however, APRPs are found in saliva of some mammal species, but not others. Several reasons have been postulated (Phillips 1996): 1) although APRPs maintain enamel, they also form an enamel pellicle that could harbor cariogenic bacteria if the animal had a high sucrose diet; 2) production of APRPs probably precludes salivary production of basic proline-rich proteins (BPRPs), which have little affinity for Ca^{2+}, but strong affinity for plant phenolic compounds (Mehansho et al. 1985; Kauffman et al. 1991). In the end, we know that production of proline-rich proteins varies widely among mammal lineages, but we also know that patterns appear to coincide with diet and types of dentition (Phillips & Tandler 1996). Thus, the proline-rich proteins are a good example of evolutionary trade-offs and interactions among dentition and other factors (Fig. 8.1).

The interactions shown in Fig. 8.1, and the previous brief summary, beg the question: what is the potential impact of a change in diet or life history strategy? Obviously, a change in diet that involves a significant decrease in dietary calcium, or demands a shift in digestive processes or metabolic rate, or alters the oral environment (as virtually any major trophic shift would do), could affect dentition in multiple ways. A shift to nectarivory is perhaps the easiest example to imagine. Under such circumstances the oral cavity and teeth would be bathed in a solution that promotes decalcification through bacterial colonization, dental erosion, and plaque formation. Teeth with extensive crushing surfaces would not be needed for lapping nectar from tropical flowers, but it is equally true that robust teeth could not be maintained for very long anyway in such an environment. An interesting sidelight of this evolutionary trend was demonstrated in one of the nectarivorous phyllostomid bats, *Leptonycteris nivalis*. In this species the intercalated ducts of the parotid salivary gland are elongate and secrete lysozyme or a lysozyme-like enzyme such as a chitinase. One explanation for this phenomenon is that salivary enzymes such as these would counteract fungal or bacterial colonization of oral epithelium and/or tooth surfaces (Phillips et al. 1998). Even so, enamel dissolution is common in these bats (Phillips 1971).

DENTAL ONTOGENY AND ERUPTION

The development (ontogeny) and eruption of teeth have been investigated in great detail in laboratory rodents and there is a large body of relevant literature. From a biological perspective, dental ontogeny is a complex subject because it involves germ layer interactions, production of three different mineralized tissues, and distinctive morphological patterns that probably reflect hierarchical genetic controls. Historically, it was not widely appreciated that morphogenesis underlies opportunities for adaptation (Thompson 1988; Gottlieb 1992). The recent discovery of conserved development genes in animals (e.g., homeobox genes) and the technical capability for studying regulation of gene expression in morphogenesis have ushered in a new era of appreciating development in an evolutionary context (Janies & DeSalle 1999; Schwartz 1999).

An understanding of dental ontogeny and tooth eruption has great potential for our attempts at understanding evolutionary mechanisms and pathways (Marshall & Butler 1966; Phillips 1971; Phillips *et al.* 1977). An illustration of this can be seen in comparative histological studies of ever-growing and non-ever-growing molars in rodents (Phillips & Oxberry 1972; Oxberry 1975; Phillips 1985). Basic information about dental ontogeny was used in conjunction with histological analyses and autoradiographic experiments with a labeled amino acid to investigate the continuous development and eruption of molars and establish a theory of their evolution. This approach enabled us to identify specific ontogenetic steps that possibly were modified to convert rodent molars into ever-growing, buttressed hypsodont teeth. When this research was conducted, little was known about gene regulation (and expression) of ontogeny. It now is practical to think in terms of genetics and there are two ways to go about this. First, because the theory is based on developmental pathways, the developmental pathway differences between ever-growing and non-ever-growing molars in rodents provide a strategic framework for identifying and isolating genes responsible for certain aspects of tooth development. Second, from an evolutionary perspective, the theory identifies steps in the developmental pathway where changes in gene regulation or expression might be investigated in order to determine the underlying basis of a major evolutionary event in rodents. Either way, the theory offers opportunities for investigating the genetics of development and elucidating mechanisms of dental evolution.

Insofar as bats are concerned, comparative ontogenetic data from dentition are few and mostly consist of incidental information. The primary ('milk') dentition of bats has attracted attention in part because of the unusual morphology of many of these teeth. Indeed, it is generally assumed that the primary dentition functions as a mechanism enabling young bats to cling to their mothers. In fact, however, the morphology of primary teeth 1) sometimes is hook-like and seemingly designed to 'hook' offspring to mother, 2) sometimes is similar to the secondary tooth that will replace it (albeit smaller in overall size), and 3) sometimes is neither hook-like nor similar to the secondary replacement, instead being similar to a presumed 'primitive' morphology (see Phillips 1971 for a review). Typically, bats have 20–22 primary teeth (Vaughan 1970), but there seems to be no way of predicting the correspondence between numbers of primary and secondary teeth. For example, in some genera (*Macrotus*) there are more primary incisors than secondary replacements (Nelson

1966; Phillips 1971), whereas in others (such as *Rhinolophus*), some primary premolars are resorbed before birth (Grassé 1955). In the overview, tooth replacement patterns and/or morphology of primary dentition has been described in a variety of microchiropteran species (e.g., Miller 1907; Reeder 1953; Stegeman 1956; Dorst 1957*a, b*; Peterson 1968; Birney & Timm 1975). Earlier comparative work by the author (Phillips 1971; Phillips *et al.* 1977) on species of glossophagine bats and by Marshall & Butler (1966) on the horseshoe bat, *Hipposideros beatus*, covered such topics as basic ontogeny, eruption and tooth replacement, developmental abnormalities, and (indirectly) genetic controls on coronal morphologies.

From an evolutionary perspective, information about the primary dentition and tooth replacement patterns in bats has considerable potential. For example, as mentioned above, the primary dentition in bats can differ morphologically and numerically from the secondary dentition. This is in contrast to many other kinds of mammals in which the primary dentition always is a small-sized version of the secondary dentition (human beings, for instance). The occurrence of two morphologies and different numbers of teeth in bats documents independence between the two sets of teeth and provides indirect evidence that although genetic control of initiation might be shared, the two sets are under separate genetic control insofar as morphology is concerned.

In bats (as in other mammals), the primary teeth usually are shed as the secondary (permanent) teeth erupt and move into the functional eruptive plane (Phillips 1971). A number of factors are involved in shedding, but basically the roots and alveolar bone sockets of primary teeth are gradually resorbed and the crown thus loses its attachment. Under some circumstances primary teeth are totally resorbed before birth (Grassé 1955) and, under others, some of the primary teeth apparently are not always shed (lower incisors in the genus *Choeronycteris,* Phillips 1971).

A comparison of shedding in three species of phyllostomid bats (a vampire bat, *Diphylla ecaudata,* a long-tongued bat, *Leptonycteris nivalis,* and the Mexican long- nosed bat, *Choeronycteris mexicana*) is shown in Fig. 8.2. Overall, the general pattern is quite similar, even though the dentitions and jaws (and function of secondary dentition) differ dramatically in these bats. For instance, in all three the upper primary (d) canine and dI2 (second incisor) are shed late, whereas lower premolars are among the earliest to be shed (Fig. 8.2). The early shedding of lower primary premolars is noteworthy, because in some species anterior lower primary premolars are resorbed prior to birth (Grassé 1955). The general similarity among these three species suggests that shedding pattern is conserved

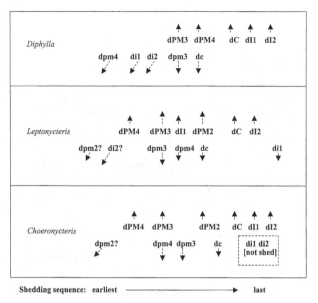

Figure 8.2. Comparisons of primary ('milk') tooth shedding in three species of phyllostomid bats (the hairy-legged vampire bat, *Diphylla ecaudata*, the long-tongued bat, *Leptonycteris nivalis*, and the Mexican long-nosed bat, *Choeronycteris mexicana*). The lower case 'd' means deciduous and the other upper and lower case letters are used to denote upper and lower teeth. The data for *Diphylla* were taken from Birney & Timm (1975) and the other two were modified from Fig. 10 in Phillips (1971).

(and thus controlled) at least in some phyllostomids. This is in contrast to examination of museum specimens of neonatal bats, which can give the impression that shedding is haphazard. Two unusual features occur in the three species. First, in *Leptonycteris nivalis* the upper, inner, primary incisor (dI1) is shed relatively early in comparison to the other two species (Fig. 8.2). Shedding of this tooth correlates with eruption of the secondary inner incisors (Fig. 8.3). Shedding of the outer, upper, primary incisors (I2) follows sometime after eruption of its secondary replacement. Second, as mentioned previously, the lower primary incisors in *C. mexicana* sometimes are not shed initially. The secondary lower incisors are initiated and more-or-less develop, but do not erupt in this bat (Phillips 1971). Retaining the primary lower incisors is a trivial matter because these teeth are mere nubs and typically are lost in adult life anyway (Phillips 1971). Indeed, the loss of lower incisors apparently is an important aspect of providing space for the tongue in this (and in some other species of) nectarivorous bats (Phillips 1971; Freeman 1995).

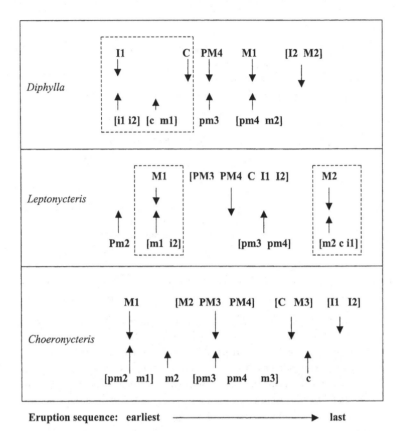

Eruption sequence: earliest ————————————➤ last

Figure 8.3. Eruption sequence of secondary ('permanent') teeth in three species of phyllostomid bats (the hairy-legged vampire bat, *Diphylla ecaudata*, the long-tongued bat, *Leptonycteris nivalis*, and the Mexican long-nosed bat, *Choeronycteris mexicana*). The data for *Diphylla* were taken from Birney & Timm (1975) and the other two were modified from Fig. 10 in Phillips (1971).

Species differences in the ontogeny of secondary teeth can be inferred from eruption patterns (Fig. 8.3). Unlike the shedding patterns, which are somewhat conserved, the eruption patterns of the secondary teeth are species-specific (and notably different) in these same bats. In the hairy-legged vampire bat, *Diphylla ecaudata*, the upper inner incisors and both lower incisors (inner and outer) erupt together followed by the lower canine and m1 and then the upper canine. Early development and eruption of this set of teeth in *Diphylla* contrasts to the two nectarivorous species. In both glossophagine bats, M1, the first lower premolar (pm2), and molar develop and erupt relatively early (Fig. 8.3). Whereas most of the anterior teeth erupt early in *Diphylla*, the homologous teeth in the two nectarivorous species erupt relatively late. The long-tongued bat, *Leptonycteris nivalis*, is interesting because development and eruption is coordinated among the upper and lower molars. The first upper and first lower molars develop and erupt together and after these molars reach the functional eruptive plane, the M2 and m2 (there are only two upper and lower molars in this species) follow them. The development and eruption of i2 and the lower canine and i1 are associated with each set of molars, respectively (Fig. 8.3). The molars in the third species, *Choeronycteris mexicana*, show the least coordination of eruption. With exception of M1/m1, the upper and lower molar teeth typically do not erupt together (Fig. 8.3).

What, if anything, is the significance of development and eruption patterns? One general statement is that relatively large teeth seem to erupt early. This is perhaps most obvious in the hairy-legged vampire bat: the upper inner incisors are large, procumbent teeth; the lower incisors are large multilobed teeth; and the upper and lower canines also are relatively large. The only exception is the modest m1, which erupts with the lower canine. Early eruption of the m1 is also typical among the three species shown in Fig. 8.3 and in a fourth species as well (*Glossophaga soricina*, Phillips 1971).

The anterior dentition is emphasized in all of the extant (and the known extinct) species of vampire bat (Ray *et al.* 1988). The enlarged razor-sharp upper incisors and canines in vampire bats are used to slice the skin of prey species (Greenhall 1988). Extremely sharp edges on these teeth are maintained through thegosis (tooth-to-tooth sharpening) and scanning electron microscopic analysis has shown that the relatively thin layer of enamel is quickly worn away (Phillips & Steinberg 1976; Vierhaus 1983; Freeman 1998). The significance of the anterior dentition in vampire bats is obvious, so in addition to correlating early development and eruption with size, one might also argue that these teeth are the direct products of natural selection. That is, the anterior dentition of vampire bats is very different from the anterior dentition of other phyllostomid species and because this difference correlates with unique feeding habits, natural selection must have focused on these teeth. If so, timing of development and eruption as well as size of the teeth reflect evolutionary adaptation to a specialized diet. Perhaps the most functionally important teeth receive a developmental priority. This theoretical conclusion leads to the testable hypothesis that timing of development and tooth eruption in other species will correlate with *functional* role. In the case of anterior teeth in the vampire bat, one can say that their function involves the way that their morphology, position, and occlusion interact to create a system of sharp teeth that can slice through the skin of mammals or birds.

Evolution of nectarivory in bats (both microchiropterans and meg-achiropterans) sometimes involves: 1) elongation of the snout; 2) reduc-tion in number of teeth (usually incisors); 3) reduction in size of cheek teeth; 4) increased spacing between teeth; 5) reduction in occlusion; 6) loss of primitive coronal morphology (W-shaped ectoloph); and 7) modifi-cation of the tongue (Phillips 1971; Greenbaum & Phillips 1974; Freeman 1995, 1998). Basically, it is easy to gain the impression that the cheek den-tition is relatively unimportant in nectarivores, although the existence of wear facets (Freeman 1998) clearly document that occlusion occurs (see illustrations in Phillips 1971). Freeman (1995) has attributed the small, narrow dentition to the need for space for the muscular, elongate tongue that these bats use to collect nectar from narrow, deep corollas.

When we examine eruption in *Leptonycteris* and *Choeronycteris* (Fig. 8.3), it is apparent that tooth size is not an obvious criterion for eruption sequence. On the other hand, the occurrence of a pattern in *Leptonycteris* begs the question of function. In this bat, I previously showed that there are significant positive correlations among coronal dimensions of upper and lower molars (Phillips 1971). Such data can be used as a measure of integration between and among the molars. The M1 and m1 develop and erupt together; 33.3% of the quantitative comparisons exhibited signifi-cant ($p < 0.05$) positive correlations. Likewise, M2 and m2 develop and erupt together (well after M1/m1, Fig. 8.3) and 20.8% of the measurements exhibited significant positive correlations (Phillips 1971). Within-jaw integration, i.e., M1 ⇔ M2 and m1 ⇔ m2, was 20% and 31.5% and inter-jaw integration among non-paired molars, i.e., M1 ⇔ m2 and m1 ⇔ M2, was 16% and 6.3%, respectively. Thus, one might postulate that the 'function' of synchronized development and eruption of upper and lower molars in *Leptonycteris* might in some way relate to the degree of morphological integration in their crown patterns. To test this idea, we can use the other glossophagine, *Choeronycteris mexicana*. In this bat there is less occlusion (based on tooth contact when the jaws are closed rather than contact during complex jaw movements, which do create wear facets in *Choeronycteris*, cf. Freeman 1998) than in *Leptonycteris*. The crown morphologies of the three upper and lower molars in *Choeronycteris* are more variable than in *Leptonycteris* (Phillips 1971). In terms of develop-ment and eruption, there is coordination between M1 and m1, but none between M2 and m2 or M3 and m3 (Fig. 8.3). No significant correlations were found between the M1 and m1 or between M3 and m3. M2 and m2, which erupt slightly out of phase, exhibited one significant positive corre-lation (6.25% of total comparisons). Thus, in this species there is little inter-jaw integration of molar coronal morphologies. Indeed, the overall inter-jaw integration (all upper and lower molars in pair-wise comparison) for *Choeronycteris* was 2.1%, whereas it was 22% for *Leptonycteris* (Phillips 1971). These data lend support to the hypothesis that there is a relationship among developmental timing, eruption sequence and synchrony, and degree of morphological integration between jaws. Moreover, this conclusion also supports the idea that patterns in developmental timing and eruption sequence might have *functional* importance.

The foregoing theoretical outline clearly needs additional testing. Such testing should be possible with available museum specimens, but the main prediction is that patterns of development and eruption of secondary teeth will reflect function and, ultimately, natural selection. This latter idea is especially important and requires additional comment in regard to the two nectarivorous species. In these two species, the amount of inter-jaw integration appears linked to degree of occlusion. The narrow molars in *C. mexicana* are widely spaced and occlusion of upper and lower molars is minimal. In *Leptonycteris* the upper and lower molars have a more complete occlusion (i.e., more inter-tooth physical contact when the jaws are closed). The model reviewed here specifically implies that for some reason(s) selection has maintained inter-jaw occlusion in *Leptonycteris*, whereas the upper and lower molars are relatively unconstrained in *Choeronycteris*.

In summary, although relatively few ontogenetic data are available in literature, the materials that have been published clearly set the stage for understanding the relationships among development, evolution, and dietary adaptation in bats.

EVOLUTION OF CHIROPTERAN DENTITION

In an ideal world we would wish to know about the dentition of the 'original' or ancestral bats and their ancestors. If we had some hard data on the mammals that were the precursors of bats, we would be able show evolutionary trends or pathways all the way back into early eutherians. Unfortunately, the fact is that the ancestors of bats have not yet been identified. Traditionally it was thought that bats shared an ancestry with insectivores (e.g., Slaughter 1970). In fact, the molar teeth of many insectivorous bats strongly resemble the molars in insectivores. This hypothesis dominated thinking for many years, and is relevant because it is the foundation of much of what has been written previously about bat teeth. For example, in his frequently cited overview of evolutionary trends in chiropteran dentition, Slaughter (1970) wrote 'there can be little doubt that bats took origin from within the Insectivora.' If one assumes that this is true, it follows that bat species with teeth similar to those in shrews can be explained by ancestry. More importantly, perhaps, if one is positive that bats evolved from insectivores, then the biology of insectivores (diet, feeding habits, physiology) must shed some light on the biology of early bats. It would be accurate to say that for many years the bat–insectivore relationship dominated thinking, and interpretation, to the extent of being a paradigm.

More recently, systematists placed bats in a group called the Archonta, which included dermopterans, scandentians (tree shrews), rodents, and primates (Pumo et al. 1998). Under these circumstances it was convenient and logical to compare bat dentitions to dentitions of dermopterans and tree shrews (Simmons & Geisler 1998). Although comparative work based on these frameworks is interesting, and perhaps valuable, molecular and other chemically based systematic analyses have produced some surprises. Indeed, genetic data cast considerable doubt on the concept of 'Archonta' as an evolutionary unit and have also failed to support the hypothesized connection between bats and shrews. In fact, one recent analysis based upon protein-coding genes in the mitochondrial genome links bats with the ferrungulates (carnivores and ungulates) (Pumo et al. 1998). If this hypothesis is supported by additional study, the origins of the dental morphology of bats might be found in fossils at the base of this large and complex group of crown orders, rather than among early insectivores or dermopterans.

Although the absolute earliest bats (i.e., true transition species that share characters with other groups) are unknown (or unidentified among fragmentary fossils), an interesting array of fossil bat species has been uncovered from Eocene deposits on several continents. These fossil specimens include some that are remarkably complete. In fact, some are in such good condition that radiographs of 50-million-year-old fossils reveal details of the dental pulp chambers, hard tissues, periodontal spaces, and compact and trabecular alveolar bone (see Fig. 27, *Paleochiropteryx*, in Simmons & Geisler 1998). An interesting comparison can be made between these radiographs of extinct ancient bats and the composite histological view of the maxillary dentition of the extant yellow-shouldered bat (*Sturnira ludovici*) shown in Fig. 10 of Phillips et al. (1977). Fundamentally, dental structure and associated tissues have been highly conserved.

Bats comprise two very distinctive groups of species. These two groups, the suborders Megachiroptera and Microchiroptera, appear to constitute two independent lines of evolution. In recent years there was a brief but intense debate about the respective origins of the suborders.

Pettigrew and colleagues (Pettigrew *et al.* 1989) argued that bats were poly-phyletic and that the megachiropterans were related to primates. This idea took origin from a paper by Smith & Madkour (1980) that was based on 'histological' data from the penis (although no actual microscopic data were presented, cf. Phillips 1994). Although this somewhat radical hypothesis was not supported by molecular genetic or morphological data sets obtained for the purpose of testing it (e.g., Baker *et al.* 1991), it is worth mentioning because it underscores the fundamental evolutionary differences between the two suborders. The extant megachiropteran species very clearly can be described as primarily 'fruit- and nectar-feeding' bats. Many (or perhaps virtually all) of the behavioral, metabolic, and morphological characteristics (including the retina and retinal pigment epithelium of the eyes) of megachiropterans relate in some way to this dietary theme. By way of comparison, the microchiropteran bats are primarily 'insectivorous' (when categorized in terms of the most common basic diet). At the same time, a notable number of microchirop-teran species feed on things other than insects. Indeed, the list of food items used by microchiropterans is truly remarkable. It includes a wide variety of arthropods (in addition to insects), vertebrate animals, blood, pollen and nectar, and fruits and flower parts (e.g., Gardner 1977). Consequently, the relationship of chiropteran evolutionary history to dental morphology and jaw mechanics, in fact to oral biology in the broad sense, is certainly obvious. The dietary diversity of bats requires a host of adaptive characters, starting with the mouth.

Recently, Simmons & Geisler (1998) have argued that flight pre-dated echolocation in the evolution of bats. They imagine the earliest bats as mammals that used vision as their primary means of orientation in the arboreal and aerial environments. In this scenario the megachirop-terans would be a branch that retained reliance on vision (and olfaction) to locate stationary food items, whereas the microchiropterans would be the branch that gradually evolved toward reliance on echolocation and aerial hawking to locate and capture flying arthropods. Although this hypothesis of early evolutionary divergence – centered on sensory systems – does not tell us what the earliest megachiropterans ate, it does suggest an early commitment to stationary food items, whereas the microchiropteran history primarily involved pursuit of moving prey.

The matter of early chiropteran divergence is important to a discus-sion of dentition because dental morphology is fundamentally different in the two suborders of bats. In most of the megachiropterans the cheek teeth are somewhat rectangular and the topology of the molars is rela-tively feature-less in comparison to most microchiropteran species.

Freeman (1998) has speculated that this aspect of megachiropteran dental morphology might be a consequence of having fused dentaries, which provide support for the tongue. Although there might be other explanations, the nearly feature-less molar topology is one reason why it is virtually impossible to make direct morphological comparisons between the dentition of megachiropterans and that of microchiropteran bats (Koopman & MacIntyre 1980). Presumably some of the low ridges and cones in megachiropteran cheek teeth do have homology (shared origin) with coronal features in the other bats and non-chiropteran mammals. Hershkovitz (1971) postulated some such homologies, whereas Hill & Smith (1984) went so far as to say that megachiropteran teeth are 'unique in structure and bear little resemblance to the teeth in any other group of mammals.'

Perhaps the most interesting questions about megachiropteran dentition are: what does the dental morphology say about the biology of these bats, and why are the cheek teeth so nearly feature-less? Neither of these interrelated questions has been addressed previously, but can be considered here. We can begin by noting that the frugivorous megachiropterans typically try to crush fruit into pulp, separating juice and fibrous components. Thus, in mechanical terms we might expect their cheek teeth to be used as crushing devices. At first thought this conclusion is inconsistent with the dental morphology. Indeed, wouldn't one expect these bats to have flat, broad cheek teeth with tall but rounded cusps? Such a morphological pattern would seem 'designed' to crush an assortment of tropical fruits. Preconceptions of what morphology should be are risky, but one would be hard pressed to present a case that cheek tooth dental morphology in megachiropteran bats is specially adapted for diet. Indeed, it happens that the crushing and processing of fruit is largely accomplished by the tongue, which pushes fruit against the ridged, heavily keratinized palate (e.g., Greet & De Vree 1984; Freeman 1998).

Adaptation of morphology to function requires genetic variation and natural selection. More importantly, perhaps, long-term (in an evolutionary sense) maintenance of a complex morphology must require something special. For instance, once a particular morphological pattern of greater survival value is acquired, it might be conserved genetically. With this thinking in mind, we can return to the question of dental morphology in megachiropteran fruit bats.

Two possibilities come to mind with regard to megachiropteran fruit bat dentition. One is the idea that the unusual dental morphologies relate to some function(s). Some evidence for this was recently reviewed by Freeman (1998), who pointed out that some megachiropteran bats are said to have a powerful bite (able to break cocoa pods according to Hill & Smith 1984). Freeman's morphometric analyses also imply functional roles for cheek teeth in megachiropterans. Another possibility (one that is not exclusive) is that the morphology of megachiropteran dentitions reflects an absence of constraint on precise morphology. A third possibility is that the ancestral megachiropteran was nectarivorous. Such a bat would have had small, relatively feature-less cheek teeth. In this scenario (which is favored by Freeman 1998), the unusual molar cusps seen in frugivorous megachiropterans would have evolved secondarily.

If one visits a museum collection and examines skulls and lower jaws of specimens of flying foxes (*Pteropus* spp.), the teeth make a startling and memorable impression. In adult bats of most species, the cheek teeth typically are eroded, abraded, or modified into stained, lump-like pegs. In other genera (e.g., *Cynopterus*, *Megaerops*), the typical adult exhibits basin-like upper and lower cheek teeth. At first glance these basins look like the products of abrasive wear, and to some extent they are due to wear. Actually, however, with trans-illumination partial decalcification of enamel is usually apparent. Decalcification is a hallmark of dental erosion. Once the apical enamel (which apparently is thin) is worn away, the exposed secondary and primary dentin are highly vulnerable to the oral environment.

Variation in dental condition due to abrasion and erosion is extreme in the average megachiropteran bat. One thing that comes to mind is the notion that maintenance of a specific dental morphology is not very important (or very practical) in these animals. In evolutionary terms, perhaps natural selection does not closely monitor, or conserve, dental morphology in megachiropteran bats. Historically, mammalogists and evolutionary biologists in general have tended to view dental phenotypes in a deterministic manner. That is, the emphasis has been on thinking in terms of directional selection. Inherent to this approach is the idea that a special dental morphology is *required* for each type of diet. The corollary is that any particular morphology is the product of selection, even if the morphological pattern does not result in a clear functional, or biological, role.

Contrary to the foregoing, I would argue that in the case of megachiropteran bats, the dentition to a large extent has been allowed to 'drift.' That is to say, there has not been intense directional selection to *obtain* a special morphology, and there has not been intense selection to *maintain* some specific, adaptive coronal pattern. One consequence is the reality that cheek teeth in megachiropteran bats bear virtually no similarity to the teeth in microchiropteran bats or other mammals either. This is an important assertion because in order to understand selection, and conservation of adaptive morphologies, it is insightful to examine the alternative.

If one accepts the premise that dentition in megachiropterans is not a product tailored by directional selection to fulfill some specific dietary purpose, then the next problem is to consider the potential biological explanation for this unusual situation. In general terms, megachiropteran fruit bats are faced with a set of fundamental physiological challenges. Most species are large-sized animals (in terms of bats), and their main problem is to acquire adequate energy and nutrition from food material that is classically low in protein and essential ions such as Na^+, but unusually high in carbohydrates. Rapid gut passage is one potential solution because it enables a bat to consume large volumes of fruit per feeding bout, in a sense searching out as much proteinaceous and ionic content as possible. Proteinases and nucleases are important, but the potentially pivotal role of gastric acid (and an acidic environment) should be readily apparent. Thus, it might not be surprising that megachiropteran bat saliva is remarkably acidic (pH ~5.5–7.0) and has a very low buffering capacity (Dumont 1997). Presumably, the acidic saliva attacks fruit as it is being chewed and the low buffering capacity is essential to both the oral environment and the food bolus in the stomach. The mechanisms that could produce such saliva are reasonably well studied and probably involve acinar and duct cells in the parotid salivary gland (Nagato et al. 1998). Insofar as megachiropteran bats are concerned, portasome- like proton pumps (which could add H^+ ions to saliva) have been described on apical membrane of striated duct cells in the parotid gland of the tailless fruit bat, Megaerops ecaudatus (Tandler et al. 1990). By way of comparison to megachiropteran fruit bats, in typical insectivorous bats the salivary pH is > 8.0 and has a very high buffering capacity.

An acidic oral environment is hardly ideal for mineralized tissues. Indeed, it leads to decalcification. Details of this process are available from extensive dental medical studies of abnormal situations in human beings, who normally have a salivary pH of ~7.6 (Shaw et al. 1978). If one examines megachiropteran dentition with this in mind, nearly all specimens reveal extensive evidence of dental erosion (especially to cheek teeth) and many have carious lesions. Indeed, the stained, rounded crowns of cheek teeth in older individuals (especially in species of Pteropus) are the result of enamel dissolution and subsequent staining of the porous underlying dentin. Upon reflection it is obvious that acidic, weakly buffered saliva is at least potentially incompatible with a dental morphology in which maintenance of complex crown patterns and sharp cutting edges is essential to survival.

Against the backdrop of the foregoing, my earlier comment (p. 251) about the significance of adequate dietary calcium and Ca^{2+} ions, and the network shown in Fig. 8.1, it is especially interesting to note that some megachiropterans supplement their fruit diet with leaves. Thomas Kunz and his colleagues (e.g., Tan et al. 1998) believe that leaves are important in the diet of megachiropteran fruit bats. Their data (and that of others as well) show that some species consume a substantial volume of leaves from a variety of tree species. The exact reasons are unknown, but the strongest hypothesis is that leaves provide protein, serve as an important source of calcium and other minerals, and possibly influence bats physiologically through plant steroids. In particular, a demand for calcium is not surprising.

Not all megachiropteran bats are frugivorous. A variety of species (e.g., especially in genera such as Eonycteris, Macroglossus, and Melonycteris) can be described as nectar- feeders and these are distinguished by having relatively less tooth area than the frugivores (Freeman 1995). Data on salivary pH are not available for any of the megachiropteran nectarivores, but the dentition in such species typically exhibits enamel erosion. However, while this erosion might be attributed to acidic saliva it also might be attributed to a combination of the bat's sugar water (nectar) diet and acid-producing, cariogenic microorganisms that flourish on the surfaces of the oral microenvironment it produces.

In the overview, with exception of a few species (e.g., in the rare genus Pteralopex and in Harpyionycteris), it appears to me that dental morphology of megachiropteran bats can best be understood in context of the oral environment. In this instance, careful examination of dental morphology can be frustrating because tooth morphology probably is a secondary matter in the origin and evolution of this entire lineage. In one interpretation the most conservative elements of dentition – initiation, development, eruption, and even the typical morphogenetic fields – have been retained, whereas the detailed coronal morphologies that might have existed in ancestral megachiropteran bats have not been vigorously maintained or directionally modified. In the alternative interpretation (based on Freeman 1998) even the ancestral megachiropterans had relatively simple coronal morphologies.

The Microchiroptera provide an interesting set of comparisons to the megachiropteran bats. Basically, most of the microchiropteran species have dental patterns in which the upper molars are characterized by a very complex W-shaped cutting ridge called an ectoloph (Slaughter

1970; Phillips 1971; Freeman 1981). This pattern can be described as the 'classic' insectivore molar. The jaw mechanics and functional aspects of this type of dental pattern have been examined in a number of studies (e.g., Freeman 1979, 1988, 1998). Basically, the upper and lower cheek teeth typically have numerous sharp cutting edges that interact to shear prey animals into fine particles. One unusual, but interesting, example is the fishing bat, *Noctilio leporinus*. This species feeds on small fish, which it chews for a brief time (~15 seconds), pauses, and then chews some more. As it chews on one side of the mouth, the fishing bat gradually deposits partly chewed fish into the pouch-like buccal cavity on the ipsilateral side (Phillips & Steinberg, unpublished). Stenson's duct from the parotid salivary gland empties in the buccal pouch, so the fish is bathed in saliva rich in digestive enzymes before the bat extracts it and continues chewing. Some of the other obvious examples are species of bats in which the prey includes beetles or other hard-bodied arthropods. Here the teeth and chewing reduce the tough chitinous exoskeleton and expose the soft tissues to salivary and gastric enzymes. Until fairly recently the textbook idea was that mammals do not produce chitinases, so comminution of insect exoskeletons seemed especially important. Recent analyses have revealed that mammals do produce chitinases and that some bats might use salivary lysozyme as a chitinase (Phillips *et al.* 1998). Even so, the teeth and chewing clearly have important roles in insectivorous bats.

The molar teeth in the insectivorous (and carnivorous) microchiropteran species can be described as conservative. That is not to imply that the morphology of these teeth is invariant. Minor differences in size and placement of the major cones and conids (paracone, protocone, and metacone on the upper molars and paraconid, protoconid, and metaconid on the lowers), shape of the ectoloph, size and placement of lingual cingular cusps, and tooth height are readily evident if one examines specimens, or uses taxonomic keys to species. However, if one takes the broad view, the molars of insectivorous species clearly share a fundamental pattern. Moreover, this specific pattern is shared with the oldest known complete bat specimens from 50 million years ago and also with a wide variety of other kinds of mammals (Simmons & Geisler 1998). From an evolutionary perspective one might say that these bats have retained a very basic dental morphology. Thus, the remarkable thing is consistency over time rather than evolutionary divergence. In the previous section I argued that cheek teeth in the megachiropterans were relatively unconstrained and, as a result, had lost the detailed complex coronal morphologies of their ancestors. In one sense, the two major evolutionary lineages of bats exemplify the results of a natural experiment. In one lineage there probably was an emphasis on selection for a type of oral environment. A secondary effect was that such an environment is not supportive of a form of dentition that provides detailed service to the animal. By using the word 'detailed' I am referring to a dental arcade in which upper and lower teeth occlude in a very finely articulated fashion, with little tolerance in the ways in which cusps and ridges interact. In the other major lineage of bats the emphasis was on selection that would maintain dentition capable of finely detailed service in conjunction with complex jaw mechanics. Indeed, morphometric analyses (e.g., Freeman 1979) have shown that subtle dental differences sometimes distinguish among species that feed on certain types of insects. In regard to insectivorous microchiropteran bats one might only speculate about the means by which the genetic underpinning of dental morphology is conserved. In these circumstances the genetic underpinnings of the structure might be physically located in highly conserved regions of one or more chromosomes and/or might be positioned among developmental genes that are used only briefly in the life of an individual. Conservation of form is an interesting evolutionary issue because success at shielding a set of genes from an excessive mutation rate[1] or expression of mutated genotypes probably is equivalent to relinquishing capability to adapt. Naturally, there also are other ways to think of genetic and morphological conservation. For instance, environmental feedback in the form of selection pressure may be necessary to maintain a finely tuned morphological system.

Yet another view of this subject was offered by Jernvall et al. (1996), who analyzed mammalian molar teeth and compared diversity in form to taxonomic diversity over time. These authors showed that diversity in molar morphology is independent of taxonomic diversity and in fact the disparity between the two was highest early in the Eocene (but was not matched by taxonomic diversity). Subsequently, when the mammalian adaptive radiation peaked (and taxonomic diversity increased), molar morphological diversity decreased (Jernvall et al. 1996). In other words, the various evolutionary lineages of mammals settled on certain morphologies and once that occurred, the basic morphologies were then conserved.

The line of reasoning shared in the foregoing paragraphs brings up a new issue. What about the dentitions in microchiropteran bats that are not insectivorous (or carnivorous)? One can readily examine the dentition in specimens of such species, but in order to understand the relationships

---

[1] Here I am thinking primarily of chromosomal events (translocations for example) and such things as transposons rather than nucleotide base substitutions.

among diet and dental morphology, it is first necessary to review the evolutionary histories and biology of such species.

In context of the suborder Microchiroptera, non-insectivorous species are unusual and restricted to one family, the Neotropical Phyllostomidae. The origin of this family is uncertain because presently none of the Eocene fossils seem to represent early phyllostomids (Simmons & Geisler 1998). Systematists most often group the family with the Noctilionidae (carnivorous/insectivorous species) and the Mormoopidae (insectivorous species). The 'frugivorous' species of phyllostomids are difficult to define because many of the species (in four of seven subfamilies) eat some fruit or other plant tissues (Gardner 1977; Fleming 1988). Thus, the impression one gains is that frugivory arose independently in various early phyllostomid lineages. If this is the case, then it seems likely that as far as phyllostomids are concerned, the term 'frugivory' connotes a broad generalization about diet rather than a specific survival strategy. With this in mind, there is ample evidence that frugivory of some type has been accomplished in different ways by phyllostomid bats. If one undertakes interspecific comparisons of the stomachs of phyllostomid frugivores, the gross anatomy, histology, and ultrastructure of the gastric mucosa is obviously variable (Forman 1972; Phillips et al. 1984). For example, the gastric mucosa in stenodermatine fruit bats is characterized by great numbers of highly active acid-producing parietal cells and few mucous cells (Phillips et al. 1984). By way of contrast, in the short-tailed fruit bat (Carollia perspicillata, Carollinae) the gastric mucosa (especially the parietal cells as determined by ultrastructural comparisons) is similar to that of insectivorous or carnivorous bats (Phillips et al. 1984). The most important point is that these differences in the gastric mucosa are not merely isolated outcomes of interspecific comparisons. Instead, such differences are signatures of system-wide differences among the various phyllostomid fruit bats. For example, in the short-tailed fruit bat (C. perspicillata) the gross anatomy of the stomach; ultrastructure of the parietal cells; ultrastructure of the parotid salivary gland acinar cell membrane and secretory product; and saliva pH ($\sim$7.5–8.5) and moderately high salivary buffering capacity, are all approximately intermediate between insectivorous bats and the stenodermatine fruit bats (Nagato et al. 1998). In a well-studied stenodermatine frugivore, the Jamaican fruit bat (Artibeus jamaicensis), the activity of the parietal cells is extreme; the pyloric gastrin- producing G-cells (which stimulate acid production in the gastric fundus) are significantly ($p < 0.05$) more abundant than in some species than in others (Mennone et al. 1986); the parotid salivary gland acinar cell membranes and product are unique; possible H^+

pumps (portasomes) are found on striated duct cell apical membranes; and the saliva is acidic (pH $\sim$5.8) and is a weak buffer (Dumont 1997).

The dentition of the short-tailed fruit bat (*C. perspicillata*) differs from that of insectivorous phyllostomids in that the teeth are slightly smaller (relative to size), slightly more widely spaced, and the molars have a less clearly defined W-shaped ectoloph. However, as Fleming (1988) noted, in a comparative sense the cheek teeth in this bat are intermediate between those of insectivorous bats and those in stenodermatines such as the Jamaican fruit bat (*A. jamaicensis*). In Freeman's recent bivariate analysis (unpublished data) of molariform area/palatal area against stylar shelf area/molariform area, *Carollia* occupies a central position among insectivorous, frugivorous, and nectarivorous bats. In these ways the dental morphology in the short-tailed fruit bat is remarkably consistent with oral environment and digestive tract. In the stenodermatines, the face is shortened (relative to the muzzle in other phyllostomids), and the cheek teeth exhibit corresponding shortness along their anterior–posterior axis. In seeming compensation for this shortening, the cheek teeth are much broader than in the short-tailed fruit bat or species in other subfamilies. Given the acidic saliva and characteristics of the digestive tract, one of the most interesting dental comparisons is between a stenodermatine such as the Jamaican fruit bat, and megachiropteran fruit bats. In the previous section I pointed out that cheek tooth dental morphology is relatively simple in most megachiropterans. Typically the teeth in these bats exhibit extensive erosion and abrasion. In *Artibeus jamaicensis* the upper molars lack a W-shaped ectoloph, but the coronal morphology of both upper and lower molars is complex and the main cones and ridges can be readily compared to homologous features in insectivorous bats. In comparison to megachiropteran fruit bats, one could say that the dentition in *A. jamaicensis* is conservative. The upper molars (particularly the large first molar) are characterized by massive enamel ridges (see Phillips *et al.* 1977 for a scanning electron microscope micrograph). Although some dental erosion and abrasion can be seen in many museum specimens of *A. jamaicensis*, cheek teeth in this species are far more resistant than are cheek teeth in any of the megachiropterans with acidic saliva. Future systematic comparisons of acid resistance of cheek tooth enamel in this bat and enamel of cheek teeth in comparable megachiropteran bats could be interesting. For example, one might use the cheek teeth of the Asian short-nosed bat, *Cynopterus sphinx*, which is physically and ecologically similar to *A. jamaicensis*, to test the hypothesis that this is an essential difference between these independently evolved lineages of fruit bats. If there is a significant difference, it would be important evidence of selection in the amelogenic system in stenodermatine fruit bats.

Although frugivory in bats has evolved independently at least twice (but probably more often), dental morphology lacks evidence of convergence. Meaningful comparisons of fruit bat dental morphology and function requires consideration of the oral environment and digestive tract.

ACKNOWLEDGMENTS

The author gratefully acknowledges the critical review and suggestions from Patricia W. Freeman, a friend and colleague at the University of Nebraska–Lincoln. Freeman's published contributions in this subject area are thought-provoking and valuable to our understanding of bat biology and the evolution of dentition. Many sources of research support have been enjoyed over the years: NIH, NSF, Hofstra University, and The University of Kansas all contributed funding that supported the author's original research articles cited herein. The Department of Biological Sciences at Texas Tech University kindly supported the writing of this essay.

REFERENCES

Andersen, K. (1912) *Catalogue of the Chiroptera in the Collections of the British Museum. I. Megachiroptera.* London: British Museum (Natural History).

Baker, R. J., Novacek, M. J. & Simmons, N. B. (1991) On the monophyly of bats. *Systematic Zoology,* **40,** 216–31.

Bennick, A. (1987) Structural and genetic aspects of proline-rich proteins. *Journal of Dental Research,* **66,** 457–61.

Birney, E. C. & Timm, R. M. (1975) Dental ontogeny and adaptation in *Diphylla ecaudata. Journal of Mammalogy,* **59,** 204–7.

Dobson, G. E. (1878) *Catalogue of the Chiroptera in the British Museum.* London: British Museum.

Dorst, J. (1957a) Considérations sur la dentition de lait des chiroptères de la famille des molossidés. *Mammalia,* **21,** 133–5.

Dorst, J. (1957b) Note sur la dentition lactéale de *Tonatia amblyotis* (Chiroptères, Phyllostomidés). *Mammalia,* **21,** 302–4.

Dumont, E. R. (1995) Enamel thickness and dietary adaptation among extant primates and chiropterans. *Journal of Mammalogy,* **76,** 1127–36.

Dumont, E. R. (1997) Salivary pH and buffering capacity in frugivorous and insectivorous bats. *Journal of Mammalogy,* **78,** 1210–19.

Fleming, T. H. (1988) *The Short-tailed Fruit Bat: A Study in Plant–Animal Associations.* Chicago: University of Chicago Press.

Forman, G. L. (1972) Comparative morphological and histochemical studies of stomachs of selected American bats. *University of Kansas Science Bulletin,* **49,** 591–729.

Freeman, P. W. (1979) Specialized insectivory: beetle-eating and moth-eating molossid bats. *Journal of Mammalogy*, **60**, 467–79.

Freeman, P. W. (1981) Correspondence of food habits and morphology in insectivorous bats. *Journal of Mammalogy*, **62**, 166–73.

Freeman, P. W. (1988) Frugivorous and animalivorous bats (Microchiroptera): dental and cranial adaptations. *Biological Journal of the Linnean Society*, **33**, 249–72.

Freeman, P. W. (1992) Canine teeth in bats (Microchiroptera): size, shape and role in crack propagation. *Biological Journal of the Linnean Society*, **45**, 97–115.

Freeman, P. W. (1995) Nectarivorous feeding mechanisms in bats. *Biological Journal of the Linnean Society*, **56**, 439–63.

Freeman, P. W. (1998) Form, function, and evolution in skulls and teeth of bats. In *Bat Biology and Conservation*, ed. T. H. Kunz & P. A. Racey, pp. 140–56. Washington: Smithsonian Institution Press.

Gardner, A. L. (1977) Feeding habits. In *The Biology of the New World Leaf-nosed Bats*, part 2, ed. R. J. Baker, J. K. Jones, Jr., & D. C. Carter, pp. 193–250. Lubbock: Texas Tech University Press.

Gottlieb, G. (1992) *Individual Development and Evolution: The Genesis of Novel Behavior*. New York: Oxford University Press.

Grassé, P.-P. (1955) Ordre des Chiroptères. In *Traité de zoologie anatomie, systématique, biologie*, ed. P.-P. Grassé, pp. 1729–1853. Paris: Presse Zéro.

Greenbaum, I. F. & Phillips, C. J. (1974) Comparative anatomy and general histology of tongues of long-nosed bats (*Leptonycteris sanborni* and *L. nivalis*) with reference to oral mites. *Journal of Mammalogy*, **55**, 489–504.

Greenhall, A. M. (1988) Feeding behavior. In *Natural History of Vampire Bats*, ed. A. M. Greenhall & U. Schmidt, pp. 111–31. Boca Raton: CRC Press.

Greet, G. De & De Vree, F. (1984) Movements of the mandibles and tongue during mastication and swallowing in *Pteropus giganteus* (Megachiroptera): a cineradiographical study. *Journal of Morphology*, **179**, 95–114.

Handley, C. O., Jr., Gardner, A. L. & Wilson, D. E. (1991a) Food habits. In *Demography and Natural History of the Common Fruit Bat, Artibeus jamaicensis, on Barro Colorado Island, Panamá*, ed. C. O. Handley, Jr., D. E. Wilson, & A. L. Gardner, pp. 141–46. Washington: Smithsonian Institution Press.

Handley, C. O., Jr., Wilson, D. E. & Gardner, A. L. (1991b) *Demography and Natural History of the Common Fruit Bat, Artibeus jamaicensis, on Barro Colorado Island, Panamá*. Washington: Smithsonian Institution Press.

Hershkovitz, P. (1971) Basic crown patterns and cusp homologies of mammalian teeth. In *Dental Morphology and Evolution*, ed. A. A. Dahlberg, pp. 95–150. Chicago: University of Chicago Press.

Hill, J. E. & Smith, J. D. (1984) *Bats: A Natural History*. Austin: University of Texas Press.

Janies, D. & DeSalle, R. (1999) Development, evolution, and corroboration. *Anatomical Record*, **257**, 6–14.

Jernvall, J., Hunter, J. P. & Fortelius, M. (1996) Molar tooth diversity, disparity, and ecology in Cenozoic radiations. *Science*, **274**, 1489–92.

Kauffman, D. L., Bennick, A., Blum, M. & Keller, P. J. (1991) Basic proline-rich proteins from human parotid gland saliva: relationships of the covalent structures of ten proteins from a single individual. *Biochemistry*, **30**, 3351–6.

Kay, R. F. & Sheine, W. S. (1979) On the relationship between chitin particle size and digestibility in the primate *Galago senegalensis*. *American Journal of Physical Anthropology*, **50**, 301–8.

Kirsch, J. A. W., Flannery, T. F., Springer, M. S. & Lapointe, F.-J. (1995) Phylogeny of the Pteropodidae (Mammalia: Chiroptera) based on DNA hybridisation, with evidence for bat monophyly. *Australian Journal of Zoology*, **43**, 395–428.

Koopman, K. F. & MacIntyre, G. T. (1980) Phylogenetic analysis of chiropteran dentition. In *Proceedings of the Fifth International Bat Research Conference*, ed. D. E. Wilson & A. L. Gardner, pp. 279–88. Lubbock: Texas Tech University Press.

Marshall, P. M. & Butler, P. M. (1966) Molar cusp development in the bat, *Hipposideros beatus*, with reference to the ontogenetic basis of occlusion. *Archives of Oral Biology*, **11**, 949–65.

Mehansho, H., Clements, S., Sheares, B. T., Smith, S. & Carlson, D. M. (1985) Induction of proline-rich glycoprotein synthesis in mouse salivary glands by isoproterenol and by tannins. *Journal of Biological Chemistry*, **260**, 4418–23.

Mennone, A., Phillips, C. J. & Pumo, D. E. (1986) Evolutionary significance of interspecific differences in gastrin-like immunoreactivity in the pylorus of phyllostomid bats. *Journal of Mammalogy*, **67**, 373–84.

Miller, G. S. (1907) The families and genera of bats. *Bulletin of the United States National Museum*, **57**, 1–282.

Nagato, T., Tandler, B. & Phillips, C. J. (1998) An unusual parotid gland in the tent-building bat, *Uroderma bilobatum*: possible correlation of interspecific differences with differences in salivary pH and buffering capacity. *Anatomical Record*, **252**, 290–300.

Nelson, C. E. (1966) The deciduous dentition of the Central American phyllostomid bats *Macrotus waterhousii* and *Pteronotus suapurensis*. *Southwestern Naturalist*, **11**, 142–53.

Oppenheim, F. G., Hay, D. I., Smith, D. J., Offner, G. D. & Troxler, R. F. (1987) Molecular basis of salivary proline-rich protein and peptide synthesis: cell-free translations and processing of human and macaque statherin mRNAs and partial amino acid sequence of their signal peptides. *Journal of Dental Research*, **66**, 462–6.

Oxberry, B. A. (1975) An anatomical, histological, and autoradiographic study of the ever-growing molar dentition of *Microtus* with comments on the role of structure in growth and eruption. *Journal of Morphology*, **147**, 337–54.

Peterson, R. L. (1968) Notes on an unusual specimen of *Scotophilus* from Vietnam. *Canadian Journal of Zoology*, **46**, 1097–8.

Pettigrew, J. D., Jamieson, B. G. M., Robson, S. K., Hall, L. S., McAnally, K. I. & Cooper, H. M. (1989) Phylogenetic relations between microbats, megabats, and primates (Mammalia: Chiroptera and Primates). *Philosophical Transactions of the Royal Society of London* B, **325**, 489–559.

Phillips, C. J. (1971) The dentition of glossophagine bats: development, morphological characteristics, variation, pathology, and evolution. *University of Kansas Museum of Natural History, Miscellaneous Publication no. 54*, 138 pp.

Phillips, C. J. (1985) Microanatomy. In *Biology of the New World Microtus*, ed. R. H. Tamarin, pp. 176–253. Special Publication no. 8, American Society of Mammalogists.

Phillips, C. J. (1994) Anatomy. In *75 Years of North American Mammalogy*, ed. J. R. Choate & E. C. Birney, pp. 234–57, Special Publication no. 11, American Society of Mammalogists.

Phillips, C. J. (1996) Cells, molecules, and adaptive radiation in mammals. In *Contributions in Mammalogy: A Memorial Volume Honoring Dr. J. K. Jones, Jr.*, ed. R. J. Baker & H. H. Genoways, pp. 1–24. Lubbock: Museum of Texas Tech University.

Phillips, C. J. & Oxberry, B. A. (1972) Comparative histology of molar dentitions of *Microtus* and *Clethrionomys*, with comments on dental evolution in microtine rodents. *Journal of Mammalogy*, **53**, 1–20.

Phillips, C. J. & Steinberg, B. (1976) Histological and scanning electron microscopic studies of tooth structure and thegosis in the common vampire bat, *Desmodus rotundus*. *Occasional Papers, Museum of Texas Tech University*, **42**, 1–12.

Phillips, C. J. & Tandler, B. (1996) Salivary glands, cellular evolution, and adaptive radiation in mammals. *European Journal of Morphology*, **34**, 155–61.

Phillips, C. J., Grimes, G. W. & Forman, G. L. (1977) Oral biology. In *The Biology of the New World Leaf-nosed Bats*, part 2, ed. R. J. Baker, J. K. Jones, Jr., & D. C. Carter, pp. 121–246. Lubbock: Texas Tech University Press.

Phillips, C. J., Steinberg, B. & Kunz, T. H. (1982) Dentin, cementum, and age determination in bats: a critical evaluation. *Journal of Mammalogy*, **63**, 197–207.

Phillips, C. J., Studholme, K. M. & Forman, G. L. (1984) Results of the Alcoa Foundation Suriname Expeditions. 8. Comparative ultrastructure of gastric mucosae in four genera of bats (Mammalia:Chiroptera), with comments on gastric evolution. *Annals of the Carnegie Museum of Natural History*, **53**, 71–117.

Phillips, C. J., Tandler, B. & Nagato, T. (1993) Evolutionary divergence of salivary gland acinar cells: a format for understanding molecular evolution. In *Biology of Salivary Glands*, ed. K. Dobrosielski-Vergona, pp. 39–80. Boca Raton: CRC Press.

Phillips, C. J., Weiss, A. & Tandler, B. (1998) Plasticity and patterns of evolution in mammalian salivary glands: comparative immunohistochemistry of lysozyme in bats. *European Journal of Morphology*, **36**, 19–26.

Pumo, D. E., Finamore, R. S., Franek, W. R., Phillips, C. J., Tarzami, S. & Balzarano, D. (1998) Complete mitochondrial genome of a Neotropical fruit bat, *Artibeus jamaicensis*, and a new hypothesis of the relationships of bats to other eutherian mammals. *Molecular Evolution*, **47**, 709–17.

Ray, C. E., Linares, O. J. & Morgan, G. S. (1988) Paleontology. In *Natural History of Vampire Bats*, ed. A. M. Greenhall & U. Schmidt, pp. 19–30. Boca Raton: CRC Press.

Reeder, W. G. (1953) The deciduous dentition of the fish-eating bat, *Pizonyx vivesi*. *Occasional Papers of the Museum of Zoology, University of Michigan*, **545**, 1–3.

Schwartz, J. H. (1999) Homeobox genes, fossils, and the origin of species. *Anatomical Record*, **257**, 15–31.

Shaw, J. H., Sweeney, E. A., Cappuccino, C. C. & Meller, S. M. (1978) *Textbook of Oral Biology*. Philadelphia: W. B. Saunders.

Simmons, N. B. & Geisler, J. H. (1998) Phylogenetic relationships of *Icaronycteris*, *Archaeonycteris*, *Hassianycteris*, and *Palaeochiropteryx* to extant bat lineages, with comments on the evolution of echolocation and foraging strategies in Microchiroptera. *Bulletin of the American Museum of Natural History*, **235**, 1–182.

Slaughter, B. H. (1970) Evolutionary trends of chiropteran dentitions. In *About Bats: A Chiropteran Symposium*, ed. B. H. Slaughter & D. W. Walton, pp. 51–83. Dallas: Southern Methodist University Press.

Smith, J. D. & Madkour, G. (1980) Penial morphology and the question of chiropteran phylogeny. In *Proceedings of the Fifth International Bat Research Conference*, ed. D. E. Wilson & A. L. Gardner, pp. 347–65. Lubbock: Texas Tech University Press.

Springer, M. S., Hollar, L. J. & Kirsch, J. A. W. (1995) Phylogeny, molecules versus morphology, and rates of character evolution among fruitbats (Chiroptera: Megachiroptera). *Australian Journal of Zoology*, **43**, 557–82.

Stegeman, L. C. (1956) Tooth development and wear in *Myotis*. *Journal of Mammalogy*, **37**, 58–63.

Strait, S. G. (1993) Molar morphology and food texture among small-bodied insectivorous mammals. *Journal of Mammalogy*, **74**, 391–402.

Studier, E. H. & Wilson, D. E. (1991) Physiology. In *Demography and Natural History of the Common Fruit Bat, Artibeus jamaicensis, on Barro Colorado Island, Panamá*, ed. C. O. Handley, Jr., D. E. Wilson, & A. L. Gardner, pp. 9–18. Washington: Smithsonian Institution Press.

Tan, K. H., Zubaid, A. & Kunz, T. H. (1998) Food habits of *Cynopterus brachyotis* (Müller)

(Chiroptera: Pteropodidae) in peninsular Malaysia. *Journal of Tropical Ecology*, **14**, 299–307.

Tandler, B. & Phillips, C. J. (1998) Microstructure of mammalian salivary glands and its relationship to diet. In *Glandular Mechanisms of Salivary Secretion*, ed. J. R. Garrett, J. Ekström, & L. C. Anderson, pp. 21–35. Basel: Karger.

Tandler, B., Nagato, T. & Phillips, C. J. (1997a) Ultrastructure of the unusual accessory submandibular gland in the fringe-lipped bat, *Trachops cirrhosus*. *Anatomical Record*, **248**, 164–75.

Tandler, B., Nagato, T. & Phillips, C. J. (1997b) A unique parotid gland in Hart's little fruit bat, *Enchisthenes hartii*. *Anatomical Record*, **249**, 349–58.

Tandler, B., Phillips, C. J., Nagato, T. & Toyoshima, K. (1990) Ultrastructural diversity in chiropteran salivary glands. In *Extraparietal Glands of the Alimentary Canal*, ed. A. Riva & P. M. Motta, pp. 31–52. Boston: Kluwer.

Thompson, K. S. (1988) *Morphogenesis and Evolution*. New York: Oxford University Press.

Vaughan, T. A. (1970) The skeletal system. In *Biology of Bats*, ed. W. A. Wimsatt, pp. 97–138. New York: Academic Press.

Vierhaus, H. (1983) Wie Vampirfledermäuse (*Desmodus rotundus*) ihre Zähne schärfen. *Zeitschrift für Säugetierkunde*, **48**, 269–77.

Whitaker, J. O., Jr., Neefus, C. & Kunz, T. H. (1996) Dietary variation in the Mexican free-tailed bat (*Tadarida brasiliensis mexicana*). *Journal of Mammalogy*, **77**, 716–24.

9

Wing ontogeny, shifting niche dimensions, and adaptive landscapes

INTRODUCTION

Growth and development of wings in bats is a unique morphogenetic process in mammals. The divergence from a 'standard issue' (ancestral) handplate to that of a handwing occurs early in skeletogenesis (Joller 1977; Adams 1992a, b; Raff 1996), and because bats begin flight on immature wings (Jones 1967; Davis 1969; Kunz 1974; Buchler 1980; Kunz & Anthony 1982; Hughes et al. 1987, 1995; Powers et al. 1991; Adams 1996, 1997, 1998; Stern et al. 1997; Hoying & Kunz 1998; Kunz & Hood in press), juveniles face energetic and functional demands confronted by few other mammals. Indeed, the aerodynamic constraints of flight are stringent for juveniles, and although lactating females provide some nutritional security for their newly volant young, this reservoir of nourishment is short-lived (Altringham 1996). In fact, juveniles of insectivorous bat species are weaned typically within two weeks of first volancy (Tuttle & Stevenson 1982; Kunz 1987) and, therefore, the capacity not only for sustained flight, but also maneuverability for pursuing and capturing evasive insects, must develop quickly. Significant changes in wing size during growth produces differential flight abilities that correlate with age (Powers et al. 1991) and foraging juveniles manifesting differing flight abilities affect population and community ecology (Adams 1996, 1997). In this chapter I describe pre- and postnatal limb morphogenesis in several bat species across suborders, and compare their developmental patterns with that of Sorex vagrans, S. cinereus, Rattus norvegicus, Mus musculus, and Acomys cahirinus in order to elucidate developmental shifts responsible in manifesting a handwing. I also contrast limb growth among bats species and compare adult morphologies encompassing seven families across the two suborders. In addition, I synthesize a hypothetical three-dimensional model integrating ontogeny, ecology, and evolution of flight. By testing

275

this model with data collected on *Myotis lucifugus*, I illustrate how integrating ontogenetic information can be useful in bridging the differential time frames between ecology and evolution, thereby illuminating the evolutionary underpinnings of ecological patterns.

ONTOGENY OF THE BAT HANDWING

Prenatal growth of the handwing: divergence from the phylotype

Studies of postnatal growth of the wing in bats are numerous (see Kunz & Stern 1995; Kunz & Hood *in press*). Description and quantification of prenatal morphogenesis, however, is comparatively inadequate. Furthermore, comparisons of prenatal limb ontogeny between bats with other mammals with less derived limbs has hardly been attempted. This is surprising considering the historical importance of developmental patterns in discerning the evolution of structures (Raff 1996; McNamara 1997). Even simple mechanistic questions concerning how the intrinsic nature of morphogenesis derives a 'standard issue' mammalian embryo (phylotype; Raff 1996) into a winged mammal still remain unsatisfactorily answered. True, in spite of the fact that changes in developmental patterns are requisite for the evolution of derived morphologies, and, in particular, key evolutionary innovations (Liem & Wake 1985; Raff 1996; McNamara 1997).

Skeletogenesis

The primary and fundamental differences in skeletogenesis of the distal forelimb between bats, shrews and rodents resides in the pre-versus postpartum appearances of the primary ossification centers (Tables 9.1, 9.2). In rodents, ossification of the carpus begins prenatally, whereas in shrews this process begins postpartum. In fact, in *Sorex* spp., the carpus, metacarpus and digits do not begin ossifying until well into the postnatal period. Sterba (1984) argued that the altricial developmental pattern exemplified by *Sorex* is ancestral relative to the precocial pattern observed in the genus *Mus*. By this reasoning, the precocial pattern of *Mus* is the result of accelerated development as compared to *Sorex* (Sterba 1984). Because *R. norvegicus* shows a similar developmental pattern to *Mus*, then the same argument concerning a relatively accelerated developmental rate would apply. Interestingly, most bats fit into this ossification scenario closer to rodents than to shrews with ossification of most of the distal ele-

Table 9.1. *Prenatal ossification sequence (in days) of primary ossification centers of the forearm bones in three terrestrial mammals (compiled from Donaldson 1924; Rugh 1968; Sterba 1977)*

Bone(s)	*Sorex vagrans*	*Mus musculus*	*Rattus norvegicus*
Humerus	17	14	17
Radius	17	14	17
Ulna	17	14	17
Carpus	PON[a]	16	18–21
Metacarpus	PON[a]	16	18–21
Digits	PON[a]	17–18	PON[a]

Note: [a] Ossification occurring postnatally

Table 9.2. *Prenatal sequence of primary ossification formation of the forearm bones (previously unpublished data) calculated as approximate percent of parturition skull length (mm) in three families of bats (Vespertilionidae, Phyllostomatidae, and Pteropodidae)*

Bone(s)	*Myotis lucifugus*	*Artibeus jamaicensis*	*Rousettus celebensis*
Humerus	70%	32%	33%
Radius	70%	32%	33%
Ulna	70%	32%	33%
Carpus	PON[a]	100% (3 of 8)	69% (all)
Metacarpus	84% (3–5)	32% (all except digit I)	69% (all)
Digits	84% (all phalanges except middle phalanx of digit I)[b]	44% (proximal phalanges), 100% (middle phalanges)[b]	60% (proximal phalanges only; 69% middle phalanges)[c]

Notes:
[a] Ossification occurs postnatally.
[b] Distal phalanges of digits III–V do not ossify in *M. lucifugus* and *A. jamaicensis*.
[c] *R. celebensis* does not posses distal phalanges on digits III–V.

ments of the forelimb occurring prenatally (Table 9.2). For *Myotis lucifugus*, all but the carpus shows ossification centers prenatally. In *Artibeus jamaicensis*, ossification of the entire carpus occurs prenatally, whereas in *Rousettus* about 70% of the carpus show ossification centers prenatally (Adams, unpublished). If the altricial developmental patterns exhibited by *Sorex* is ancestral in mammals (Sterba 1984), then it reasons that rodents

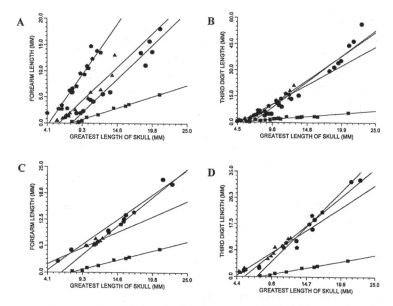

Figure 9.1. Growth in length of the forearm and third digit in several species of Microchiroptera, Megachiroptera, and *Rattus norvegicus* relative to skull growth. For plots A and B, circles – *Artibeus jamaicensis*, triangles – *Eptesicus fuscus*, pentagons – *Myotis lucifugus*, and squares – *Rattus norvegicus*. In plots C and D, dots – *Rousettus celebensis*, triangles – *Syconycteris australis*, pentagons – *Megaloglossus woermanni*, and squares – *Rattus norvegicus*.

and bats show similarly derived developmental organizations, but surely the selection pressures for these convergent patterns were independent. That is, in rodents, divergence from the ancestral developmental pattern is correlated with the evolution of precocial young (Sterba 1984), whereas in bats, the derived developmental pattern of the forelimb is likely due to unique functional and developmental integrations necessary for the evolution of wings in mammals (Adams 1998).

Growth rates

Despite some similarities in ossification sequences, bats do differ significantly both developmentally and anatomically from both rodents and shrews (and most other mammals) and these differences can be illustrated by plotting growth trajectories that give insight into the relative rates of growth for specific anatomical structures. Growth trajectories of the forearm and third digit of the vespertilionids *M. lucifugus* and *Eptesicus fuscus*, the phyllostomid, *Artibeus jamaicensis* (Fig. 9.1A, B), and three pteropodids, *Megaloglossus woermanni, Syconycteris australis,* and *Rousettus celeben-sis* (Fig. 9.1C, D) differ from the trajectory exhibited by *Rattus norvegicus,* the latter of which is representative a proportionately ancestral mammal-ian morphology of the limb (Wood 1962; Adams 1992a). The forearm and third digit of bats grow in length at rates faster relative to skull growth than observed in *R. norvegicus.* (Fig. 9.1). Among the microchiropterans, *M. lucifugus* has a faster growing forearm, even faster than another vespertil-ionid, *E. fuscus. E. fuscus* has, however, a slightly faster growing third digit than either *M. lucifugus* or *A. jamaicensis* (Fig. 9.1A, B). Of the three micro-chiropteran species depicted, none shows characteristics of forelimb growth similar to *R. norvegicus.* Joller (1977) documented similar differ-ences in forearm growth between the spiny rat, *Acomys cahirinus,* and *Myotis myotis.*

For two megachiropterans, relative growth trajectories for forelimb length are more variable across species than observed in the Micro-chiroptera (Fig. 9.1C, D). Curiously, *Syconycteris australis* tracts similarly with *R. norvegicus,* whereas *Rousettus celebensis* and *Megaloglossus woermanni* share a similar growth pattern that is distinctive from *Rattus norvegicus* and *S. aus-tralis. M. woermanni* shows the steepest slopes and, therefore, the highest rate of growth, for forearm length (Fig. 9.1C) and third digit length (Fig. 9.1D) among the three megachiropterans (Adams, unpublished).

Bone fusions

Besides the obvious differences in lengths of the forearm bones in bats relative to shrews and rodents, the functional significance of the derived forelimb in bats lies not only in its elongation, but in the fusions of bones and divergent diametrical growth between the ulna and radius (Fig. 9.2; Adams 1992b). In fact, it is through distinctive developmental fusions between forearm bones that gives rise to three distinctive forearm types (bowstring, half-strut and strut configurations) among the 40 extant species of bats across six families of microchiropterans (Table 9.3) and the Pteropodidae examined for this study.

Despite the fact that there are three distinctive adult forearm morphologies among bats, all types begins with chondrogenesis of a 'stan-dard issue' mammalian forelimb, delineating the vertebrate developmen-tal and evolutionary underpinnings of the handwing. The morphology of the vespertilionid forearm I describe as 'bowstring' in that the lateral cur-vature of the radius resembles a bow and the ulnar diaphysis, which trans-forms from cartilage to ligament during morphogenesis, resembles a string (Adams 1992a). Differential growth between the midshaft diameters

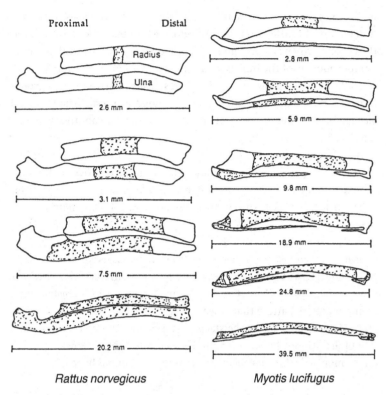

Figure 9.2. Camera lucida drawings illustrating growth and ossification patterns of the radius and the ulna in *Rattus norvegicus* and *Myotis lucifugus*. (From Adams 1992a.)

of the radius and of the ulna results in the radius becoming the dominant bone of the forearm in size and, apparently, function (Fig. 9.3A, B; Vaughan 1959). Fusion between the distal tips of the forearm bones results in a unique structure termed the radio-ulnar bridge that forms the articulation surface with the carpus (Adams 1992a). The radius grows to form most of the proximal articulation surface with the humerus, and the ulna, although fully fused with the radius, retains a proximal extension of the olecranon process for attachment of the m. triceps brachii bundle (Fig. 9.4). Curiously, although the ulnar shaft appears to be 'missing' in examinations of adult gross anatomy (Vaughan 1959), cleared and stained, wholemount fetuses and histological analyses show that the mid-diaphyseal shaft is not lost, but transforms into dense connective tissue (ligament) that comprises the string component of the bowstring style forearm (Fig. 9.5A; Adams 1992a).

Development of the ulna in vespertilionid bats, then, is unique in three fundamental ways: 1) diametrical growth of the ulnar diaphysis is

Table 9.3. *Forearm types (see text for descriptions) as distributed throughout microchiropterans based upon analysis of postcranial skeletons of adult bats housed at the University of Kansas Natural History Museum, Lawrence (previously unpublished data)*

Bowstring	Half-strut	Strut
Vespertilionidae	Molossidae	Phyllostomidae
Myotis lucifugus	*Eumops perotis*	*Anora geoffroyi*
Myotis volans	*Molossus molossus*	*Artibeus literatus*
Antrozous pallidus	*Molossus ater*	*Artibeus jamaicensis*
Lasiurus cinereus	*Molossus sinalae*	*Brachyphylla cavernarum*
Eptesicus fuscus	*Nyctinomops macrotus*	*Carollia perspicillata*
	Tadarida brasiliensis	*Choeronycteris mexicana*
	Tadarida mexicana	*Dermanura cinereus*
	Tadarida elegans	*Glossophaga sorcina*
	Tadarida pumila	*Leptonycteris curasae*
		Lonchorhina aurita
		Lonchorhina robustus
		Macrotus californicus
		Monophyllus plethodon
		Phylloderma stenops
		Phyllostomus discolor
		Sturnira ludovici
		Vampyrops vittatus
		Desmodus rotundus
		Mormoopidae
		Mormoops megalophylla
		Pteronotus davyi
		Pteronotus parnellii
		Rhinolophidae
		Rhinolophus ferrumequinum
		Nycteridae
		Nycterus thebiaca

Note: Sample sizes vary per species, but none was less than three.

truncated early (developmental stage 3, Adams 1992b) in skeletogenesis, resulting in a thread-like, vestigial appearance of the diaphysis in adults; 2) the central third of the diaphysis of the ulna differentiates from cartilage to ligament (the ulnar ligament, Adams 1992a) and stretches between the ossified proximal portion of the diaphysis and the ossified distal epiphysis; 3) unlike most other mammals, the ulna does not develop a

A

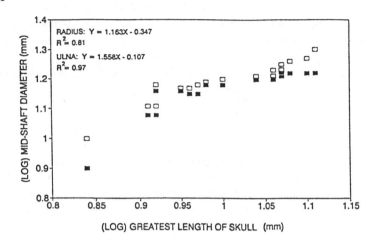

B

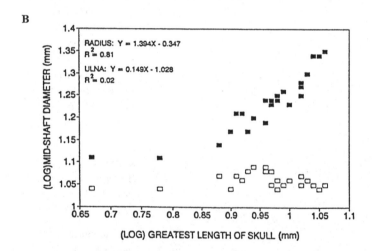

Figure 9.3. Scatter plots showing log midshaft diameter vs. log greatest length of skull for A) *Rattus norvegicus* and B) *Myotis lucifugus*. R^2 values generated by Spearman rank correlation analysis. (From Adams 1992a.)

trochlear notch, and very little of the ulna comes to articulate with the humerus proximally. However, the olecranon process does provide the insertion point for the m. triceps brachii (Vaughan 1959). Formation of the radio-ulnar bridge is consistent among vespertilionids as a derived articulation surface in contact with the carpus. Although variation in the size and shape of the bridge occurs across species, the overall developmental pattern is consistent, as is general morphology (Fig. 9.6) and,

PROXIMAL FOREARM

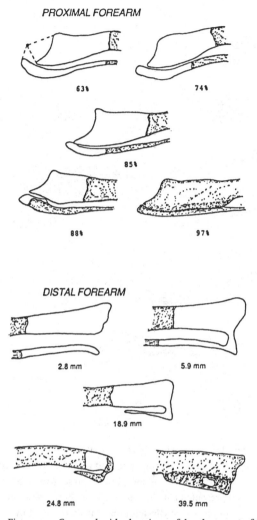

DISTAL FOREARM

Figure 9.4. Camera lucida drawings of development of the proximal radius (above) and ulna (below) in *Myotis lucifugus*, illustrating radio-ulnar fusion and the overtaking of the elbow joint by the radius, whereas the distal radius and ulna fuse forming the radio-ulnar bridge. (From Adams 1992a.)

therefore, so likely is function. The derived ulnar shaft of vespertilionids is unique and its specific function has yet to be determined, much less fully appreciated. Indeed, it is plausible that the ulnar ligament is a flight adaptation that serves as an energy-storage unit. Strain-gage measurements taken from the megachiropteran *Pteropus poliocephalis* (which curiously exhibits a similar bowstring morphology to, but not identical with,

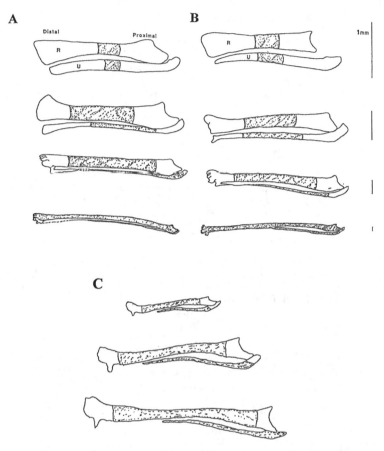

Figure 9.5. Camera lucida drawings illustrating two of the three forearm morphologies present in microchiropterans and the singe forearm type in megachiropterans. A) The bowstring morphology present in all examined vespertilionids is represented here by *Eptesicus fuscus*. B) The strut morphology present in all examined phyllostomid bats is represented by *Artibeus jamaicensis*. C) Megachiroptera exhibit a forearm morphology most similar to vespertilionids, but there are distinctive characteristics of the megachiropteran style represented here by *Syconycterus australis*. R – radius, U – ulna. (After Adams 1992a.)

vespertilionids, see below), showed that, at mid-downstroke, its forearm underwent dorsal compressive forces and ventral tensile forces that cause the radius to bend in an upward, concave manner (Swartz *et al.* 1992; Swartz 1998). If these forces are similar to what is experienced by vespertilionid bats, concave bending of the dorsal radius would create tensile forces on the ulnar diaphysis, predictably resulting in stretching of the

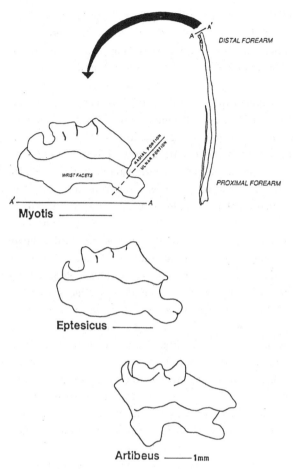

Myotis ————

Eptesicus ————

Artibeus ———— 1mm

Figure 9.6. Camera lucida drawings of an end-on view of the distal radial–ulnar complex for *Myotis lucifugus, Eptesicus fuscus,* and *Artibeus jamaicensis.* The distal ulna contributes some portion of the wrist articulation surface in all bats examined, although in the vespertilionid condition the ulnar epiphysis contributes more surface area than in the representative phyllostomid. Drawings not to scale. (After Adams 1992a.)

ulnar ligament. Hypothetically, at the bottom of the downstroke, tension created in the ulnar ligament by the upward bending of the radius would facilitate recoil of the radius to its relaxed, ventrolaterally curved position. Thus, the bowstring-style forearm may allow not only for greater flight efficiency, but also increased maneuverability. Kinetic storage mechanisms have been noted in other flying vertebrates such as birds (Jenkins *et al.* 1988).

Although proximo-distal fusion of the radius and ulna is consistent among bats, subtle differences in the details of morphology may give clues to the natural history of any particular group. As mentioned, the bowstring morphology described for vespertilionids is not ubiquitous among the Microchiroptera. As exemplified in *Artibeus jamaicensis* (Fig. 9.5B), phyllostomids exhibits a 'strut-style' forearm morphology (Table 9.3). The strut-styled forearm, although passing through a developmental sequence similar to the bowstring style, involves fusion of the tip of the reduced ulnar diaphysis to a ventrolaterally positioned radial tubercle located at about the midpoint of the radial diaphysis (Fig. 9.5B). The functional significance of this unique forearm morphology is unknown, but the 'design' suggests that it would be more sound structurally and less elastic than the bowstring type. Strut morphology of the forearm is present in species representing the families Mormoopidae, Phyllostomidae, Rhinolophidae, and Nycteridae (Table 9.3). Whether or not the developmental patterns resulting in a strut-style morphology among these families are similar is unknown, but is under investigation.

A third forearm morphology present in microchiropterans is presumably present only in the free-tailed bats (Molossidae, Table 9.3). The molossid forearm exhibits an intermediate appearance between bowstring and strut morphologies and, therefore, I term this condition the 'half-strut.' A half-strut forearm involves the ulna contacting the radius at about the midpoint of the radial diaphysis as in the strut-style, however, there is no radial tubercle present and nor is there fusion between midshafts of the ulna and radius. Instead, the distal portion of the ulna is tightly bound to the radius by connective tissue along approximately half its length from mid-diaphysis to distal tip (Adams, unpublished). The developmental events giving rise to this forearm morphology are under investigation. In comparing bowstring and strut morphologies in terms of the proportion of the distal ulna that contributes to the carpal joint, the former exhibits a larger ulnar component than the latter (Fig. 9.6). However, in all cases thus far investigated, the ulna does contribute to a portion of the forearm articulation with the carpus.

Curiously, the forearm of megachiropterans (Fig. 9.5C, *Syconycteris australis*), as mentioned above, is most similar to that of vespertilionids. The diaphysis of the ulna becomes thread-like during development and the radius dominates both the elbow and carpal joint. Whether or not the ulnar diaphysis becomes ligamentous has not yet been determined. The size and shape of the radio-ulnar bridge in pteropodids is distinctive from that observed in vespertilionids thus far. Most salient among these differences is that there is no ossification of the ulna proximally to the radio-ulnar bridge in megachiropterans, but instead a peg-like, lateral projection from the bridge is present (Fig. 9.5C). The functional significance of this arrangement is unknown.

The manus

Prenatal development of the manus also shares similarities and differences across bats. Few attempts have been made to establish stages of fetal bat development, except in *Myotis lucifugus* where changes in external anatomy were correlated with skeletogenesis (Adams 1992*b*). Development of the manus in *M. lucifugus* begins in similar fashion to that of other mammals (Fig. 9.7). Presence of limb-buds define stage 1 in which no cartilage model (chondrogenesis) is apparent in whole-mount embryos (Adams 1992*b*). As the handplate of the forelimb forms, chondrogenic elements (stage 2) of the humerus, radius, ulna, and five metacarpal elements appear (stage 3, Fig. 9.7A). Stages 4 (Fig. 9.7B) and 5 are marked by elongation of the limb-bud and the external appearance of a bend approximating the position of the elbow joint. Ossification of the long bones is sequenced proximodistally with the humerus, radius, and ulna showing diaphyseal (primary) centers of ossification foremost. Stage 6 (Fig. 9.7C) is marked by relative elongation of the forearm, metacarpus, and digits. The metacarpus and phalanges begin to ossify in early and late stage 6 respectively and to grow and ossify through stage 7 (Fig. 9.7D).

Growth of the wing membrane, or patagium, parallels growth of the digits increasing in size (Fig. 9.8). Adams (1998) gives evidence that length increases of the bony elements of the wing are strongly influenced by growth of the patagium throughout flight ontogeny, rather than vice versa. Unfortunately, because data presented herein were gathered from specimens that were fluid-preserved (which deforms soft tissues), analyses of shape changes in the wing membranes within and across developmental series was seriously confounded. However, some general similarities and differences in prenatal wing development between mega- and microchiropterans can be advanced. In both suborders, the patagium stretches between all digits except digits I and II. In both mega- and microchiropterans, claw formation begins prenatally. In the Microchiroptera, however, only digit I is clawed, whereas in Megachiroptera, digits I and II are clawed. Curiously, prenatal morphogenesis of the claw of digits I and II occurs simultaneously and is correlated with loss of the patagial coverings of the distal tips of digits I and II.

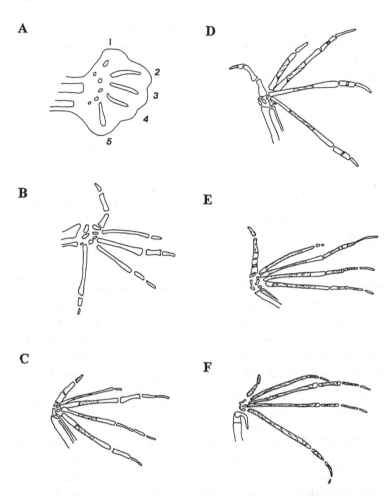

Figure 9.7. Growth of the hand of *Myotis lucifugus*. A) First chondrogenesis within the handplate of the limb-bud. B)–E) Prenatal growth. F) Juvenile. Drawings not to scale.

Postnatal growth of the wing

Ward (1904) provided one of the earliest studies to quantify postnatal growth in bats. He measured relative forelimb growth and proportional changes in body dimensions for several species in the families Mormoopidae, Vespertilionidae, Phyllostomatidae, and Molossidae. Since then, postnatal growth for many other species also has been described: *Nycticeius humeralis* (Jones 1967); *Nyctalus noctula*, *Pipistrellus pipistrellus*, and *Eptesicus serotinus* (Kleiman 1969); *Myotis thysanodes*

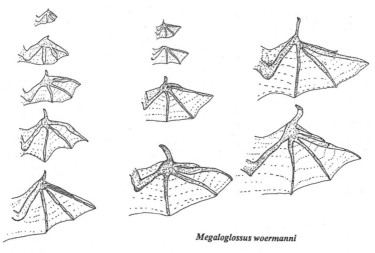

Megaloglossus woermanni

Artibeus jamaicensis

Figure 9.8. Camera lucida drawings from preserved specimens illustrating prenatal growth of the digits and patagium for *Artibeus jamaicensis* and *Megaloglossus woermanni*. Drawings not to scale.

(O'Farrell & Studier 1973) and *M. lucifugus* (O'Farrell & Studier 1973; Buchler 1980; Kunz & Anthony 1982; Powers *et al.* 1991; Adams 1992*a*, *b*, 1996, 1997, 1998; Stern & Kunz 1998; Hoying & Kunz 1998; Kunz & Hood *in press*); *Eptesicus fuscus* (Kunz 1974); *Lasiurus borealis* (Stangl *et al.* 1996); *Rhinolophus ferrumequinum* (Hughes *et al.* 1987); *Pipistrellus pipistrellus* (Hughes *et al.* 1995); *Miniopterus schreibersii* (Serra-Cobo 1989); *Myotis myotis* (Joller 1977); *Pipistrellus mimus* (Isaac & Marimuthu 1997).

Patterns of ossification in the wing bones for known-aged juvenile *M. lucifugus* (Table 9.4) show that although primary ossification centers of the radius and ulna begin to develop during the prenatal period, ossification of secondary centers do not arise until 16 days postnatal age. For the carpus, ossification does not begin until 26 days postnatal age. Full ossification of the carpus corresponds with approximate age of first flight (24 days postnatal age on average) in this species (Powers *et al.* 1991; Adams 1996). Postnatal development of the patagium in *M. lucifugus* involves large-scale increases in surface area; however, there are no significant changes in wing shape relative to area (Adams 1998). Presumably, the functional integrity of the growing wing is not compromised during postnatal morphogenesis because one parameter (wing shape) is held stable while, simultaneously, another (wing area) is changing markedly (Adams 1998).

Table 9.4. *Ossification patterns for primary (1°) and secondary (2°) centers for the forearm bones and carpus in known-age juvenile* Myotis lucifugus

Age	Radius 1°	Radius 2°	Ulna 1°	Ulna 2°	Carpus
2 hours	yes	no	yes	no	no
3 days	yes	no	yes	no	no
5 days	yes	no	yes	no	no
16 days	yes	yes	yes	yes	lunar
					trapezium
					magnum
24 days	yes	yes	yes	yes	scaphoid
					pisiform
					unciform
					cuneiform
					trapezoid

DEVELOPMENTAL ECOLOGY OF FLIGHT

Onset of flight

Interpretations of postnatal development of the wing in bats are most compelling when couched in terms of their ecological and evolutionary significance (for review see Kunz & Hood *in press*). Powers *et al.* (1991), working with *M. lucifugus*, illustrate how juveniles exhibit 'flopping behavior' (falls without wing flapping) at 0–5 days postnatal age. 'Drop-evoked fluttering' occurs at 10 days, whereas 'flapping behavior' begins at 17 days. 'Flight' is not achieved until 24 days postnatal age on average. In addition, *M. lucifugus* exhibits mature flight muscles capable of delivering adult-level flapping rate prior to017 days postnatal, indicating that neuromuscular control becomes established before the onset of flight. In Chapter 11 (this volume), Hermanson notes that the pectoralis muscles of *M. lucifugus* contain the myosin isoforms necessary to produce adult-level contractile activity as early as 20 days before the onset of flight.

Dimensions of the juvenile wing

Aspect ratio (=wing span^2/wing area) and wing loading (=body weight/wing area) are important and predictive parameters defining flight ability in adult bats (Aldridge 1986; Aldridge & Rautenbach 1987; Norberg 1994). Although aspect ratio is typically described as a measure of 'wing shape' (Norberg 1990), it is highly inaccurate in this regard because it is not a measure of 'size-free' shape (Adams 1996, 1998). Measurements of aspect ratio do provide, however, a criterion of wing morphology known to correlate with flight style and habitat utilization in adults (Norberg 1990, 1994). Less obvious is how changes in aspect ratio and wing loading affect flight ability during ontogeny. General growth of the wing has been described in several bat species. For juvenile evening bats, *Nycticeius humeralis*, accelerated growth of the forearm and digit V (Jones 1967) occurs between days 0 and 35, but levels off thereafter. The wings of horseshoe bats, *Rhinolophus ferrumequinum*, show the typical pattern of increasing area and span between 0 and 30 days, this trend persisting throughout the time of first flight and sustained flight (Hughes *et al.* 1987). Similar data are available for *Phyllostomus hastatus* (Stern *et al.* 1997). Little brown bats also show the typical pattern of accelerated growth in wing span and wing area (Powers *et al.* 1991) between 0 and 30 days postnatal age as they proceed from flopping behavior to sustained flight (Fig. 9.9). It appears a common trait among insectivorous bats that large changes in body mass (Kunz *et al.* 1998) and wing proportions persist throughout the learning-to-fly period. For example, in *M. lucifugus*, aspect ratio and flapping rate increase (Fig. 9.9A, C respectively), whereas wing loading decreases (Fig. 9.9B) as juveniles age.

What also decreases dramatically as young bats take to the wing is survivorship (Humphrey & Cope 1976). Awkward flying juveniles often do not make it back to the maternity roost, falling victim to the myriad dangers facing newly volant young (Adams & Pedersen 1994). Survivorship in juveniles is clearly influenced, among other attributes, by maturation of the wing. Postnatal growth of the wing is correlated with changes in resource use (i.e., niche dimensions associated with habitat utilization and diet) that change throughout flight ontogeny (Adams 1996, 1997). In the next section I discuss how changing niche dimensions associated with juvenile ontogeny (the ontogenetic niche) have significant effects on population and community dynamics in animals as well as in bats (Adams 1996, 1997).

Defining the ontogenetic niche

Resource partitioning by age-class, reflecting differences in morphology between juvenile and adult animals, is important in structuring populations, assemblages, and communities (Polis 1984). Patterns of resource use relative to changes in age of a juvenile from birth to maximum size define its *ontogenetic niche* (Werner & Gilliam 1984). The

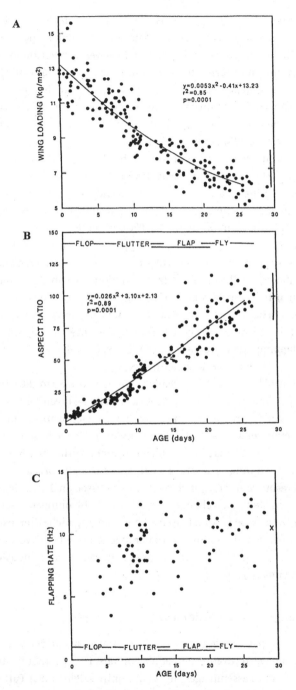

Figure 9.9. Relationship between A) wing loading, B) aspect ratio, and C) flapping rate and age in known-age juvenile *Myotis lucifugus*. (From Powers *et al.* 1991.)

ontogenetic component of the niche refers to the age/size differences of niche breadth that potentially 1) reduce competition among age groups within a population and 2) increase the overall niche breadth of a species (Christiansen & Fenchel 1977; Polis 1984, 1988, 1991; Adams 1996). Age-dependent segregation may reflect behavioral shifts, such as differential use of habitats or food resources between juveniles and adults, as well as among juveniles of different ages. Segregation among age-classes may also be an inherent consequence of disparate morphologies across age groups (Tschumy 1982; Polis 1984). In most animals, juveniles are smaller than their parents (Ellstrand 1983), and it is size itself that physically restricts use of certain resources from certain age groups. Because most studies on developmental ecology (ontoecology) are on animals that undergo complete metamorphosis and, thus, exhibit dramatic and obvious morphogenetic states, relating specific developmental characters with resource use is readily achievable (Snodgrass 1954; Martof & Scott 1957; Anderson 1967; Anderson & Graham 1967; Chapman 1969; Bell 1975; Marangio 1975; Wassersug & Sperry 1977; Wilbur 1980; Fox 1984; Duellman & Trueb 1986). However, ontoecological shifts are documented for nonmetamorphic animals as well, including holometabolous insects (With 1994), arachnids (Polis 1984, 1988), fish (Stoner 1980; Gilliam 1982; Stoner & Livingston 1984; Werner & Hall 1984), reptiles (Pough 1973, 1978; Ballinger *et al.* 1977; Schoener 1977; Kjaergard 1981; Voris & Moffit 1981; Mushinisky *et al.* 1982; Seigel *et al.* 1987; Congdon *et al.* 1992), birds (Skutch 1945; Van Tyne 1951; Stewart 1956; Kalmbach 1958; Welty 1979), and mammals (Van Horne 1982; Adams 1996).

Coppinger & Smith (1990) developed a model to illustrate changes in juvenile behavior throughout postnatal ontogeny (Fig. 9.10). They suggest that the attributes of juveniles are not simply the precursors of adult traits, but rather, that juvenile behaviors manifest their own discernable adaptive values. As juvenile behavior wanes and adult behavior waxes, juveniles progress from maternal dependency to independence (Fig. 9.10). Indeed, it is this transitional period from dependency to independence that has considerable evolutionary significance (Coppinger & Smith 1990). As juveniles develop, their physiological and behavioral needs, as well as their morphology, are adaptive to specific ages (Bateson 1987). Mammalian ontogeny, therefore, is anything but linear. In fact, many aspects of juvenile development are not precursors of adult behavior and morphology, but instead are age-specific adaptations with their own evolutionary histories and trajectories.

For bats, wings are an adaptation shared by neonate, juvenile, and adult alike. However, it is juveniles that must learn to fly under the conditions of changing wing area, and an argument can be made that the

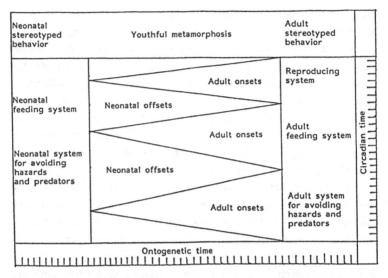

Figure 9.10. Model describing the timing of offset of neonatal versus onset of adult behaviors in mammals throughout ontogeny. (From Coppinger & Smith 1990.)

juvenile stage is when selection is most intense (Humphrey & Cope 1976; Adams 1997). Interest in interpreting wing ontogeny relative to flight ability and resource use has led to efforts to integrate development, functional morphology, behavior, and ecology in bats (see Hermanson's and also Jones's chapters, this volume). Powers et al. (1992) explored the early stages of juvenile flight in M. lucifugus and correlated changes in morphology with flight ability and estimated age (Kunz & Anthony 1982). Young pups (1–9.8 days ± 4.9 days postnatal age) 'flopped' from the launching pad without exhibiting flapping behavior and either kept their wings close to the body or extended them during free-fall. Spontaneous evoked 'fluttering' behavior was first observed at 9.8 ± 4.2 days of age, whereas some horizontal 'flapping' flight was achieved at 17.3 ± 4.9 days of age. Juveniles were capable of flying about the room at 23.8 ± 4.1 days of age.

Once flight is achieved by juvenile M. lucifugus and they enter the foraging arena, interspecific interactions among conspecifics molds, in an apparently important way, population dynamics. For example, by age-grouping juvenile M. lucifugus based upon the amount of cartilage present in the distal epiphysis of the third metacarpal, Buchler (1976, 1980) calculated the ratio of mean flight time to percentage track time in flight and found them significantly correlated. Young bats flew reasonably well, but

Table 9.5. *Percentages of captures per clutter index of class A (CAA) and class B adults (CBA) and juveniles (J) of* Myotis lucifugus *in 1990 versus 1991*

Year	Clutter index[a]	Class A adults	Class B adults	Juveniles
1990		$n = 88$	$n = 73$	$n = 53$
	0	4.5	8.2	43.4
	1	59.1	4.1	20.8
	2	0	9.6	0
	3	30.7	21.9	11.3
	4	2.3	23.3	15.1
	5	3.4	32.9	9.4
1991		$n = 71$	$n = 57$	$n = 45$
	0	8.1	13.1	57.7
	1	40.2	35.2	24.1
	2	0	4.0	0
	3	14.3	17.1	6.7
	4	20.1	21.4	4.9
	5	17.3	9.2	6.6

Note: [a] Clutter index grades are: CI-o = open habitat > 3 m above ground, 1 = open habitat within 3m of ground, 2 = semi-clutter habitat over vegetation 2–3 m in height, 3 = semi-clutter, canopy present, no undergrowth, flight paths > 3 m in diameter, 4 = heavy clutter, canopy present, undergrowth present, flight paths between 1 and 3 m in diameter, 5 = heavy clutter, canopy present, undergrowth present, flight path < 1 m in diameter.
Source: After Adams (1997).

were slow and made wide, gentle turns while avoiding cluttered habitats and other bats, whereas adults were more maneuverable and used more cluttered habitats within which to forage. In addition, emergence times differed according to relative age with juveniles typically leaving the roost later in the evening than did adults.

Juvenile *M. lucifugus* also exhibit differential resource utilization, not only relative to adults, but also to other juveniles differing in age. The degree of niche separation between juveniles and adults is related strongly to population density, whereas niche segregation among juveniles is a construct of wing ontogeny and flight ability (Adams 1996, 1997). At a Wyoming maternity colony of *M. lucifugus*, Adams (1996, 1997), using well-defined microhabitats (Table 9.5), found that at high population density, adults dispersed from their preferred feeding areas once juveniles of the colony became volant and entered the foraging arena. In May and June, adults foraged predominately in open areas, within 3 m of the ground. However, once juveniles became volant in July and began to forage in the open areas, adults shifted their foraging mostly to heavily cluttered forests (Table 9.5). In 1991, when only about one half of the adult population returned to the site, adults did not change foraging patterns throughout the summer and, instead, continued to forage close to the ground in open areas when juveniles became volant. Independent of year and population density, newly volant juveniles foraged predominately in the least cluttered microhabitat >3 m above the ground in the open (Table 9.5). In 1990 there were significant differences among class A adults (CAA), class B adults (CBA) and juveniles (J) in microhabitat use determined by χ^2 analysis for general overlap (GOadj; Ludwig & Reynolds 1988): GOadj = 0.616, V = 119.406, $p < 0.001$. Specific overlap (SO), a pairwise comparison among groups, shows that CAA and CBA overlap more with J than they do with each other ($SO_{CAA:J}$ = 50%; $SO_{CBA:J}$ = 49.6%; $SO_{CAA:CBA}$ = 37%) indicating that adults shift their foraging pattern after juveniles are present in the foraging arena. In 1991, approximately 20% more overlap occurred between CBA and J than in 1990 and pairwise comparison indicates that CAA and CBA overlap more with each other than either does with J: GOadj = 0.745; $SO_{CAA:CBA}$ = 62%; $SO_{CAA:J}$ = 48%; $SO_{CBA:J}$ = 58%. Curiously, dietary differences between adults and juveniles under low population density were not significant and all age groups consumed a large proportion of Diptera (Table 9.6). Under high population density, and greater sympatry between juveniles and adults while foraging, a significant difference in diet was detected (Table 9.6), although both age-groups showed an increased consumption of Hemiptera (Corixidae) in 1991 (Adams 1993). For 1990, high general overlap among groups CAA, CBA and J occurred, indicating no significant differences among groups (GOadj = 0.9400, $p > 0.05$). In 1991, there were significant differences among groups (GOadj = 0.804, $p < 0.001$) and specific overlap pairwise test showed that the diet of juveniles overlapped most with CBA ($SO_{J:CBA}$ = 80.6%; $SO_{J:CAA}$ = 51.4%). Interestingly, consumption of primary food resources such as trichopterans differed across age groups depending upon degree of adult/juvenile habitat overlap. When overlap in foraging habitat was low (high population density), adult consumption of trichopterans dropped from 35.4% to 8.3% of diet when juveniles became volant. Simultaneously, trichopterans contributed 25% of the overall food intake of juveniles. When overlap in foraging habitat was higher between juveniles and adults (low population density), adults continued to consume a high percentage of trichopterans (20% of diet) after juveniles became volant, whereas juveniles' diets consisted of a mere 5.7% trichopterans. This represents an almost 23% decrease in consumption of trichopterans

Table 9.6. *Percentages (calculated as percent frequency of individuals) of orders of insects occurring in the diet of juvenile and adult* Myotis lucifugus *in 1990 and 1991*

	1990			1991		
Insects	Juveniles ($n = 12$)	Class A adults ($n = 17$)	Class B adults ($n = 8$)	Juveniles ($n = 17$)	Class A adults ($n = 13$)	Class B adults ($n = 23$)
Coleoptera	7.7	8.3	11.1	13.2	12.1	10.9
Diptera	32.7	35.4	38.1	24.5	21.1	23.4
Lepidoptera	9.6	16.7	22.2	22.6	5.5	7.8
Trichoptera	25.1	35.4	8.3	5.7	50.5	20.3
Hymenoptera	1.9	0.0	2.8	5.7	3.2	1.6
Homoptera	1.9	0.0	5.7	0.0	1.1	1.6
Neuroptera	1.9	0.0	0.0	0.0	0.0	0.0
Ephemeroptera	11.5	0.0	2.8	1.9	0.0	0.0
Hemiptera	7.7	4.2	8.3	26.4	6.6	34.4

Source: After Adams (1997).

by juveniles when foraging sympatry with adults was high (Adams 1997). Although this study was rather short in duration, and did not provide enough data from independent insect sampling for statistically determining if fluctuations in insect populations at the site contributed to the dietary patterns of M. lucifugus, the reporting of significant differences between the diets of adults and juveniles of other species, such as hoary bats, Lasiurus cinereus (Rolseth et al. 1994), support the assertion of age-specific resource partitioning. Harptrap captures at the exit holes of the Wyoming colony showed that, although a few adults always left the colony first, juveniles follow them and emerged simultaneously with the remaining adults. This differs from data gathered by Buchler (1980) who showed age-specific temporal differences in emergence at his study colonies. Kunz & Anthony (1996) illustrated that age of volant juveniles correlated strongly with emergence time at maternity sites in New Hampshire. Very young juveniles left the roost later than did older juveniles and adults, who together had similar emergence times. Whether or not the Wyoming colony adheres to the Kunz & Anthony (1996) age-specific emergence model remains to be tested.

Among juveniles foraging at Fort Laramie, age-specific foraging patterns could be discerned, but these patterns did not correlate with population density. Juvenile foraging patterns, however, did correlate with

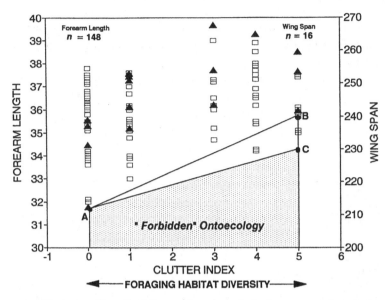

Figure 9.11. Plot of 148 forearm lengths and 16 wing spans against clutter indices as a measure of foraging habitat diversity in juvenile *Myotis lucifugus*. Individuals with shorter forearms are found only to utilize microhabitats of least clutter (clutter indices 1 and 2), whereas those with longer forearms forage in both open and more cluttered microhabitats (clutter indices 3 through 5), thereby increasing microhabitat diversity with increasing size and age. 'Forbidden' ontoecology refers to incongruence between morphospace and ecospace in that juveniles with less developed wings are restricted to forage in less cluttered habitats due to lack of maneuverability, the result of mechanical constraints. The stippled area is capped by a line drawn between points A and C and indicates where there is incongruence between forearm length and foraging ecology. The line drawn between A and B caps a larger area illustrating where wing span is incongruent with foraging ecology, both due to the physical constraints associated with underdeveloped wings that handicaps flight ability in more obstacle-laden microhabitats.

ontogenetic changes in wing area (Adams 1996). Indeed, juveniles show shifts in resource use throughout flight ontogeny that relate strongly to limb morphogenesis as well as changes in wing area. For example, as forearm length increases, so does the diversity of microhabitats used by juvenile *M. lucifugus* (Fig. 9.11). Juveniles with forearms measuring 32 mm or less in length were not captured flying in any microhabitat except for the least cluttered (clutter index 0). Growth increases of a mere 2.5 mm in forearm length is associated with greater habitat utilization. The digits

Table 9.7. *Summary of discriminant function analysis of bony elements making up wing length*

Discriminant function	Eigenvalue	Relative %	Canonical correlation
1	1.334	89.99	0.756
2	0.150	10.11	0.361

	Standard discriminant function coefficient	
Element lengths	1	2
Forearm	−0.052	−0.552
Third metacarpal	0.605	0.521
Proximal phalanx	−0.667	−0.141
Middle phalanx	0.936	−0.462
Distal phalanx	−0.0006	0.715

Source: After Adams (1996).

and the patagium, of course, also increase in size with growth, and correlate with increased habitat diversity. Factor analysis on individual bony elements of digit III shows that relative growth of the proximal phalanx (standard discriminant function coefficient = 0.936, Table 9.7) is the most discriminatory element of the handwing during flight ontogeny and corresponds with shifts in habitat use by juveniles. Pairwise analysis indicates no significant differences between early- and mid-July juveniles (SO = 93%, $p > 0.05$). There was, however, significant difference between mid- and late-July juveniles (SO = 75.3%, $p < 0.05$) and highest significant difference between early- and late-July juveniles (SO = 70.6%, $p < 0.001$). Discriminant function analysis groups open-area juveniles and differentiates them from cluttered-area juveniles based upon the relative growth of the middle phalanx of digit III (Fig. 9.12).

Juvenile *M. lucifugus* with smaller wings consume different diets than do larger juveniles as well as adults (Table 9.8). Juveniles captured in early July consume more Diptera and Hemiptera (67.5% of diet) than consumed by individuals in mid- (56.3%) or late July (22.8%). Lepidopterans also decrease in consumption relative to age (35% early July–2.8% late July). Most significant increased consumption by juveniles as they grow is in trichopterans (12.5% to 71.4%). Late July juveniles eat a dietary diversity not dissimilar to that observed in Class A adults captured before juveniles become old enough to fly. Although it is difficult to think of poor flight skills as adaptive, at the populational level differential flight skills among different-age juveniles serves to help alleviate the potential for

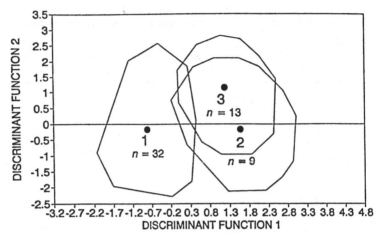

Figure 9.12. Plot representing discriminant function analysis of bony elements contributing to length of wing (forearm, radius, metacarpal III, and phalanges of digit III) of *Myotis lucifugus*. Open-area juveniles are easily discriminated from those captured in either semi- or heavy-clutter habitats based upon the lengths of the long bones of digit III. Discriminant function 1 explains 89.99% of sample variation (Eigenvalue = 1.334) and discrimination among groups is driven predominately by length of the second phalanx of digit III exhibiting a standard discriminant function coefficient of 0.936. 1 = open habitat, 2 = semi-clutter habitat, 3 = heavy-clutter habitat. Dots represent centroids of dispersion clouds for group data points, lines trace outer perimeter of data clouds for each group. (From Adams 1996.)

Table 9.8. *Percentages from pooled data for 1990 and 1991 of insect orders consumed by juvenile* Myotis lucifugus *in July.*

	Diptera	Lepidoptera	Hemiptera	Trichoptera
July 1–10	32.5	20.0	35.0	12.5
July 11–20	37.6	18.7	18.7	25.0
July 21–31	20.0	5.7	2.8	71.4

Source: After Adams (1996).

competition for food and foraging space among age groups. Flight ability tends to segregate age groups from competing for similar resources since some individuals are physically restricted from consuming some foods and flying in some microhabitats. For *M. lucifugus*, segregation was more intense under higher population density, a fact that gives credence to the shifting ontogenetic niche hypothesis because one would predict a

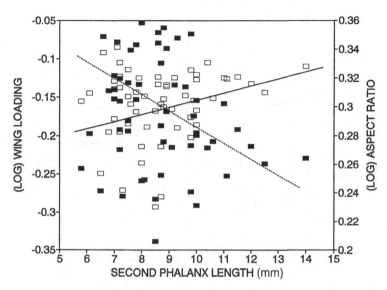

Figure 9.13. Plot of second phalanx length of the third digit against wing loading (solid rectangles, $r = -0.21$) and aspect ratio (open rectangles, $r = 0.24$) for 54 juvenile *Myotis lucifugus*. Solid line demarcates the trend in aspect ratio and the dashed line demarcates the trend in wing loading relative to growth of the second phalanx (most discriminating element among groups, see Figure 9.12). As flight begins, aspect ratio and wing loading are inversely proportional as compared with these same parameters in adults, but converge and reverse relationship at approximately the mid-point of flight ontogeny.

greater effect when resources become more limited due to greater numbers of individuals.

Juvenile bats adhere to similar pattern of ontoecology from year to year regardless of population density. The gradual change from juvenile to adult morphology and behavior in bats requires proportional changes of wing growth versus body growth, calculated as aspect ratio and wing loading. As the wing grows, wing loading decreases and aspect ratio increases (Fig. 9.13) and as these parameters change in concert with growth and development, juveniles shift and add more microhabitats to their foraging repertoire (Fig. 9.14). In this stage, on average, young bats cannot fly in the second most cluttered microhabitat (clutter index 1) until parameters for wing loading and aspect ratio have converged which occurs on average at wing length 88.5 mm (Fig. 9.14). Ability to forage in forested areas (clutter indices 3–5) occurs after wings have grown more than 88.5 mm in length and the relationship of wing loading and aspect

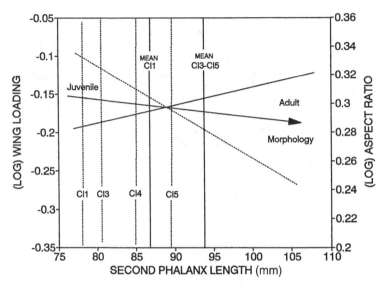

Figure 9.14. Plot combining data plotted in Fig. 9.13 with changes in habitat utilization among juvenile *Myotis lucifugus*. Dotted vertical lines represent minimum length of second phalanx of digit III captured in each microhabitat (defined as clutter indices, CI1–CI5), whereas solid vertical lines demarcate mean phalanx length captured in each microhabitat. Although juveniles with short second phalanges and becoming volant before wing loading and aspect ratio have reached adult proportions are occasionally captured in semi- and heavy cluttered microhabitats (CI3–4), on average only juveniles representing a proportionateley adult-like relationship between aspect ratio and wing loading are captured in microhabitats represented by CI3–CI5.

ratio have become inversely proportional to that observed in juveniles with smaller wings relative to body size (Adams 1996). The capture of a few juveniles flying in habitats predictably too difficult according to measures of wing loading and aspect ratio (<mean), suggest that young bats may attempt to enter more cluttered habitats to test their flight ability, but likely return to less cluttered habitats fairly quickly. Shifts in niche space during flight ontogeny increase the niche dimensions of young bats and provide a greater resource base for the population overall. In addition, because smaller juveniles (forearm length<32 mm, wing loading <0.17, aspect ratio<0.29) cannot forage in cluttered habitats effectively, nor eat a comparable assortment of insects, they are segregated in resource use from middle-age juveniles (forearm length>32 mm, <33.5 mm; wing loading>0.17, <0.20; aspect ratio>0.29, <0.30) that are segregated from older juveniles (forearm length>34.0; wing loading>0.20, aspect ratio>0.30) and also adults. Resources, then, are partitioned according to age and owing to physical constraints accompanying wing growth, thereby alleviating much of the potential for competition among juveniles. Effective measures to alleviate potential intraspecific competition among conspecifics is important because older juveniles likely would outcompete younger juveniles for food and foraging space, leading to problems in age-recruitment (Tschumy 1982). Niche partitioning among age-/size-cohorts allows for groups of bats to grow, develop flight skills, and forage together, alleviating competitive interactions among different-age juveniles. (Actually, this is not unlike the philosophical underpinnings in designing children's education systems where students proceed through grades in cohorts of similar age and size, thereby minimizing competition among different age groups in the classroom, although age-integrated interactions may certainly occur on the playground!)

Using juvenile wing morphology to predict habitat use differs fundamentally from the predictions associated with interspecific patterns of resource use as correlated with adult wing morphology (Norberg & Rayner 1987). For juveniles, shorter, underdeveloped wings command a greater turning radius, whereas the opposite is true for adults. The functional differences in adult versus juvenile wings during flight resides in the relatively rapid changes in wing area during growth that alters the relationship between wing loading and aspect ratio as shown in my ontoecological model (Fig. 9.14). In a sense, newly volant young are flying 'upside-down,' because the relationship between wing loading and aspect ratio is inversely proportional to that of adults (Adams 1996). Although the wing area of newly volant young is sufficient to create the necessary lift required for flight, it apparently is insufficient in providing adept turning control and overall maneuverability. Decreases in wing loading and increases in aspect ratio during flight ontogeny correspond with stepwise changes in foraging patterns that lead to increased utilization of cluttered habitats and available prey (Fig. 9.14).

Wings, natural selection, and evolution

The use of adaptive landscapes as metaphors for how natural selection works on populations began with Sewall Wright (1931). Attempting to derive a multivariate technique for understanding the distribution of allele combinations within populations, Wright depicted 'harmonious' (adaptive) allele aggregates within populations as congregating at the peaks of a landscape. Valleys between peaks represented the positions of less adaptive allelic combinations as well as delineating areas on the topographic map where selection works to sift out less fit forms (Eldridge 1989). Although in its original manifestation the model was portrayed as being static, Wright eventually generated his shifting-balance theory where adaptive peaks were dynamic and movable within the landscape. The Wrightean landscape was co-opted first by Dobzhansky (1937), who visualized peaks as representing highly adapted populations of species in a field of potential gene combinations, and later by Simpson (1944) to illustrate horse evolution based upon morphological character divergence across geological time. Simpson replaced Wright's allelic-combination theory with adaptive morphology as the playground of evolutionary change and by doing so adjusted the resolution of the landscape into macroevolutionary representation. Simpson (1944, 1953) used the landscape metaphor to develop his theory of quantum evolution that characterized causal pathways enabling populations of species to move from one peak to another. Recasting landscape theory into the realm of developmental biology and morphogenesis, Waddington (1940, 1957) depicted a ball rolling across an 'epigenetic landscape' to illustrate metaphorically the ontogenetic fate (canalization) of a zygote during morphogenesis. His model depicted totipotent cells becoming channeled as they travel along certain irreversible morphogenetic pathways with chance playing less and less a role in ontogeny throughout developmental time.

If ontogeny can be considered an important component in the ecology and evolution of organisms, the next logical echelon is to superimpose ontoecology upon landscape theory. In fact Werner & Gilliam (1984) suggest that the next step in understanding the fundamental role of ontogeny in the ecology of organisms should be cast as a three-dimensional approach. To do this one must integrate ontogeny, functional ecology, and evolutionary theory in a way not previously synthesized. One must build an ontoecological landscape.

THE ONTOECOLOGICAL LANDSCAPE

A model of flight ontogeny, ecology, and evolution

Pedersen (1995) built a developmental landscape model to illustrate ontogenetic migration from a rudimentary, nasal-emitting morphospace toward an oral-emitting morphospace in bats (see Chapter 6, this volume). Similarly to Waddington, Pedersen illustrated the canalization of ontogenetic patterns of derived morphologies as building upon ancestral growth trajectories. In his case, the derived head posture of nasal-emit-ting bats is the consequence of derived ontogeny that begins with a devel-opmentally ancestral head posture shared by all mammals. The adult bats' morphology is the consequence of divergent developmental events that appear as discrete peaks on an adaptive landscape, strongly linking divergent evolution in bats with shifts in developmental timing.

However, can a generalized model be drawn to illustrate the rela-tionships among morphogenesis, not only with evolutionary patterns, but also involving ecological ones in order to pinpoint times of intense selective pressure during development? If changes in wing loading and aspect ratio correlate with shifts in niche dimensions, then times of intense selection pressure should be implicit and recognizable in the model. In all studies of wing ontogeny in bats, wing loading decreases and aspect ratio increases as the individual grows. When changes in micro-habitat use are tightly coupled with wing ontogeny as shown for *M. lucifugus*, then the challenges involved with increasing niche breadth necessary for juvenile survival may result in selection for certain wing dimensions, and/or rate of wing growth, at specific times during ontog-eny. By casting a three-dimensional model of the dynamic relationship between wing loading, aspect ratio, and their relevance to age-specific niche partitioning during flight ontogeny, evolutionary interpretations based upon Darwinian selection on natural populations should be revealed. Indeed, Werner & Gilliam (1984) state: 'A cohort of such organ-isms can be viewed as navigating a landscape of ecological niches, and one of the challenging problems is to develop theory that will help us predict the course a species will take and the ecological consequences.'

Following through on this idea accords a three-dimensional land-scape model that investigates the synthetic relationship between the con-tributions of development and ecology to the evolutionary patterns of wing form in bats (Fig. 9.15). In a metaphorical sense, the peaks of this hypothetical model represent points at which flight is adaptive enough to allow for survival in that particular microhabitat. Those that are not com-petent fliers in a particular microhabitat are selected out and these areas are demarcated by valleys in the landscape. Hypothetically, valleys impli-cate critical stages of juvenile morphogenesis, that is, periods of highest selective pressures associated with making the transitions towards adult-hood. This balancing act between developmental rate, flight ability, habitat utilization, and dietary diversity all contribute to survivorship ability, which is severely challenged early in the ontogeny of flight and, therefore, the valleys, representing most intense selective pressures, are deepest in the landscape model initially.

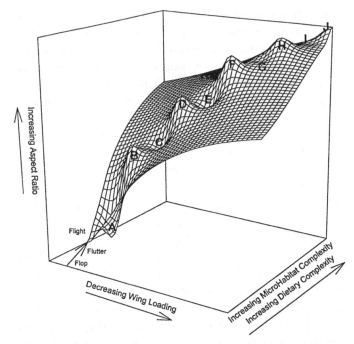

Figure 9.15. Hypothetical model illustrating a three-dimensional adaptive landscape for wing morphology transition relative to foraging ecology in the form of habitat and dietary associations. Beginning with the onset of flopping behavior proceeding to fluttering flight, newly volant young bats venture from a nonvolant condition through various volant life stages. The first few days of flight are highly dangerous and the likelihood of death is high. See text for full explanation.

Beginning with the onset of flight (Powers *et al.* 1991), juveniles exhibiting flopping and fluttering flight become newly volant young that venture away from the maternity roost. Juveniles manifest high mortality within the first few days of volancy (Humphrey & Cope 1976). Point A on the landscape model represents this potential pitfall of survivorship as a deep hole in the mesh within which many young bats will fall and never recover. Some individuals, however, will survive this initial ordeal and begin climbing the first adaptive peak of the landscape, the pinnacle (B) of which represents competent flight in open areas and the ability to capture enough insects for survival and further growth and development. As more flight competence is gained as a result of changing wing loading and aspect ratio during growth, young bats begin making the next transition to flying in semi-clutter habitats and capturing more evasive insects, thereby increasing both foraging habitat diversity and dietary diversity. The climb to peak (D) requires the traversing of another deep hole in the landscape (C) which selects out more individuals from the population. Survivors of this transition find the next series of adaptive peaks to require even greater flight ability leading to increased habitat utilization and increased dietary complexity. The possibility of survivorship increases as individuals progress up and across the landscape and, therefore, valleys are less deep (G, I) and peaks less tall (F, H, J) as one traverses the model. Decreases in wing loading accounts for the left to right shift of 'successful' individuals as they ascend the landscape, the tilt of which is driven by increases in aspect ratio. The exact numbers of metaphorical adaptive peaks required to account for the ontoecology of any given species will vary depending upon the mosaic nature of the habitat and diversity of the insect fauna available. This hypothetical model is intended to illustrate the integration of development, flight ability, foraging ecology, and evolution of wing form for aerial insectivores based upon data gathered from many studies of vespertilionid bats. Perhaps different models will need generating for understanding bat species having significantly different natural histories and contrasting ecologies.

Do data fit the model?

For little brown bats (*Myotis lucifugus*), increase in use of foraging habitats has been documented and correlated with increasing aspect ratio and decreasing wing loading during juvenile morphogenesis (Buchler 1980; Adams 1996, 1997). A landscape model built from raw data ($n = 56$) collected on age-specific ontoecology in *M. lucifugus* (Fig. 9.16A) supports and tracks the hypothetical model (Fig. 9.15) in several important ways. (Because older juveniles are capable of foraging in all microhabitats, in order to represent the individuals restricted to flying in lesser-cluttered microhabitats or less complex diets, peaks in the landscape represent one standard deviation below the mean length of wing of individuals captured in that particular microhabitat.) 1) As wing loading and aspect ratio change due to wing growth, juveniles are distributed along the landscape and generate peaks that correspond to microhabitat affiliations. 2) The formation of valleys correspond with transitions between open, semi-clutter, and heavy-clutter habitats. 3) The peaks shift from left to right along the landscape. 4) The steepest 'climb' for juveniles is from first volancy to competent flight (A to B) in open areas and is followed by a deep depression in the mesh. This initial steepness followed by a deep depression corresponds with harsh selective pressures on the population as predicted by the hypothetical model. In addition, the rigor of the landscape lessens as it ascends, and closely manifests the Type III survivorship curve known for this species (Humphrey & Cope 1976).

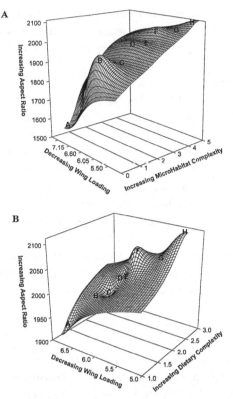

Figure 9.16. Raw data on aspect ratio and wing loading plotted against A) microhabitat complexity and B) dietary complexity. See text for full explanation.

Dietary diversity measured at the ordinal level of insect classification also adheres to the generalized landscape model (Fig. 9.16B) with transition along the rising landscape gradient consisting of hills and valleys. Valleys are steepest at the beginning of the landscape which illustrates high selective pressures on the population relative to wing morphology and ecology. As aspect ratio increases and wing loading decreases, dietary diversity increases presumably due to increased flight skill and ability to capture a diversity of insect prey.

That valleys in the landscape depict selection pressures is supported by what is known of juvenile survivorship in bats. Before juveniles become volant, survivorship is high, with mortality averaging 3% (Tuttle & Stevenson 1982; Ransome 1990). Early flight and weaning are associated with weight loss in juveniles (Tuttle & Stevenson 1982; Kunz 1987; Kunz & Stern 1995), most likely due to the stresses imposed by perfecting flight and echolocation skills as well as learning to catch evasive insects.

Survivorship during early flight is low (Humphrey & Cope 1976), but increases as flight ability betters with age. The initial high mortality and then increase in survivorship with age is illustrated by deeper valleys and steeper peaks present at the foot of the landscape (representing early juvenile volancy) and then attenuating as individuals climb, thereby acquiring developmental and ecological maturity. Data from *M. lucifugus* follow this prediction for both habitat use and dietary complexity as niche dimensions shift with ontogeny. As above mentioned, selection works predominately at transition periods where juveniles, previously constrained by immature wings, increase foraging niche space by attempting to fly in more challenging habitats and chasing more evasive insect species. The risks established at each transition phase are of two factors: 1) as juveniles attempt to forage through more complex habitats, the chances of fatal or injurious collision with inanimate objects increases, and 2) as juveniles attempt to forage on more evasive insect species to establish a 'balanced' diet, the chances of failure in securing enough food to grow and survive increases. Those that can 'traverse the ontoecological landscape' more quickly stand to gain the most energy for growth and development afforded by a wider range of resource availability, whereas those that flounder likely are engulfed by selection's grip, thereby becoming unresolvably stranded in death's valleys.

CONCLUSIONS

Wing ontogeny begins *in utero* as a 'standard issue' mammalian template, but diverges from the ancestral pattern due to differential developmental patterns (heterochrony via regional peramorphosis; McNamara 1997). Forearm, metacarpus, and finger growth for both Mega- and Microchiroptera outpace skull growth in a way not observed in insectivores nor rodents. Among bats, Microchiroptera have similar growth trajectories in both the forearm and digit III, however, in the Megachiroptera growth trajectories are variable among species, and do not necessarily track similarly. Fundamental differences between suborders of bats involves retention of the third phalanx in microchiropterans and loss of this element in digits III through V in megachiropterans thus far studied. In addition, ossification sequences differ between suborders, but are also varied among families within Microchiroptera. Working from the premise that groups most closely related should show highest similarity in developmental patterns, gathering of data on morphogenesis, above and beyond those illustrated herein, would predictably give greater insight into arguments concerning the mono- versus diphyletic origins of bats.

Flight ability and foraging ecology is tightly coupled with postnatal growth of the wings in *M. lucifugus* and, predictably, in other aerial insectivorous bats as well. Age- specific resource partitioning occurs, not only between adults and juveniles within a population, but also among juveniles of different ages, sizes, and flight abilities (Adams 1996, 1997). Shifts in niche space (microhabitat utilization and dietary complexity) are tightly coupled with changes in morphospace (i.e., wing loading and aspect ratio) as juveniles grow and develop. The dynamic continuum of differential growth and ecology of juvenile bats can be illustrated by an ontoecological landscape that depicts peaks of flying competence (relative to habitat clutter and dietary diversity) surrounded by transitional valleys where selection operates most strongly. The landscape model integrates ontogeny, ecology (ontoecology), functional morphology, and potential populational selection effects, providing a compelling scenario for the influence of juveniles on population dynamics and on the evolution of a species.

ACKNOWLEDGMENTS

I would like to thank David M. Armstrong, Thomas H. Kunz and Scott C. Pedersen for providing valuable comments on an earlier draft of this paper. I would also like to thank the American Museum of Natural History, New York, the Harrison Zoological Museum, Sevenoaks, England, and the University of Wisconsin Zoological Museum, Madison, for loaning me specimens. Funding was provided by the University of Wisconsin Research Grant and the Department of Biological Sciences at University of Wisconsin–Whitewater.

REFERENCES

Adams, R. A. (1992*a*) Comparative skeletogenesis of the forearm of the little brown bat *(Myotis lucifugus)* and the Norway rat *(Rattus norvegicus)*. *Journal of Morphology*, **214**, 251–60.

Adams, R. A. (1992*b*) Stages of development and sequence of bone formation in the little brown bat, *Myotis lucifugus. Journal of Mammalogy*, **73**,160–7.

Adams, R. A. (1993) Consumption of water boatmen (Hemiptera: Corixidae) by little brown bats, *Myotis lucifugus. Bat Research News*, **34**, 66–7.

Adams, R. A. (1996) Size-specific resource use in juvenile little brown bats, *Myotis lucifugus* (Chiroptera: Vespertilionidae): is there an ontogenetic niche? *Canadian Journal of Zoology*, **74**, 1204–10.

Adams, R. A. (1997) Onset of volancy and foraging patterns of juvenile little brown bats. *Journal of Mammalogy*, **78**, 239–46.

Adams, R. A. (1998) Evolutionary implication of developmental and functional integration in bat wings. *Journal of Zoology (London)*, **246**,165–74.

Adams, R. A. & Pedersen, S. C. (1994) Wings on their fingers. *Natural History*, **103**, 48–55.

Aldridge, H. D. J. N. (1986) Manoeuvrability and ecological segregation in the little brown bat (*Myotis lucifugus*) and yuma (*M. yumanensis*) bat (Chiroptera: Vespertilionidae). *Canadian Journal of Zoology*, **64**, 1878–82.

Aldridge, H. D. J. N. & Rautenbach, I. L. (1987) Morphology, echolocation and resource partitioning in insectivorous bats. *Journal of Animal Ecology*, **56**, 763–78.

Altringham, J. D. (1996) *Bats: Biology and Behaviour*. Oxford: Oxford University Press.

Anderson, J. D. (1967) A comparison of the life histories of costal and montane populations of *Ambystoma macrodactylum* in California. *American Midland Naturalist*, **77**, 323–55.

Anderson, J. D. & Graham, R. E. (1967) Vertical migration and stratification of larvae of *Ambistoma t. tigrinum*. *Ecology*, **52**, 1107–12.

Anthony, E. L. P. & Kunz, T. H. (1977) Feeding strategies of the little brown bat, *Myotis lucifugus*, in southern New Hampshire. *Ecology*, **58**, 775–86.

Ballinger, R. M., Newlin, M. & Newlin, S. (1977) Age-specific shifts in the diet of the crevice spiny lizard, *Sceloporus poinsetti*, in southwestern New Mexico. *American Midland Naturalist*, **97**, 482–4.

Bateson, P. (1987) Biological approaches to the study of behavioral development. *International Journal of Behavior and Development*, **10**, 1–22.

Bell, G. (1975) The diet and dentition of the smooth newt larvae, *Triturus vulgaris*. *Journal of Zoology (London)*, **176**, 411–24.

Buchler, E. R. (1976) Prey selection by *Myotis lucifugus* (Chiroptera: Vespertilionidae). *American Naturalist*, **110**, 619–28.

Buchler, E. R. (1980) The development of flight, foraging and echolocation in the little brown bat (*Myotis lucifugus*). *Behavioral Ecology and Sociobiology*, **6**, 211–18.

Chapman, R. F. (1969) *The Insects: Structure and Function*. New York: Elsevier.

Christiansen, F. & Fenchel, T. (1977) *Theories of Populations in Biological Communities*. New York: Springer-Verlag.

Congdon, J. D., Gotte, S. W. & McDiarmid, R. W. (1992) Ontogenetic changes in habitat use by juvenile turtles, *Chelydra serpentina* and *Chrysemys picta*. *Canadian Field Naturalist*, **106**, 241–8.

Coppinger, R. P. & Smith, C. K. (1990) A model for understanding the evolution of mammalian behavior. In *Current Mammalogy*, ed. H. Genoways, pp. 335–74. New York: Plenum Press.

Dalgleish, A. E. (1964) Development of the limbs of the mouse. PhD dissertation, Stanford University.

Davis, R. (1969) Growth and development of young pallid bats *Anthozous pallidus*. *Journal of Mammalogy*, **50**, 729–36.

Dobzansky, T. (1937) *Genetics and the Origin of Species*. New York: Columbia University Press.

Donaldson, H. H. (1924) *The Rat*. Philadelphia: Memoirs of the Wistar Institute of Anatomy and Biology, no. 6.

Duellman, W. E. & Trueb, L. (1986) *Biology of Amphibia*. New York: McGraw-Hill.

Eldridge, N. (1989) *Macroevolutionary Dynamics*. New York: McGraw-Hill.

Ellstrand, N. C. (1983) Why are juveniles smaller than their parents? *Evolution*, **37**, 1091–4.

Findley, J. S. (1993) *Bats: A Community Perspective*. Cambridge: Cambridge University Press.

Foresman, K. R. (1994) Comparative embryonic development of the Soricidae. *Special Publication, Carnegie Museum of Natural History*, **18**, 241–57.

Fox, H. (1984) *Amphibian Morphogenesis*. Clifton: Humana Press.

Gilliam, J. F. (1982) Habitat use and competition bottlenecks in size-structured fish populations. PhD dissertation, Michigan State University, East Lansing.

Hoying, K. M. & Kunz, T. H. (1998) Variation in size at birth and post-natal growth in th insectivorous bat *Pipstrellus subflavus* (Chiroptera: Vespertilionidae). *Journal of Zoology (London)*, **245**, 15–27.

Hughes, P. M., Ransome, R. D. & Jones, G. (1987) Aerodynamic constraints on the flight ontogeny in free-living greater horseshoe bats, *Rhinolophus ferrumequinum*. In *European Bat Research*, ed. V. Hanák, J. Horácek, & J. Gaisler, pp. 255–62. Prague: Charles University Press.

Hughes, P. M., Rayner, J. M. V. & Jones, G. (1995) Ontogeny of true flight and other aspects of growth in the bat *Pipstrellus pipstrellus*. *Journal of Zoology (London)*, **235**, 291–318.

Humphrey, P. M. & Cope, J. B. (1976) Population ecology of the little brown bat, *Myotis lucifugus*, in Indiana and north-central Kentucky. *Special Publication, American Society of Mammalogists*, **4**, 1–80.

Isaac, S. S & Marimuthu, G. (1997) Development of the wing in the Indian pygmy bat *Pipistrellus mimus*. *Journal of Bioscience*, **22**, 193–202.

Jenkins, F. A., Jr., Dial, K. P. & Goslow, G. E., Jr. (1988) A cineradiographic analysis of bird flight: the wishbone in starlings is a spring. *Science*, **242**, 1495–8.

Joller, V. H. (1977) Zur Ontogenese von *Myotis myotis* Borkhausen (verglichen mit jener von *Acomys cahirinus dimidiatus*). *Natureorschende Gesellschaft, Basel*, **86**, 87–151.

Jones, C. (1967) Growth development, and wing loading in the evening bat, *Nycticeius humeralis* (Rafinesque). *Journal of Mammalogy*, **48**, 1–19.

Kalmbach, E. R. (1958) Life histroies of North American blackbirds, orioles, tanagers and their allies. *Bulletin of the United States National Museum*, **221**, 1–212.

Kjaergard, J. (1981) A methods for examination of stomach content in live snakes and some information of the feeding habits in common viper (*Vipera berus*) in Denmark. *Natura Jutlandica*, **19**, 16–37.

Kleiman, D. G. (1969) Maternal care, growth rate and development in the noctule (*Nyctalus noctula*), pipistrelle (*Pipistrellus pipistrellus*) and serotine (*Eptesicus serotinus*). *Journal of Zoology*, **157**, 187–211.

Kunz, T. H. (1974) Reproduction, growth and mortality of the vespertilionid bat, *Eptesicus fuscus*, in Kansas. *Journal of Mammalogy*, **55**, 1–13.

Kunz, T. H. (1987) Size of bats at birth and maternal investment during pregnancy. *Symposium of the Zoological Society of London*, **57**, 79–106.

Kunz, T. H. & Anthony, E. L. P. (1982) Age estimation and postnatal growth in the bat, *Myotis lucifugus*. *Journal of Mammalogy*, **63**, 23–32.

Kunz, T. H. & Anthony, E. L. P. (1996) Variation in the timing of nightly emergence behaviour in the little brown bat, *Myotis lucifugus* (Chiroptera: Vespertilionidae). In *Contributions in Mammalogy: A Memorial Volume Honoring Dr. J. Knox Jones, Jr.*, ed. H. H. Genoways & R. J. Baker, pp. 225–35. Lubbock: Texas Tech University Press.

Kunz, T. H. & Hood, W. R. (in press) Parental care and postnatal growth in the Chiroptera. In *The Reproductive Biology of Bats*, ed. E. G. Crichton & P. H. Krutzsch. London: Academic Press.

Kunz, T. H. & Stern, A. A. (1995) Maternal investment and post-natal growth in bats. In *Ecology, Evolution and Behaviour of Bats*, ed. P. Racey & S. Swift, pp. 123–38. Oxford: Oxford University Press.

Kunz, T. H., Wrazen, J. A. & Burnett, C. D. (1998) Changes in body mass and fat reserves in prehibernating little brown bats (*Myotis lucifugus*). *Ecoscience*, **5**, 8–17.

Liem, K. F. & Wake, D. B. (1985) Morphology: current approaches and concepts. In *Functional Vertebrate Morphology*, ed. M. Hildebrand, D. M. Bramble, K. F. Leim

& D. B. Wake, pp. 366–77. Cambridge: Belknap Press of Harvard University Press.

Ludwig, J. A. & Reynolds, J. F. (1988) *Statistical Ecology: A Primer on Methods and Computing*. New York: John Wiley.

Marangio, M. S. (1975) Phototaxis in larvae and adults of the marbled lizard, *Ambystoma opacum. Journal of Herpetology*, **9**, 293–7.

Martof, B. S. & Scott, D. C. (1957) The food of the salamander *Leurognathus. Ecology*, **38**, 494–501.

McNamara, K. J. (1997) *Shapes of Time: The Evolution of Growth and Development*. Baltimore: Johns Hopkins University Press.

Moss, M. L. (1962) The functional matrix. *Vistas in Orthopedics*, **35**, 85–95.

Mushinisky, H. R., Hebrard, J. J. & Vodopich, D. S. (1982) Ontogeny of water snake foraging ecology. *Ecology*, **63**, 1624–9.

Norberg, U. M. (1990) *Vertebrate Flight*. Berlin: Springer-Verlag.

Norberg, U. M. (1994) Wing design, flight performance, and habitat use in bats. In *Ecological Morphology*, ed. P. Wainwright & S. Reilly, pp. 170–205. Chicago: University of Chicago Press.

Norberg, U. M. & Rayner, J. M. V. (1987) Ecological morphology and flight in bats (Mammalia: Chiroptera): wing adaptations, flight performance, foraging strategy and echolocation. *Philosophical Transactions of the Royal Society of London B*, **182**, 207–27.

O'Farrell, M. J. & Studier, E. H. (1973) Reproduction, growth and development in *Myotis thysanodes* and *M. lucifugus* (Chiroptera: Vespertilionidae). *Ecology*, **54**, 118–30.

Pedersen, S. C. (1995) Cephalometric correlates of echolocation in the Chiroptera: II. Fetal development. *Journal of Morphology*, **225**, 107–123.

Polis, G. A. (1984) Age structure component of niche width and intraspecific resource partitioning: can age groups function as ecological species? *American Naturalist*, **123**, 541–64.

Polis, G. A. (1988) Exploitation competition and the evolution of interference, cannibalism and intraguild predation in age/size structured populations. In *Size Structured Populations: Ecology and Evolution*, ed. L. Perrson, & B. Ebenmann, pp. 185–202. New York: Springer-Verlag.

Polis, G. A. (1991) Complex interactions in deserts: an empirical critique of food web theory. *American Naturalist*, **138**, 123–55.

Pough, F. H. (1973) Lizard energetics and diet. *Ecology*, **54**, 837–44.

Pough, F. H. (1978) Ontogenetic changes in endurance in water snakes (*Natrix sipedon*): physiological correlates and ecological consequences. *Copeia*, **1978**, 69–75.

Powers, L. V., Kandarian, S. C. & Kunz, T. H. (1991) Ontogeny of flight in the little brown bat, *Myotis lucifugus*: behavior, morphology and muscle histochemistry. *Journal of Comparative Physiology A*, **168**, 675–85.

Raff, R. R. (1997) *The Shape of Life*. Chicago: University of Chicago Press.

Ransome, R. D. (1990) *The Natural History of Hibernating Bats*. London: Christopher Helm.

Rolseth, S. L., Koehler, C. E. & Barclay, R. M. R. (1994) Differences in diets of juvenile and adult hoary bats, *Lasiurus cinereus. Journal of Mammalogy*, **75**, 394–8.

Rugh, R. (1968) *The Mouse*. Minneapolis: Burgess.

Schoener, T. W. (1977) Competition and the niche. In *Biology of the Reptilia*, ed. C. Gans & D. Tinkle, pp. 35–136. New York: Academic Press.

Seigel, R. A., Collins, J. T. & Novak, S. S. (1987) *Snakes: Ecology and Evolutionary Biology*. New York: MacMillan.

Serra-Cobo, J. (1989) Primary results of the study on *Miniopterus schreibersi* growth.

In *European Bat Research*, ed. V. Hanák, I. Horácek & J. Gaisler, pp. 163–73. Prague: Charles University Press.

Skutch, A. F. (1945) Incubation and nesting periods of Central American birds. *Auk*, **62**, 8–37.

Simpson, G. G. (1944) *Tempo and Mode in Evolution*. New York: Columbia University Press.

Simpson, G. G. (1953) *The Major Features of Evolution*. New York: Columbia University Press.

Snodgrass, R. E. (1954) Insect Metamorphosis. *Smithsonian Miscellaneous Collections*, **122**, 1–124.

Stangl, F. B., Jr., Dalquest, W. W. & Grimes, J. V. (1996) Observations on the early life history, growth, and development of the red bat, *Lasiurus borealis* (Chiroptera: Vespertilionidae), in north Texas. In *Contributions in Mammalogy: A Memorial Volume Honoring Dr. J. Knox Jones, Jr.*, ed. H. H. Genoways & R. J. Baker, pp. 139–48. Lubbock: Museum of Texas Tech University.

Sterba, O. (1977) Prenatal development of central European insectivores. *Folia Zoologica*, **22**, 27–44.

Sterba, O. (1984) Ontogenetic patterns and reproductive strategies in mammals. *Folia Zoologica*, **33**, 65–72.

Stern, A. A. & Kunz, T. H. (1998) Intraspecific variation in postnatal growth in the greater spear-nosed bat. *Journal of Mammalogy*, **79**, 755–63.

Stern, A. A., Kunz, T. H. & Bhatt, S. S. (1997) Seasonal wing loading and ontogeny of flight in *Phyllostomus hastatus* (Chiroptera: Phyllostomidae). *Journal of Mammalogy*, **78**, 1199–1209.

Stewart, R. E. (1956) Ecological study of the ruffed grouse broods in Virginia. *Auk*, **73**, 33–41.

Stoner, A. W. (1980) Feeding ecology of *Lagodon rhomboides* (Pisces: Sparidae): variation and functional response. *Fisheries Bulletin*, **78**, 337–52.

Stoner, A. W. & Livingston, R. J. (1984) Ontogenetic patterns in diet and feeding morphology in sympatric sparid fishes from seagrass meadows. *Copeia*, *1984*, 174–87.

Swartz, S. M.1998) Skin and bones: functional, architectural, and mechanical differentiation in the bat wing. In *Bat Biology and Conservation*, ed. T. H. Kunz & P. A. Racey, pp. 94–109. Washington: Smithsonian Institution Press.

Swartz, S. M., Bennet, M. B. & Carrier, D. R. (1992) Wing bone stresses in free flying bats and the evolution of skeletal design for flight. *Nature*, **359**, 726–9.

Tuttle, M. D. & Stevenson, D. (1982) Growth and survival in bats. In *The Ecology of Bats*, ed. T. H. Kunz, pp. 105–50. New York: Plenum Press.

Tschumy, W. O. (1982) Competition between juveniles and adults in age-structured populations. *Theoretical Population Biology*, **21**, 255–68.

Van Horne, B. (1982) Niches of adult and juvenile deer mice (*Peromyscus maniculatus*) in seral stages of coniferous forest. *Ecology*, **63**, 992–1003.

Van Tyne, J. (1951) A cardinal's, *Richmondena cardinalis*, choice of food for adult and for young. *Auk*, **68**, 110.

Vaughan, T. A. (1959) Functional morphology of three bats: Eumops, Myotis, Macrotus. *University of Kansas Publications, Museum of Natural History*, **12**, 1–153.

Voris, H. K. & Moffit, M. W. (1981) Size and proportional relationships between the beaked sea snake and its prey. *Biotropica*, **13**, 15–19.

Waddington, C. H. (1940) *Organizers and Genes*. New York: Cambridge University Press.

Waddington, C. H. (1957) *The Strategy of Genes*. London: Allen & Unwin.

Ward, H. L. (1904) A study in the variations of proportions in bats, with brief notes on some of the species mentioned. *Transactions of the Wisconsin Academy of Science*, **14**, 630–54.

Wassersug, R. J. & Sperry, D. G. (1977) The relationship of locomotion to differential predation on *Pseuacris triseriata* (Anura: Hylidae). *Ecology*, **58**, 830–9.

Welty, J. C. (1979) *The Life of Birds*. Philadelphia: W. B. Saunders.

Werner, E. E. & Gilliam, J. F. (1984) The ontogenetic niche and species interactions in size-structured populations. *Annual Review of Ecology and Systematics*, **15**, 393–425.

Werner, E. E. & Hall, D. J. (1984) Ontogenetic habitat shifts in bluegill: the foraging rate–predation risk trade-off. *Ecology*, **69**, 1352–66.

With, K. A. (1994) Ontogenetic shifts in how grasshoppers interact with landscape structure: an analysis of movement patterns. *Functional Ecology*, **8**, 477–85.

Wilbur, H. H. (1980) Complex life cycles. *Annual Review of Ecology and Systematics*, **11**, 67–93.

Wood, A. E. (1962) Tertiary rodents of the family Paramyidae. *Transactions of the American Philosophical Society*, **53**, 1–210.

Wright, S. (1931) Evolution in Mendelian populations. *Genetics*, **1**, 356–66.

RICK A. ADAMS & KATHERINE M. THIBAULT

10

Ontogeny and evolution of the hindlimb and calcar: assessing phylogenetic trends

INTRODUCTION

Comparative development can be used to understand the derivation of form and also to understand homology versus the convergent evolution of structures. Indeed, Shubin (1994) states: 'Ontogenetic data can greatly aid in the formation of hypotheses of homology . . . comparative studies of ontogenetic sequences, and the analysis of interspecific patterns of variability can all provide information about potential homologies.' Phylogenetic homology (Roth 1984, 1994) refers to those structures derived from the same feature in a common ancestor and shared by two or more taxa (synapomorphies). Biological homology, however, refers to characters that share a set of specific developmental constraints (Wagner 1989). Understanding both phylogenetic and biological homologies is necessary to fully discern the evolutionary origins of characters. In this chapter, we examine the developmental patterns of the hindlimb and calcar and their implications for the phylogenetic relationships among bats.

THE HINDLIMB: A DERIVED STRUCTURAL COMPOSITE

The skeletal system of bats is unique in form and function, predominately due to suites of characters derived for flight. For example, the hindlimb of bats is unlike that of most other mammals and its fundamental morphology is a compromise between adequate structural support and reduced weight for flight (Howell & Pylka 1977). The bat hindlimb has evolved specializations involving bone reductions and joint rotations (MacAlister 1872; Vaughan 1959), as well as highly derived characteristics such as the passive digital lock (Schaffer 1905; Bennet 1993; Quinn & Baumel 1993; Schutt 1993, 1998) and the calcar (Vaughan 1959, 1970a, b; Koopman 1988; Schutt & Altenbach 1997; Schutt & Simmons 1998).

Reduced diameters (relative to length) of the long-bones of the hindlimb have been argued to be a weight-reducing measure, adaptive in the evolution of flight in bats. Hypothetically, selection for reduced diameters of hindlimb bones resulted in limbs unable to support the compressive forces of body weight, and, therefore, selection for hanging posture in bats occurred (Howell & Pylka 1977). This reasoning, however, is inconsistent with biomechanics in that the compressive strength of bone is much greater (165×10^6 Pa) than its tensile strength (110×10^6 Pa; Kardong 1998). Consequently, hanging posture is more likely to fracture thin long-bones than are the compressive forces generated by quadrupedal posture. A more probable explanation for evolutionary reduction in diameter of long-bones in bats is that selection favored lightening of the skeleton for flight, but the extent of this reduction was actually constrained, not necessitated, by the evolution of hanging posture in bats.

Although the selective advantages, or evolutionary trade-offs, associated with hanging posture in bats remain under debate, the hindlimb of bats is unique among mammals, not only in its adult condition, but also in its growth and development. In this chapter we describe the underlying growth and skeletogenic patterns associated with the derived characters of the lower hindlimb and calcar. We compare patterns of growth and development of the skeleton among families and across the suborders Microchiroptera and Megachiroptera.

ONTOGENETIC PATTERNS OF THE LOWER HINDLIMB

Morphogenesis among the bones comprising the forelimb (see Chapter 9, this volume) and hindlimb in bats is asymmetrical. The developmental sequence of the hand bones (digits II–V) in *Myotis sodalis*, *M. lucifugus*, and *M. myotis* exhibits a proximodistal bias (Bader & Hall 1960; Joller 1977; Adams 1992, 1998), whereas digits of the foot have a virtually random developmental sequence across species (Bader & Hall 1960). Those digits of the manus that are associated with the flight membranes proceed through a nonrandom proximodistal pattern of morphogenesis and increasing length variation. The random pattern of ontogeny observed in the hindfoot of bats is very similar to that reported for nonvolant mammals such as humans (Lewenz &Whitney 1902) and rats (Olson & Miller 1958; Fritz & Hess 1970; Joller 1977). Hence, differences in developmental sequencing between the hand bones and foot bones in bats are likely driven by differences in the functional nature of each (Bader & Hall 1960).

Since the hindlimb of most, if not all, bats is adapted for supporting a hanging posture, one would predict general similarities in morphology

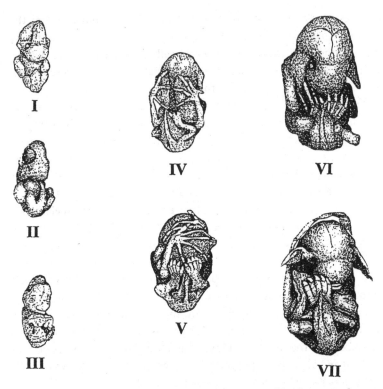

Figure 10.1. Camera lucida drawings of stages 1–7 of development of prenatal *Myotis lucifugus*. (After Adams 1992.)

and morphogenesis to be present across groups, regardless of phylogeny. Although much more work involving many more species is needed to build strong conclusions concerning phylogenetic relationships based upon developmental patterns in bats, we provide some compelling data illustrating the utility of integrating ontogeny into such investigations.

Microchiroptera

Detailed analyses of skeletogenesis have been carried out in very few species of bats. Joller (1977) described general skeletogenesis in *Myotis myotis* by staining the bones of fetuses with alizarin red. Unfortunately, because he did not stain for cartilage precursors to bone, much information remained undocumented concerning the developmental origin of skeletal components. Using cleared and stained specimens, Adams (1992) characterized prenatal development in *M. lucifugus* into seven stages by using relative changes in body proportions (Fig. 10.1) as compared with skeletogenic events. In summary, chondrogenic models of the tibia, fibula, three of the tarsal bones, all of the metatarsus, and the first phalanx of each digit were apparent beginning in stage 3 (mean crown–rump length [MCRL] = 8.7 mm). In stage 4 (MCRL = 10.8 mm), chondrogenic models of all lower leg and hindfoot bones are present. Skeletogenesis of the cartilage precursors of the tibia and fibula appeared in stage 5 (MCRL = 10.9 mm), whereas phalanges began ossifying in stage 7 (MCRL = 12.8 mm). Postnatal development of the hindlimb proceeds with little change in ossification between a late-stage 7 fetus (Fig. 10.1 & 10.2A) and a 5-day-old juvenile (Fig. 10.3). At 16 days after birth, the tarsus was fully ossified, with the exception of the calcaneus, which showed approximately two-thirds ossification. Full ossification of the calcaneus occurred by day 24 and was accompanied by the formation of secondary ossification centers in the distal tibia and fibula. In the phyllostomid *Artibeus jamaicensis*, the sequence of ossification is essentially the same as in *M. lucifugus*. However, complete ossification of the lower hindlimb elements occurs before birth as does formation of secondary ossification centers in the tibia and fibula (Fig. 10.2B). Truncated relative growth of the proximal fibula occurred between 16 and 26 days postnatal in *M. lucifugus*, resulting in the tibia outgrowing the fibula in length. Asymmetry in form between the lower hindlimb bones is retained in adults and the proximal end of the fibula is attached to the tibia via a ligament (Fig. 10.3).

In phyllostomids (Adams & Thibault, unpublished), differential relative growth between tibia and fibula begins prenatally, but unlike in *M. lucifugus* the length of the fibula in *A. jamaicensis* never equals that of the tibia, not even during chondrogenesis.

Megachiroptera

For two megachiropterans (Fig. 10.2C, D; *Megaloglossus woermanni* and *Rousettus celebensis*), general skeletogenesis of the lower hindlimb and foot is similar to that of microbats (Adams & Thibault, unpublished). However, since complete, or nearly complete, ossification of the tarsus, metatarsus, and phalanges occurs prenatally, these two pteropodids show an ossification rate similar to that observed in *A. jamaicensis*. Lengths of the fibula and tibia are never equal once ossification begins, and the fibula remains approximately two-thirds the length of the tibia throughout skeletogenesis. Secondary ossification centers of the tibia and fibula are not present in the prenatal specimens of these species.

Similarities in developmental patterns are consistent with similarities of form and function in the hindlimb across groups of bats. The

A

B

C

D

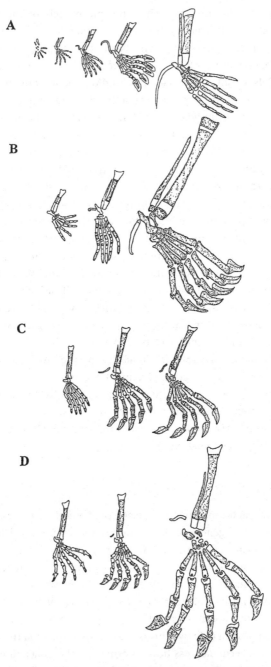

Figure 10.2. Camera lucida drawings of lower hindlimb and calcar development for A) *Myotis lucifugus*, B) *Artibeus jamaicensis*, C) *Megaloglossus woermanni*, and D) *Rousettus celebensis*.

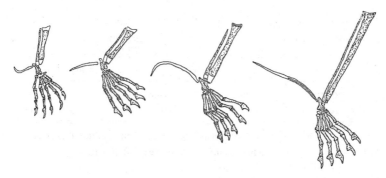

Figure 10.3. Postnatal growth of the lower hindlimb and calcar of known-age (left to right, 5, 16, and 24 days) juveniles and an adult *Myotis lucifugus*.

overall size of the foot relative to the lower leg is consistent among groups, with foot length contributing 50% to 70% of lower hindlimb length during prenatal development and at birth (Tables 10.1, 10.2). The predominance of the hindfoot in terms of size and state of development during prenatal morphogenesis suggests its large importance to the newborn young immediately after birth (Ward 1904; Vaughan 1959). The proportional dominance of the foot during hindlimb morphogenesis recedes during postnatal growth in *Myotis lucifugus*, and in adults the foot contributes 40% of the lower hindlimb length (Table 10.3). Growth trajectories of the tibia and fibula, as plotted relative to growth of the skull (Fig. 10.5), are similar throughout bats, although the megachiropteran, *Megaloglossus woermanni*, does show the slowest relative growth rate of the lower leg bones.

ONTOGENETIC PATTERNS OF THE CALCAR

The calcar is a structure unique to bats. Its description varies depending upon the species, but is estimated to be composed of one or more of the following: cartilage, bone, or calcified cartilage (Vaughan 1959; Schutt & Simmons 1998). The calcar has not been investigated histologically, and, therefore, its actual composition, or variability in composition among bats, remains uncertain. Calcars occur in most, but not all, species throughout both suborders of bats, presumably performing the same primary function of supporting the uropatagium (Vaughan 1959, 1970a, b; Norberg 1990). In some species of microchiropterans, the calcar is elaborate, in some cases forming a keeled structure, and is used in supporting the edges of the uropatagium in making a basket for catching insects (Vaughan 1959; Simmons & Geisler 1988). Presence of the calcar in

Table 10.1. *Prenatal growth proportions (percentage of length) of the hindfoot relative to total length of the lower hindlimb in two Microchiroptera*

	Myotis lucifugus			Artibeus jamaicensis		
Ossification index[a]	0.39	0.55	0.68	0.33	0.55	0.89
Percentage of length	64	59	60	58	59	53

Note: [a] Ossification index is calculated as total ossified length of the tibia divided by tibial length and is used here to indicate relative developmental state.

Table 10.2. *Prenatal growth proportions (percentage of length) of the hindfoot relative to total length of the lower hindlimb in two Megachiroptera*

	Rousettus celebensis			Megaloglossus woermanni		
Ossification index[a]	0.62	0.71	0.78	0.70	0.76	0.80
Percentage of length	60	54	61	54	56	62

Note: [a] Ossification index is calculated as total ossified length of the tibia divided by tibial length and is used here to indicate relative developmental state.

Table 10.3. *Growth proportions (percentage of length) of the hindfoot relative to total length of the lower hindlimb in three juvenile* Myotis lucifugus

	Juvenile	Juvenile	Juvenile	Adult
Ossification index[a]	0.74	0.79	0.88	1.0
Percentage of length	58	58	50	39

Notes: [a] Ossification index is calculated as total ossified length of the tibia divided by tibial length and is used here to indicate relative developmental state.

both suborders of bats has resulted in some investigators designating it a synapomorphy (Van Valen 1979; Koopman 1988; Simmons 1994, 1995, 1998), thereby compelling systematists to incorporate this character into phylogenetic analyses (Simmons & Geisler 1998). Striking differences between the gross structure, position, and fossil history of the calcar in Microchiroptera and Megachiroptera, however, suggests to some that the calcar might be independently derived in each suborder (Schutt & Simmons 1998). So far, only characteristics of the adult calcar have been utilized in phylogenetic schemes. In this section we describe and compare development of the calcar (previously unpublished data) using whole-mount cleared and differentially stained (alcian blue for cartilage and alizarin red for calcium) fetuses and juveniles of representative micro- and megachiropterans in hopes of further assessing origins of this unique structure.

Microchiroptera

The relative position of the developing calcar in microchiropterans, as depicted in *Myotis lucifugus* and *Artibeus jamaicensis*, is in proximity to the calcaneal tuberosity (Fig. 10.2A, B). In both species, the calcar begins development as a relatively small cartilaginous structure adjacent to, but not in contact with, the calcaneus. The cartilage precursor of the calcar is first apparent in whole-mount specimens cleared and stained after the entire chondrogenic model of the hindlimb is already in place, and the primary ossification centers have appeared in the tibia and fibula. This occurs at approximately stage 4 in *M. lucifugus*. In *M. lucifugus*, the calcar grows extensively throughout pre- and postnatal development (Fig. 10.2A, Fig. 10.4) and late in postnatal development, at approximately the onset of volancy, part of the calcar apparently becomes calcified. Calcification of the proximal tip and inner core, for approximately one-half its proximo-distal length, is indicated by the uptake of alizarin red stain marking the infusion of calcium salts into the structure (Fig. 10.4). There appears to develop a hinge-type articulation between the calcar and the calcaneal tuberosity (Fig. 10.4A), but this can only be confirmed histologically (in progress). This developmental pattern is consistent, and representative, of other vespertilionids so far investigated (unpublished data). For the phyllostomid *A. jamaicensis*, initial developmental position of the calcar is similar to that in *M. lucifugus* with the proximal tip adjacent to the calcaneal tuberosity. However, chondrogenesis occurs much later in development, after much of the diaphyses of the tibia and fibula are ossified. Calcification of the calcar occurs only at the most proximal end where it articulates with the calcaneal tuberosity (Fig. 10.2B), and calcification begins, and is completed, before birth.

Megachiroptera

Whereas the calcar of adult microchiroptrans is located in articulation with the calcaneal bone, the calcar of adult megachiropterans is positioned lateral to, and about four-fifths from, the fibular proximus (Schutt & Simmons 1998). Based upon our analyses (Adams & Thibault,unpublished) of development in *Megaloglossus woermanni* and *Rousettus celebensis*,

A

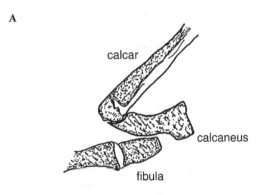

B

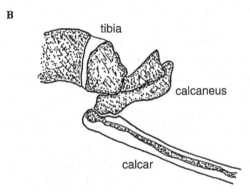

Figure 10.4. Camera lucida drawings of the A) dorsal and B) ventral views of the calcar–calcaneal complex in a subadult *Myotis lucifugus*.

chondrification of the calcar occurs lateral to the fibula where it imbeds in the tissue of m. gastrocnemius, never contacting bone. No calcium deposits are observed in the calcars of these two megachiropterans and apparently it remains cartilaginous in adults. As mentioned, the relative position of the calcar in adults differs between Microchiroptera and Megachiroptera and this differential positioning is consistent with the differential developmental patterns and positions between suborders.

SIMILARITIES AND DIFFERENCES ACROSS SUBORDERS

The hindlimb morphology of bats has been used as a character in helping decipher phylogenetic relationships among groups and is generally described as a synapomorphic structure among bats (Walton & Walton 1970). Wide variation occurs in the details of hindlimb structure

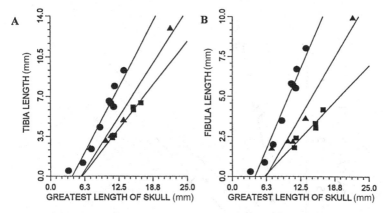

Figure 10.5. Growth trajectories for lengths of A) tibia and B) fibula for *Myotis lucifugus* (circles), *Artbeus jamaicensis* (triangles), and *Megaloglossus woermanni* (squares).

among bats, and these distinctive specializations are the result of adaptations to specific ecologies (Findley 1972; Schutt & Altenbach 1997; Schutt 1998). Developmental patterns of the tibia/fibula complex demonstrate high similarity across suborders and support the monophyletic origin of bats. Reduction in the proximal end of the fibula is consistent among bats (although there is full development of the fibula in two of the three vampire bat species), with megachiropterans showing a greater proportional loss (almost 50%) in fibular length during growth, relative to the tibia. General ossification patterns are consistent across groups. Tibial development is comparable and shows similar growth trajectories among species (Fig. 10.5A). Megachiroptera, however, do show slower relative growth of the fibula relative to skull growth compared to microchiropterans (Fig. 10.5B).

In all species so far examined, calcar formation in whole-mount specimens is marked by a chondrogenic condensation. Beyond this, there are clear distinctions between calcar developmental patterns in micro- and megachiropterans. In fact, differences in calcar position and structure between the two suborders are significant enough to compel some investigators to rename the structure in megachiropterans to highlight its uniqueness versus the microchiropteran condition: i.e., 'styliform bone' (Humphrey 1869; Andersen 1912), the 'tarsal spine' (Mori 1961), or 'uropatagial spur' (Schutt & Simmons 1998). Renaming the 'calcar' in megachiropterans would inherently imply nonhomology with this structure in microchiropterans (Schutt & Simmons 1988). Differences in the megachiropteran calcar in terms of developmental position, apparent

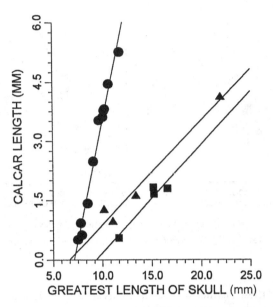

Figure 10.6. Growth trajectories of the calcars for *Myotis lucifugus* (circles), *Artibeus jamaicensis* (triangles), and *Megaloglossus woermanni* (squares).

lack of calcium investment during development, and its intimate nature with the soft tissue of the gastrocnemius bundle instead of a bony element, support the hypothesis of independent origins for the micro-chiropteran calcar and the megachiropteran 'uropatagial spur.' Hence, the calcar, as currently understood, does not meet the assumptions of biological homology (Wagner 1989). On the other hand, growth trajectories of the calcar (Fig. 10.6) do not differentiate microchiropterans from mega-chiropterans, nor do form and function in any significant way. Indeed, *Myotis lucifugus* shows the highest exponential growth of the calcar, whereas *A. jamaicensis* and *Megaloglossus woermanni* share a similar growth trajectory relative to skull growth. If the calcar in Micro- and Megachiroptera lack common ancestry, then the apparent similarity in size, shape and function of this structure across suborders must be the result of convergence.

FOSSILS, PHYLOGENY, AND THE CALCAR

As described above, the calcar and uropatagial spur show highly distinctive patterns in terms of developmental position and relative timing of appearance. The heterotopic nature of the calcar of micro-chiropterans and the uropatageal spur of megachiropterans in adults is the result of differential patterns of morphogenesis between suborders. For example, the calcar and uropatagial spur do not share a common position at chondrogenesis, but instead, at least what can be discerned from investigations using whole-mount specimens, the position and gross anatomy of the microchiropteran and megachiropteran structures are unique developmentally, as well as in adults.

Further compelling support for convergent evolution can be found in the distinctive nature of the muscles associated with the calcar in microchiropterans and the uropatagial spur in megachiropterans (Fig. 10.7). The m. calcaneocutaneous is absent in extant pteropodids, but is present in all microchiropterans examined, as is the m. calcaneotibialis in most cases, whereas the opposite taxonomic distribution is true for the m. uropatagium (Schutt & Simmons 1998). Comparative morphology of extant groups establishes a case for the independent origins of the calcar and uropatagial spur within each suborder. But what can history tell us? Simmons & Geisler (1998), using fossil bats, provide further evidence of an independent origin of the calcar. Neither *Icaronycteris index* nor *Archeonycteris trigonodon* apparently possessed either calcars or uropatagial spurs. However, *Archaeopteropus transiens*, which was considered by many to be the ancestor of the megachiropterans (Meschinelli 1903; Andersen 1912; Simpson 1945; Romer 1966; Koopman & Jones 1970), did manifest a calcar that lay in articulation with the calcaneus (Schutt & Simmons 1998). Does *A. transiens* support the assertion of common ancestry for both the calcar and uropatagial spur, and, therefore, a monophyletic origin of Mega- and Microchiroptera, or is *A. transiens* not a megabat at all? Schutt & Simmons (1998) argue the latter, even though *A. transiens* does possess characteristics that are clearly megachiropteran. However, Schutt & Simmons (1998) argue that characteristics such as possession of a claw on the second digit (known in all extant megachiropterans and no microchiropterans) can be considered primitive for all mammals and therefore that character is inappropriate for use in analyses of the relationship between micro- and megachiropterans.

A corollary of the homology concept is that: for those structures deemed to originate in common ancestry, they would predictably share both a phylogenetic history and shared developmental similarity (see Goodwin 1984; Roth 1984 for counter argument). Within the purview of homology known as phylogenetic homology (based largely upon evidence from the fossil record), the evolution of the calcar and the uropatagial spur is evidently a result of parallelism or convergence. Under the concept of homology known as biological (ontogenetic) homology, the picture remains less clear. Ontogenetic data presented herein demonstrate that

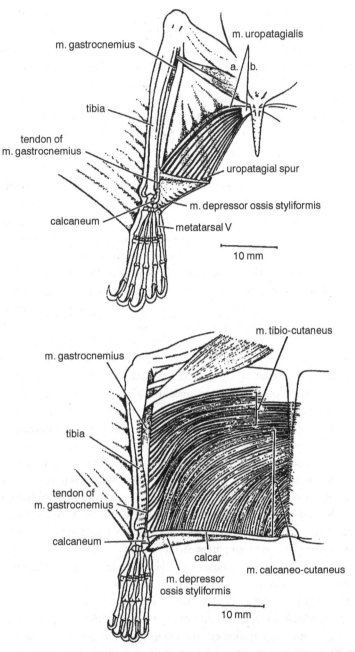

Figure 10.7. Illustrations of the calcar/calcaneal complex of *Phyllostomus hastatus* (upper drawing) and the tibia/uropatagial spur complex of *Rousettus amplexicaudatus* (lower drawing). (From Schutt & Simmons 1998.)

the developmental origins of the calcar (at least as evidenced in whole-mount, cleared and stained fetuses) and the uropatagial spur are topologically distinctive in their formations. If homology were the case, one might expect homotopy of cartilage condensations to be evident in both micro- and megachiropterans, with later relocation of the calcar during growth and development. This in itself would, at least, support the hypothesis and requisites of developmental, and therefore, evolutionary, homology. However, the cartilage condensations of the calcar and the uropatagial spur apparently initiate development in different relative positions and these positions, as represented in the adult morphology of each group, are also likely representative of the position in which this structure evolved.

The oldest known ancestor to present-day bats, *Icaronycteris index*, apparently did not possess a calcar or spur as evidenced by fossil remains (Simmons & Geisler 1998). Although it is possible that the calcar, or spur, of this specimen was present, but never fossilized, close examination of the calcaneus shows no evidence of a calcaneal tuberosity, known to articulate with the calcar in microchiropterans. This, of course, does not rule out the possibility that *I. index* had a uropatagial spur as in extant megachiropterans. Curiously, the Eocene bats *I. index* and *Archaeonycteris trigonodon* are both surmised to have had uropatagial anatomy consistent with extant free-tailed species (Molossidae) (Simmons & Geisler 1998). Examination of free-tailed bat anatomy shows that these species possess calcars that support the trailing edge of their relatively small uropatagia (Schutt & Simmons 1998). In addition, although they are aerial insectivores, molossids 'filter-feed' on insects by ingesting prey directly into the mouth as opposed to using the wings and uropatagium as a catchment like other species of insectivorous microchiropterans. Curiously, of all known extant species, only *Craseonycteris thonglongyia* (Craseonycteridae), the smallest known bat in the world, has a 'broad-type' uropatagium, but lacks a calcar or spur that would support the trailing edge. All other species that do not possess a calcar or strut have either absent, rudimentary, or V-cut uropatagia (Schutt & Simmons 1998). In a functional sense, it is hard to imagine a uropatagium of any measure not having a structure that prevents flapping of its trailing-edge during flight. In fact, Schutt & Simmons (1998) have recast previous determinations of membrane morphology in *Archaeopteropus transiens* based upon calcar morphology suggesting that this species possessed a 'broad' uropatagium, supported along its trailing-edge by a calcar.

In extant bat species (and apparently evidenced by some extinct species), calcar size and shape correlate with surface area and function of the uropatagium (Simmons 1998). In the present study we show that the megachiropterans, *Rousettus celebensis* and *M. woermanni*, both known to have shortened tails and reduced uropatagia, have small uropatagial spurs relative to foot size. In microchiropterans, the frugivorous *Artibeus jamaicensis* has a reduced calcar and surface area of the uropatagium relative to *Myotis lucifugus* and other vespertilionids that use their membranes to catch insects in flight, this supporting the assertion that calcar shape and size is consistent with the extent of its function (Simmons 1998). Developmental data presented for the first time herein support the hypothesis that the calcar is not homologous within the Chiroptera, but is a convergent structure, evolving independently within each suborder. Specialization of the vespertilionid calcar is apparently a product of its accelerated growth rate resulting in a structure adaptive in support of the uropatagium as an insect catchment during aerial foraging. Relative growth rate similarities noted between the calcar in *A. jamaicensis* and the uropatagial spur in *Megaloglossus woermanni* and *R. celebensis* is apparently a product of similar growth trajectories (Fig. 10.6) resulting in a less elaborate structure (compared to most vespertilionids), whose primary function is to stabilize the trailing-edge of the uropatagium during flight.

In closing, it should be mentioned, however, that the developmental events described herein are at a relatively coarse scale because they are based upon whole-mount, cleared and stained fetuses. In order to truly address the question of homology, investigations into formation of the primordia of the calcar would be highly insightful. Independent of differential positioning, if the calcar (as well as other related structures of the distal hindlimb) is formed from a homologous stream of mesenchyme that generates these structures in both groups, then an argument can be made for homology. If the calcar condensations prove independent of limb formation in either both, or in one, of the subordinal groups, then either convergent or parallel evolution may be invoked. What these data confer relative to arguments concerning mono- versus diphyletic origins of bats currently remains unclear, but, hopefully, with further investigations into developmental patterns, the evolution of the calcar as representing either a chiropteran synapomorphy, or a convergent/parallel adaptation, will finally be discerned.

ACKNOWLEDGMENTS

I would like to thank William Schutt and an anonymous reviewer for providing valuable comments on an earlier draft of this paper. I would also like to thank the American Museum of Natural History, New York, the

Harrison Zoological Museum, Sevenoaks, England, and the University of Wisconsin Zoological Museum, Madison, for loaning me specimens. Funding was provided by the University of Wisconsin Research Grant and the Department of Biological Sciences at University of Wisconsin–Whitewater.

REFERENCES

Adams, R. A. (1992) Stages of development and sequence of bone formation in the little brown bat, *Myotis lucifugus*. *Journal of Mammalogy*, **73**, 160–7.

Adams, R. A. (1998) Developmental and functional integration in bat wings. *Journal of Zoology (London)*, **246**, 165–74.

Andersen, K. (1912) *Catalogue of the Chiroptera in Collections of the British Museum*, vol. 1. *Megachiroptera*, 2nd edn. London: British Museum. (Repr. New York: Johnson Reprint Corporation.)

Bader, R. S. & Hall, J. S. (1960) Osteometric variation and function in bats. *Evolution*, **14**, 8–17.

Bennet, M. B. (1993) Structural modification involved in the fore and hind limb grip of some flying foxes (Chiroptera: Pteropodidae). *Journal of Zoology*, **229**, 237–48.

Findley, J. S. (1972) Phenetic relationships among bats of the genus *Myotis*. *Systematic Zoology*, **21**, 31–52.

Fritz, H. & Hess, R. (1970) Ossification of the rat and mouse skeleton in the perinatal period. *Teratology*, **3**, 331–8.

Goodwin, B. (1984) Homology, development and heredity. In *Homology: The Hierarchical Basis of Comparative Anatomy*, ed. B. K. Hall, pp. 230–47. New York: Academic Press.

Howell, D. J. & Pylka, J. (1977) Why bats hang upside down: a biomechanical hypothesis. *Journal of Theoretical Biology*, **69**, 625–31.

Humphrey, G. M. (1869) The myology of the limbs in *Pteropus*. *Journal of Anatomy and Physiology*, **3**, 294–319.

Joller, V. H. (1977) Zur Ontogenese von *Myotis myotis* Borkhausen (verglichen mit jener von *Acomys cahirinus dimidiatus*). *Naturforschende Gesellschaft, Basel*, **86**, 87–151.

Kardong, K. V. (1998) *Vertebrates*. Dubuque: Wm. C. Brown.

Koopman, K. F. (1988) Systematics and distribution. In *Natural History of Vampire Bats*, ed. A. M. Greenhill & U. Schmidt, pp. 7–17. Boca Raton: CRC Press.

Koopman, K. F. & Jones, J. K. (1970) Classification of bats. In *About Bats*, ed. B. H. Slaughter & W. D. Walton, pp. 22–8. Dallas: Southern Methodist University Press.

Lewenz, M. A. & Whitney, M. A. (1902) A second study on the variability and correlation of the hand. *Biometrika*, **1**, 345–630.

MacAlister, A. (1872) The myology of the Cheiroptera. *Philosophical Transactions of the Royal Society of London*, **162**, 125–71.

Meschinelli, L. (1903) Un nuovo chiroptero fossile (*Archeopteropus transiens* Mesch.), *Della ligniti di arti*, **62**, 1329–44.

Mori, M. (1961) Muskulatur des *Pteropus edulis*. *Okajamas Folia Anatomica*, **36**, 253–307.

Norberg, U. M. (1990) *Vertebrate Flight*. Berlin: Springer-Verlag.

Olson, E. C. & Miller, R. L. (1958) *Morphological Integration*. Chicago: University of Chicago Press.

Quinn, T. H. & Baumel, J. J. (1993) Chiropteran tendon locking mechanism. *Journal of Morphology*, **35**, 197–208.

Romer, A. S. (1966) *Vertebrate Paleontology*, 3rd edn. Chicago: University of Chicago Press.

Roth, V. L. (1984) On homology. *Biological Journal of the Linnaean Society*, **22**, 13–29.

Roth, V. L. (1994) Within and between organisms: replicators, lineages, and homologues. In *Homology: The Hierarchical Basis of Comparative Anatomy*, ed. B. K. Hall, pp. 302–33. New York: Academic Press.

Schaffer, J. (1905) Anatomische-histologische Untersuchunge über den Bau der Zehen bei Fledermäusen und einigen kletternden Säugetieren. *Zeitschrift für wissenshaftliche Zoologie*, **83**, 231–84.

Schutt, W. A., Jr., (1993) Digital morphology in the Chiroptera: the passive digital lock. *Acta Anatomica*, **148**, 219–27.

Schutt, W. A., Jr., (1998) Chiropteran hindlimb morphology and the origin of blood feeding in bats. In *Bat Biology and Conservation*, ed. T. H. Kunz & P. A. Racey, pp. 157–68. Washington: Smithsonian Institution Press.

Schutt, W. A., Jr. & Altenbach, J. S. (1997) A sixth digit in *Dyphylla ecaudata*, the hairy legged vampire bat (Chiroptera, Phyllostomatidae). *Mammalia*, **61**, 280–5.

Schutt, W. A., Jr. & Simmons, N. B. (1998) Morphology and homology of the chiropteran calcar, with comments on the phylogenetic relationship of *Archaeopteropus*. *Journal of Mammalian Evolution*, **5**, 1–31.

Shubin, N. H. (1994) History, ontogeny, and evolution of the archetype. In *Homology: The Hierarchical Basis of Comparative Anatomy*, ed. B. K. Hall, pp. 250–73. New York: Academic Press.

Simmons, N. B. (1994) The case for chiropteran monophyly. *American Museum Novitates*, **3103**, 1–54.

Simmons, N. B. (1995) Bat relationships and the origin of flight. In *Ecology, Evolution and Behaviour of Bats*, ed. P. A. Racey & S. M. Swift, *Symposium of the Zoological Society of London*, **67**, 27–43.

Simmons, N. B. (1998) A reappraisal of interfamiliar relationships of bats. In *Bat Biology and Conservation*, ed. T. H. Kunz & P. A. Racey, pp. 3–26. Washington: Smithsonian Institution Press.

Simmons, N. B. & Geisler, J. H. (1998) Phylogenetic relationships of *Icaronycteris, Archaeonycteris, Hassianycteris*, and *Paleochiropteryx* to extant bat lineages, with comments on the evolution of echolocation and foraging strategies in Microchiroptera. *Bulletin of the American Museum of Natural History*, **235**, 1–82.

Simpson, G. G. (1945) The principles of classification and a classification of mammals. *Bulletin of the American Museum of Natural History*, **85**, 1–350.

Van Valen, L. (1979) The evolution of bats. *Evolutionary Theory*, **4**, 104–21.

Vaughan, T. A. (1959) Functional morphology of three bats: Eumops, Myotis, Macrotus. *Publications of the Museum of Natural History of Kansas*, **12**, 1–153.

Vaughan, T. A. (1970a) The skeletal system. In *Biology of Bats*, ed. W. A. Wimsatt, pp. 97–138. New York: Academic Press.

Vaughan, T. A. (1970b) Adaptations for flight in bats. In *About Bats*, ed. R. Slaughter & D. Walton, pp. 127–43. Dallas: Southern Methodist University Press.

Wagner, G. P. (1989) The biological homology concept. *Annual Review of Ecology and Systematics*, **20**, 51–69.

Walton, D. W. & Walton, G. M. (1970) Post-cranial osteology of bats. In *About Bats*, ed. R. Slaughter & D. Walton, pp. 93–125. Dallas: Southern Methodist University Press.

Ward., H. L. (1904) A study in the variations of proportions in bats, with brief notes on some of the species mentioned. *Transactions of the Wisconsin Academy of Sciences*, **14**, 630–54.

11

A comparative perspective on the ontogeny of flight muscles in bats

INTRODUCTION

Flight muscles, and the pectoralis muscle in particular, are the 'motors' behind flight in bats. Whereas the locomotory muscles of most mammals are composed of multiple fiber types (discussed below) which facilitate different kinds of movements (slow versus fast movements, for example), the flight muscles of several bats have been classified as either unitypic or bitypic (Hermanson & Foehring 1988; Hermanson 1998). For example, in the unitypic condition, the pectoralis muscle is made of only a single (fast-twitch) fiber type and is highly specialized for a specific pattern of contraction. In contrast, bats with bitypic muscles have pectoralis muscles containing two fiber types which are both presumably fast-twitch, but which probably differ in underlying metabolic specializations. These observations about the pectoralis can be extended to other muscles in the shoulder girdle, most notably the serratus anterior ('serratus ven-tralis' of Nomina Anatomica Veterinaria 1994), subscapularis, and short head of biceps brachii. Inclusion of these muscles as primary downstroke muscles was hypothesized by Vaughan (1959) based on his careful dissec-tion studies of bats, and was substantiated by histochemical (Foehring & Hermanson 1984; Hermanson & Foehring 1988; and others) and electro-myographic studies of these muscles (Hermanson & Altenbach 1981, 1985; Altenbach & Hermanson 1987). Most studies of bat flight muscles and anatomy were based on analysis of adult specimens. Thus, it is not surpris-ing that we have learned little about the normal ontogeny of bat muscles and about the problems faced by growing bats, nor about the comparative ecology of juvenile bats from different chiropteran species.

A large body of literature exists about the normal ontogeny of loco-motory muscles of terrestrial mammals. The biomedical literature is replete with information about muscle development as it relates to developmental biology, neuromuscular disease, and environmentally triggered muscular deficits. The agricultural literature, largely under the heading of 'meat science,' contains many studies pertinent to the growth and productivity of domestic species used in human food production. The purpose of this chapter is to provide reference to relevant literature pertaining to other animals that highlights our understanding of the common principles of muscle development to review current knowledge about the normal ontogeny of flight muscles in bats, and finally, to provide questions that should provide fruitful study with respect to chiropteran evolution and ecology.

MUSCLE STRUCTURE AND FUNCTION IN TERRESTRIAL MAMMALS

The general design of mammalian muscles provides for several behavioral options. That is, the muscles are composed of two or three fiber types, each of which is best suited to facilitate divergent, although perhaps overlapping levels of force production. Fiber types of various bird and mammal species have been reviewed elsewhere (Hermanson 1998; Pette & Staron 1990) but will be briefly summarized here (and see Table 11.1, from Sypert & Munson 1981). The classification system of Brooke & Kaiser (1970; see also Guth & Samaha 1969, 1970) was based on myosin ATPase properties and often is expressed as type I (slow), type IIa (fast), and type IIb (fastest). A parallel classification based on the work of Peter et al. (1972) is broadly used by physiologists and accounts for myosin ATPase activity as well as the levels of activity of several metabolic enzymes. The latter classification correlates fast and slow myosin activity with aerobic potential. Fast fibers can correlate with high glycolytic or anaerobic activity as FG fibers, meaning fast and glycolytic (anaerobic but little aerobic or fatigue resistant properties) or, fast fibers can exhibit high glycolytic and oxidative activity in FOG fibers, meaning fast, oxidative (fatigue resistant), and glycolytic (potential to use anaerobic pathways for 'sprinting' activity). The two classifications are comparable but do not necessarily always agree. Identification of type IIa fibers based on myosin ATPase over a range of pH preincubations, as in the original Brooke & Kaiser (1970) methodology, is not always synonymous with the type FOG of the Peter et al. (1972) classification (Nemeth & Pette 1981). All students of muscle biology should be fluent in both classifications because they are both used extensively in the contemporary literature. The selection of one or the other classification may reflect the heritage of a particular laboratory reporting their work, or may reflect an important aspect of the question being studied in the labor-

Table 11.1. *Physiological and histochemical properties of muscle units medial gastrocnemius and soleus muscles in the domestic cat; these values should suggest general properties (such as high, low, or intermediate (inter.) enzyme activity levels) of the several fiber types observed in adult terrestrial mammals*

Histochemical fiber type	Medial gastrocnemius muscle					Soleus
	Type IIb (FG)	Type IIc (FI)	Type IIa (FOG)	Type I (S)	Type I (S)	Type I (S)
Percent in motor unit pool	45%	5%	25%	25%	25%	100%
Twitch tension (g)	8	1.7	0.3	0.4	0.4	
Twitch contraction time (ms)	24	25	15	63	63	98
Resistance to fatigue	low	inter	high	high	high	high
Mean fiber area (μm²)	5290	?	2890	1730	1730	4320
Mean fiber size	large	?	inter	small	small	small
Fiber capillary supply	sparse	?	rich	rich	rich	rich
mATPase (alkaline preincubation)	high	high	high	low	low	low
Oxidative enzyme activity	low	inter	inter/high	high	high	high
Glycolytic enzyme activity	high	high	high	low	low	low

Source: Modified from Sypert & Munson (1981).

atory. For instance, a laboratory studying the effect of exercise on muscle metabolism is likely to report their results based on the Peter *et al.* (1972) classification. Another laboratory examining the impact of exercise on the presumed alteration of myosin heavy chain isoforms will tend to focus on the Brooke & Kaiser (1970) classification.

Terrestrial animals such as the cat generally have hindlimb muscles composed of varying proportions of the three main fiber types, the proportion being dependent on the activation history of that muscle or how it is used in posture and/or locomotion. A muscle such as the soleus of domestic rats, a stout muscle in the lower leg that holds the ankle extended against gravity, is composed of 80 to 90% type I (slow oxidative: SO) fibers (Kugelberg 1973; Armstrong & Phelps 1984). Thus, this muscle is ideally suited not to produce work (work implies muscle shortening), but to maintain a near-isometric contraction for long periods while the animal stands still. The adjacent lateral gastrocnemius muscle of the cat contains far fewer type I fibers (in the range of 2 to 24% depending on the muscle compartment sampled: English & Letbetter 1982) but does contain about 58 to 82% type IIb (FG) fibers. The lateral gastrocnemius will facilitate long periods of quiet standing by recruiting only about 25% of its total muscle fibers, the type I and IIa fibers, while the IIb fibers remain quiescent. The same muscle is suitably designed to facilitate a range of higher speed movements by recruiting the fatiguable type IIb fibers for short duration bouts of activity (Walmsley *et al.* 1978). The most forceful actions, such as leaping to a higher perch, will require recruitment of many of the muscle's type IIb fibers (expressed as FG fibers in the studies by Taylor 1978; Walmsley *et al.* 1978). These type IIb fibers are poorly endowed with aerobic potential, as evidenced by weak histological staining for nicotinamide dinucleotide dehydrogenase or succinic dehydrogenase, enzymatic markers specific for the aerobic pathways in metabolism. As such, the type IIb components of the lateral gastrocnemius cannot sustain forceful contractile activity for long periods of time (Table 11.1 for comparable data from medial gastrocnemius). This type of fatigue can be seen in the high-speed pursuit of prey by a cheetah. Failure to capture prey during the initial pursuit, lasting only a few seconds, usually results in the cheetah breaking off the chase due to muscle fatigue (Schaller 1972). Heat overload also has been invoked as a limiting factor in the maximum pursuit time used by cheetahs (Taylor & Rowntree 1973). This fatigue would be predicted if the major locomotory muscles of the cheetah consist mostly type IIb fibers (high power producing but fatiguable) as suggested by the above-mentioned studies on domestic cats. In fact, the vastus lateralis contained 76% FG (type IIb) and 9% FOG (type IIa), and gastrocnemius (head not specified) contained 59% FG and 3% FOG fibers (Williams *et al.* 1997). Clearly both the type I and IIa fibers are important in the locomotion of the cat. Type IIa fibers, in particular, demonstrate a compromise between good aerobic, as well as anaerobic potential (thus, these are the FOG fibers of the Peter *et al.* (1972) classification). The latter is measured by histochemical staining for α-GPD, an enzyme of significance in the basic glycolysis pathway of metabolism. The anaerobic scope in cat muscles facilitates high force production on demand, such as escaping dogs and cars, or simply jumping on a couch (domestic cats can exhibit a vertical jump of about 1.4 m: Zomlefer *et al.* 1977). The aerobic scope facilitates long-term recruitment of a population of muscle fibers: their rich endowment of mitochondria, blood capillaries, and smaller fiber size permits the muscle fiber to be activated and reactivated many times before the onset of fatigue becomes a factor. Dogs tend to have locomotory muscles that are well suited to aerobic function (Armstrong *et al.* 1982; Snow *et al.* 1982). The wild dogs of Africa (genus *Lycaon*) are noted for their ability to run down prey for long periods of time at high speeds and over the course of several kilometers (Estes 1991). North American wolves generally can pursue prey (deer, for example) for distances of up to 1 km or more without undue fatigue (Mech 1970). One study of domestic dogs (*Canis familiaris*) showed that they could run for up to 17 hours with only 5-minute rest periods every 30 minutes (Dill *et al.* 1932), covering a distance as much as 132 km during that time. Thus, the muscles of several canid species are highly specialized for aerobic potential.

Recent studies have elucidated a type IIx type of muscle fiber, correlated with a unique isoform of the myosin heavy chain (LaFramboise *et al.* 1990; Talmadge *et al.* 1996). In the cat, many of the fibers previously described as type IIb have been reassigned to this type IIx category based on the biochemical or isoform properties. For the present discussion, no attempt is made to revise earlier characterizations of developing muscle fibers as type IIb into the new type IIx category. More work is ongoing in many laboratories and will appear in the literature over the next several years. With respect to the bats studied, there is no common opinion that the myosin isoforms have the type IIx isoforms. As techniques for studying muscle improve there will be certain revisions in our interpretations of the muscle biology of many organisms.

ONTOGENY OF MUSCLES IN TERRESTRIAL MAMMALS

A characteristic of most terrestrial mammals is that muscles go through a period of enormous transition during their development. The muscles of neonatal cats and rats are uniformly slow in contraction time, regardless of their adult myosin ATPase properties (Denny-Brown 1929). This slow contraction speed corresponds with histochemical properties of neonatal cat muscle (Nyström 1968): the soleus muscles of 3-week-old kittens show a mosaic pattern of histochemical differentiation that is largely obscured by 6 weeks of age as the muscle matures towards its uniformly slow-twitch adult phenotype. In contrast, the gastrocnemius muscle (both lateral and medial heads) continues its differentiation and retains a mosaic pattern of histochemical fiber types in adults (see English & Letbetter 1982). In most muscles studied, the neonatal muscles include a number of muscle fibers that are quite difficult to classify using traditional myosin ATPase studies: the fibers exhibit properties of fast- and slow-twitch muscle when studied under alkaline and acidic preincubation conditions (Engel & Karpati 1968). The latter category is often referred to as type IIc, a kind of 'waste basket' for fibers we don't quite know how to classify. As an example of ontogeny in a fast muscle, the extensor digitorum longus muscle of neonatal laboratory rats includes about 4% slow fibers (type I myosin), most of which are constrained to one part, or compartment, of the muscle (Thompson et al. 1989). Some of the fibers may be transformed to fast-twitch types (type II myosin) during the early postnatal period (Kugelberg 1973; Thompson et al. 1989). In the extensor digitorum longus muscle of mice, Whalen et al. (1984) demonstrated a decrease in the percentage of slow fibers from about 10% of muscle fibers in the newborn, to about 0.8% in the adult. Thus, for the murine digital extensors there was a developmentally programmed loss of slow myosin that occurred after birth. Similar replacement or transformation of type I (slow) fibers is observed in neonatal little brown bat (*Myotis lucifugus*) pectoralis muscles (Schutt et al. 1994).

In a functional sense, neonatal muscles perform quite differently than their adult counterparts. Neonatal muscle tends to exhibit slower contractile speed (Close 1964; Buller & Lewis 1965), possibly because the myosin heavy chain isoforms, one of the important determinants of the speed of shortening of a muscle (Reiser et al. 1985; and see recent discussion by Bottinelli et al. 1994), are different from those seen in adults. For example, there are several transitional myosin heavy chain isoforms reported in the literature, each of which can be detected by monoclonal antibodies in histological study or by electrophoretic separation (Whalen et al. 1981; LaFramboise et al. 1991). Why go through these ontogenetic stages with multiple transitional isoforms? It might seem energetically wasteful to create a large number of transitional protein molecules only to deconstruct them at a later time and to replace them with new proteins. But, these transitions seem to reflect the hormonal and behavioral environment of the fetus or the neonate. Experimental alteration of thyroid function, for example, alters the normal expression of type II myosin heavy chains (Whalen et al. 1985). Experimental reduction of movement can inhibit normal muscle growth as well as alter the expression of myosin and metabolic proteins (Fitts et al. 1985, 1989). Indeed, transitional myosin protein isoforms and both fetal and neonatal fiber types are present in most mammals and birds that have been studied. These transitional forms of proteins and fiber types are an important part of the normal cycle of constructing an adult muscle.

Myogenesis or the development of muscle has been intensely studied in several species of laboratory rodents (reviewed by Hauschka 1994). Based on the analysis of muscle development in rats (Rattus norvegicus) and mice (Mus musculus), early researchers concluded that there were at least two waves of myofiber development in each muscle during gestation (cf. Kelly & Zacks 1969; Ontell & Kozeka 1984). In both rats and mice, gestation lasts about 21 days. In the first 12 days of gestation, the general framework of the limbs is laid down. At 12 days gestational age primary myotubes can be observed with either transmission electron microscopy or traditional light microscopy. These myotubes appear as widely spaced, undifferentiated cellular structures within the developing muscle masses. The primary myotubes contain central nuclei and vacuoles, and disorganized and short muscle contractile proteins: actin and myosin. The myotubes develop into myofibers when the nuclei migrate peripherally and when myofilaments become more highly organized (the actin and myosin become organized parallel to the long axis of the muscle fiber: Ontell & Kozeka 1984). Secondary myotubes appear in the muscle matrix at about 14 to 16 days gestation age and are associated with individual primary myofibers (Ontell & Kozeka 1984). It is believed that the secondary myotubes use the primary myofiber as a scaffold along which they develop and elongate (Kelly & Zacks 1969; Ontell & Dunn 1978).

Exactly how are the myosin isoform differences between different fibers established during ontogeny? This question has been debated by many researchers for several decades. One line of evidence suggests that myosin isoform differences are established at the time of formation of the individual primary and secondary myotubes (for chicken muscle, see Crow & Stockdale 1986; for rat muscle, see Condon et al. 1990) and that these myotubes can undergo differentiation into adult fast and slow myofibers in the absence of extrinsic cues such as innervation. In any case, two categories of primary myosin heavy chain isoforms, slow myosin heavy chain isoform, and neonatal myosin heavy chain isoform (which will serve as a precursor to the expression of adult fast myosin heavy chains), are expressed in specific regions of the muscles correlated with the histochemical pattern seen in the adult muscle. Furthermore, patterns of myosin heavy chain expression are correlated with the birth date of the myofibers and thus with the generation (primary or secondary) of myotubes from which they arise (Narusawa *et al.* 1987; Condon *et al.* 1990). The position of myotubes within a muscle and the birthdate (primary vs. secondary myotube) both determine the distribution of fiber types within a muscle (Zhang & McLennan 1998). Of interest in the Zhang & McLennan study was that primary myotubes differentiated into adult type I or type IIb fibers, but not into type IIa fibers. The type I and IIb fibers are functionally most divergent with type IIa fibers occupying a more intermediate position in a functional continuum (high force vs. low force production; high glycolytic vs. low glycolytic potential).

Large animals, sheep (*Ovis aries*) in particular, have been shown to have several generations of developing muscle fibers, perhaps to accommodate the immense number of muscle fibers that will eventually compose the cross-sectional area of a muscle (Wilson *et al.* 1992). This complicates the relatively simple idea that myogenesis could be accomplished by two waves or by two generations of myotubes. Domestic sheep, *O. aries*, generally have a gestational period of about 155 days with muscle formation commencing at about 32 days gestation age and primary myotubes generated through about 38 days. Secondary myotube formation occurs in multiple waves and continues from about 80 to 140 days gestation age, depending on the muscle type and the species studied (Ashmore *et al.* 1972; Maier *et al.* 1992). While muscle biologists have learned much about the phenotypic and functional ontogeny of muscle in terrestrial animals, this area remains open for continued investigation.

ONTOGENY OF FLIGHT MUSCLES IN BATS

What is unique about flight muscle ontogeny that should be relevant to the adaptive responses of bats to their environment? Everything! Many North American bats have a well-scripted schedule of ontogeny that plays out in about three to four weeks. During this period, deciduous teeth are shed and replaced by insect-shearing permanent teeth. In parallel, the young bats grow anywhere from 0.2 g per day (during the first 5 postnatal days) to 0.3 g per day (from postnatal days 10 through 15), reaching 90% of the adult body mass by the ages of about 24 to 29 days (Jones 1967 for *Nycticeius humeralis*; Powers *et al.* 1991 for *Myotis lucifugus*). The bats are weaned sometime in the fourth week of postnatal life and must adapt to volancy at this same time. The price of neuromuscular failure at this time can be a catastrophic crash that can injure or kill the bat, or that can make the young bat accessible to predators. Compared to kittens which will experiment with their developing locomotory skills over a period of eight weeks (Peters 1983), little brown bats (*M. lucifugus*) achieve volancy relatively quickly, a condition similar to that seen in certain avian species that nest in elevated but rather inaccessible locations (for example, barn swallows or cliff swallows: Welty 1975; common swifts: Martins 1997). These birds, as with insectivorous bats at weaning, must pursue vigorous locomotor strategies shortly after fledging. Flight muscles appear to develop in a highly channelized fashion to ensure that they are suitably developed and are capable of producing sufficient power to enable flight in the fourth week of life. Muscle and flight ontogeny has been well studied in certain insectivorous North American bats, particularly little brown bats (*M. lucifugus*: Kunz & Anthony 1982; Powers *et al.* 1991; Schutt *et al.* 1994). Relevant literature regarding flight muscle development in bats (*M. lucifugus*) will be presented, followed by prospective contrasts with adaptively different bats from other families and other parts of the world.

The end result of neuromuscular ontogeny in little brown bats is a suite of flight muscles that are homogeneous: all muscle fibers are adapted for fast-twitch contractile performance, all are highly oxidative, and all fibers appear to be poorly suited for glycolytic (anaerobic) special-ization (Table 11.2; Brigham *et al.* 1990; Powers *et al.* 1991; Schutt *et al.* 1994). The pectoralis muscle has been better studied than any other muscle in bats. In general, however, the other primary downstroke muscles exhibit the same fundamental characters described above. It is unusual in mammals to see a single muscle, or several muscles as in bats, so important to an animal's locomotion and exhibiting such a singular focus in design as does pectoralis (see reviews of muscle composition in dogs by Armstrong *et al.* 1982; in rats, see Armstrong & Phelps 1984; for an old review of other mammals see Ariano *et al.* 1973).

Returning to the subject of bat flight muscles provides contrast to the above-mentioned terrestrial mammals. The flight muscles of adult little brown bats may, at first, appear uninteresting to a physiologist because of the remarkable homogeneity of the pectoralis and its focus on powering flight. I have referred to these flight muscles, and the pectoralis muscle in particular, with the 'toggle switch' hypothesis. As one possibil-ity related to this hypothesis, the little brown bat does not use the pector-alis much to modulate flight styles or to maneuver. Instead, much like the design of a twin-engine aircraft, the pectoralis functions simply to provide the sustained power to develop thrust (parallel to the jet in this example),

Table 11.2. *Histochemical properties of adult muscle fibers in pectoralis, acromiodeltoideus, biceps brachii, and semitendinosus muscles of adult little brown bats (Myotis lucifugus); these values approximate general properties of fiber types in several vespertilionid species*

Histochemical fiber type	Pectoralis	Acromiodeltoideus[a]			Semimembranosus		
	IIa (FO)[b]	IIa (FOm)	IIa (FOh)	IIa (FOl)	IIb (FG)	IIa (FOG)	I (SO)
Percent in muscle	100%	19%	45%	36%	–	55%	45%
Fiber diameter (μm)	19	–	–	–	–	33	29
Fiber area[c]	408	–	618	–	–	–	–
mATPase (alkaline preincubation)	high	high	high	low	–	high	low
Oxidative enzyme activity	high	high	high	lower	–	high	high
Glycolytic enzyme activity	high	low	low	low	–	high	low

Notes:

[a] Data for acromiodeltoideus from Armstrong et al. (1977). These data illustrate another classification scheme for bat muscle that was based heavily on oxidative potential.

[b] The terms FO (fast and oxidative), FOG (fast, oxidative and glycolytic), and SO (slow and oxidative) represent another classification system often used in physiology.

[c] Fiber area (μm²) data estimated in transverse sections and adapted from Powers et al. (1991).

Data adapted from Hermanson et al. (1991) and from Schutt et al. (1994) unless indicated otherwise.

while the distal parts of the wing and, thus, the muscles of the forearm may be primarily responsible for maneuvering. As an example, a simple jet aircraft design presents two jet engines mounted in pods close to the fuselage of the aircraft. The jets provide for forward thrust. The relatively small wing flaps, located farther along the wing than the jets, capitalize on basic physics to provide the upward or downward moments to the wing necessary to initiate a turn. The tail, similarly, presents a relatively small surface area but is sufficient to effect lateral control of the flight path. Although the tail of a little brown bat bears no resemblance to that of a jet aircraft, it is necessary in the side-to-side maneuverability of bats as well as in adjusting overall pitch stability of the flying bat. In summary, the pectoralis muscles could produce much of the power required for lift and thrust generation, whereas the power required for maneuverability may derive from smaller muscles distributed along the forearm (Fig 11.1). How does this relate to the ontogeny of the pectoralis muscle? Recall that the pectoralis of adult *M. lucifugus* become highly focused and contain only one muscle fiber type, one physiological potential. Yet the newborn pectoralis of *M. lucifugus* contain at least two identifiable fiber types and perhaps as many as four transitional myosin heavy chain isoforms. More work remains to understand why this transformational sequence occurs. Finally, as described in the next paragraph, our understanding of the functional attributes of neonatal and adult muscle is also equivocal, and the 'toggle switch' hypothesis may represent an oversimplification.

Criticism of the 'toggle switch' hypothesis comes from two studies of turning flight in bats and birds. Norberg (1976) reported two maneuvers of bats that create a rapid loss of height: a 180° roll followed by downward 'lift' generated by the now upside-down airfoils; and sideslip which used a large component of drag to cause downward motion. In both of these maneuvers one wing was pronated (the leading edge rotated downward) to create reduced angle of attack and, therefore, a reduction of lift on one side of the bat's body. It is likely that the wing pronation was initiated at the base of the wing (resulting from a pectoralis contraction). In birds, Dial (1992) demonstrated that rock doves (*Columba livia*) could sustain steady-state forward flight subsequent to denervation of some or most of the forearm muscles (controlling the distal wing). However, these birds were incapable of performing a 'controlled' landing and were evidently incapable of turning behaviors. The recent study by Warrick & Dial (1998) provides the first kinematic evidence that the pectoralis is the prime controller of turning flight. These authors demonstrated that banking was elicited not from shape changes (increase or decrease of lift surface) or angle of attack alterations, but rather from wingtip velocity

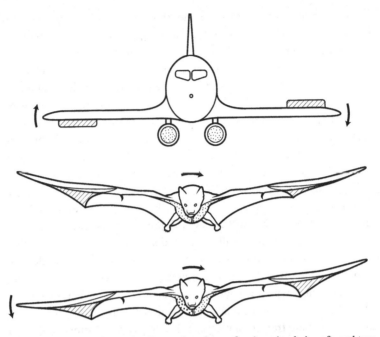

Figure 11.1. Schematic of control surfaces of an imagined aircraft, and two versions of a stylized bat viewed from the front. In the aircraft, the jets may modulate thrust production but do not alter the direction of thrust. Turning is controlled by specific control surfaces or flaps located on the wings and some distance from the center of mass. In the bats, the 'toggle switch' idea suggests that the left and right pectoralis produce equal force (equal stippling, both sides) to power the downstroke, but are not capable of controlling maneuvers. Turns might be effected by changes in wing conformation caused by muscles of the forearm. A likely scenario, however, includes pectoralis modulation such that the pectoralis of one side produces more force faster than the contralateral muscle. This results in accelerated downstroke movements on one side relative to the other, coupled with conformation changes in the distal wing. These models are presented because the unitypic (all homogeneous fast-twitch fibers) composition of the pectoralis of vespertilionid bats challenges how or even whether the pectoralis can be modulated to contribute to maneuvers.

asymmetries between the inside and outside wing (relative to the direction of a turn). Warrick & Dial concluded from these data that the pectoralis was functioning as more than just a powerful motor, but was critical in producing large force asymmetry in the downstroke, asymmetries that needed to be countered by active upstroke muscle force asymmetries on the ensuing upstroke to stabilize the bird's flight. More study is needed to

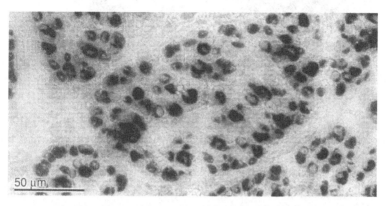

Figure 11.2. Transverse section of late gestational age pectoralis muscle of *Myotis lucifugus*, probably about 10 days prior to expected birth (based on developmental state and mean time of parturition within the colony). This section was stained for myosin ATPase following acidic preincubation (pH 4.3), a protocol that stains type I, potential slow-twitch myotubes darkly. Note extensive connective tissue accumulations between myotubes. Many myotubes contain central nuclei (unstained clear inclusions), the primary determinant to call them myotubes as opposed to myofibers at this stage. Small myotubes located in apposition to the larger myotubes likely represent secondary generation myotubes. The pectoralis is very crudely organized at this stage.

clarify the muscular basis of turning behavior in bats. At present, my 'toggle switch' hypothesis is not supported by the literature. A majority of the downstroke work for straight horizontal flight is provided bilaterally by the pectoralis and by the accessory downstroke muscles, including serratus anterior, subscapularis, and short head of biceps brachii while some component of thrust (drag-based) might also be produced during the active upstroke (Altenbach & Hermanson 1987).

Similar to the transitional stages observed during muscle ontogeny in terrestrial mammals, bats express transitional histochemical fiber types and myosin isoforms during their development. In prenatal (Fig. 11.2) or early postnatal bats (less than four days age), type I fibers are distributed sporadically throughout the deepest parts of the pectoralis muscle (Figs. 11.3, 11.4; and see Schutt *et al.* 1994). No evidence of type I fibers is retained in the adult muscle, nor can type I fibers be detected in bats showing spontaneous flapping behavior (aged 10 days or older: Fig. 11.5). Comparison with hindlimb muscles (knee flexors such as the biceps femoris, semimembranosus and semitendinosus) of *M. lucifugus* shows expression of type I myosin in about 10% of all hindlimb muscle fibers

346

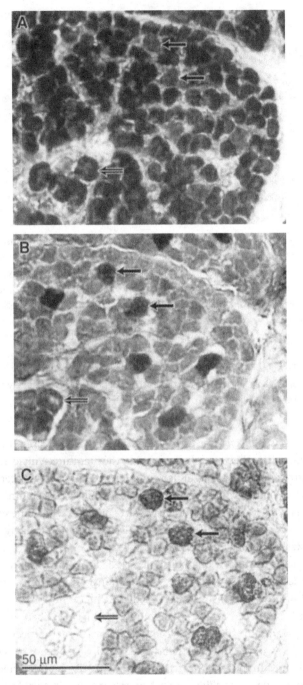

Figure 11.3. Transverse sections of 12-hour-old *Myotis lucifugus* pectoralis muscle stained for myosin ATPase following A) alkaline preincubation (pH 10.3), B) acidic preincubation at pH 4.4, and C) after reaction with anti-slow (S58) myosin antibody. Type II, presumed fast-twitch fibers stain darkly and

sampled. These type I fibers are not lost through development and may play a significant role in postural maintenance during roosting in adult bats. Thus, type I myosin heavy chain isoform is expressed throughout the development stages and adult life in certain muscles of little brown bats. In the pectoralis of little brown bats, however, type I myosin disappears from all myofibers during the normal course of ontogeny and before the onset of flight. We do not know if the type I fibers of neonatal bats are lost during ontogeny (see McClearn *et al.* 1995) or if they transform to a fast-twitch phenotype.

Transitional forms of myosin heavy chain isoforms exist in developing little brown bats, and in particular, in the flight muscles which will develop into unitypic muscles. Schutt *et al.* (1994) noted multiple transitional myosin heavy chain isoforms in prenatal and neonatal little brown bats (Fig. 11.6). These transitional isoforms did not persist at older postnatal ages. By about 6 days postnatal age, SDS polyacrylamide gel electrophoresis revealed the expression of a myosin heavy chain isoform that comigrated at the level of the adult type II myosin heavy chain isoform, and that showed increasing levels of protein concentration with increasing age during the preflight period. During the period from 6 to 10 days postnatal, the transitional isoforms disappeared from the SDS-PAGE results. Thus, nearly 20 days before the onset of true volancy, and 4 days before the onset of spontaneous bilateral wing flapping behavior (stage I of Powers *et al.* 1991), the pectoralis muscle of little brown bats contained the myosin isoforms that appear necessary to facilitate the contractile activity of adult powered flight. This contrasts with a report by Yokoyama & Uchida (1979) who concluded that although adult fiber number is fixed at the time of birth, fiber type differentiation occurs at the time that juveniles become volant (approximate age 20 days). These authors defined myofiber differentiation on the basis of three characters: t-tubule orientation, lipid

Figure 11.3 (*cont.*)
type I should not stain following alkaline preincubation. However, at this age, transitional isoforms appear to cause all fibers to react positively in this myosin ATPase reaction. In the acidic preincubation myosin ATPase study, all presumed slow-twitch (type I) fibers stain darkly and correlate with serial sections reacted against anti-slow antibodies include many type I myosin positive fibers. Two type I fibers are indicated with arrows. Histochemistry needs to be correlated with other approaches in newborn bats and most other mammals (see Hermanson 1998). Note that significant connective tissue continues to infiltrate the space between fibers and muscle fascicles. Some central nuclei are still present (double arrow).

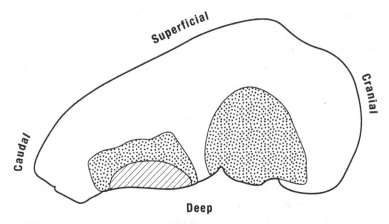

Figure 11.4. Schematic transverse section through the midbelly of pectoralis of a 1-day-old *Myotis lucifugus*. The stippled and hatched areas contain type I fibers, identified through staining with myosin ATPase following acidic preincubation or through reactions against myosin type I specific antibodies. The hatched area represents a separate part of the pectoralis complex, the pectoralis abdominalis. Unshaded areas of the muscle contain exclusively type II fibers, identified through histochemical and immunocytochemical means. (After Schutt *et al.* 1994.)

density, and mitochondrial content. Yokoyama & Uchida were not using the same histochemical studies that we (Hermanson *et al.* 1991; Schutt *et al.* 1994) used for the study of *Myotis lucifugus*. The complexity and size of the chiropteran pectoralis has been a factor underlying the dearth of data regarding quantification of fiber number in the pectoralis of neonatal and newly volant bats.

No modification of the myosin heavy chain isoform complement is observed once volancy is practiced, suggesting that the translation of myosin proteins is tightly controlled by genetic factors and is not dependent on exercise-related feedback from increasing levels of flight experienced by 20- to 40-day-old *M. lucifugus*. Other studies, notably those of O'Farrell & Studier (1973) described flight in little brown bats at 15 days of age in Nevada. It is difficult to reconcile those early flight observations with our studies, which have been based largely on studies of little brown bats in eastern North America. However, the early maturation of the contractile proteins may provide an element of safety for the juvenile bats in cases where early flight is dictated by disruption of the maternity roost by external factors, or simply to protect against the consequences of accidental dislodgement from the roost. In the latter case, the ability to fly, even briefly, would prevent a young bat from falling to the floor of a roost or may facilitate avoidance of predators. The floor of a maternity roost can

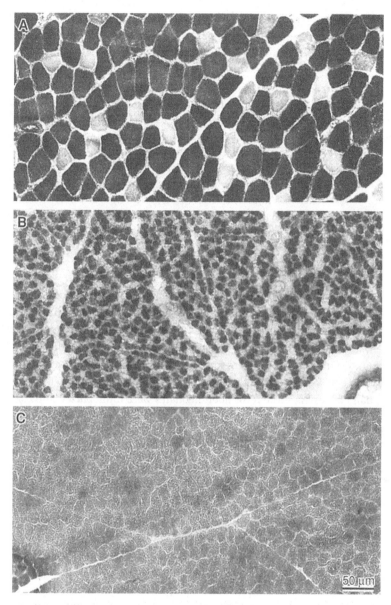

Figure 11.5. Transverse sections of 9-day-old *Myotis lucifugus* semitendinosus (A) and pectoralis (B) muscles stained for myosin ATPase following alkaline preincubation (pH 10.3). Pectoralis fibers and fascicles are well differentiated although substantial connective tissue still lies between fascicles. This connective tissue is less apparent in adult pectoralis (C) and the space between fascicles is primarily defined by perimysium, blood vessels, and nerve bundles. Note that semitendinosus fibers are larger in diameter than age-matched pectoralis, and that numerous type I (unstained) fibers are present in the hindlimb muscle.

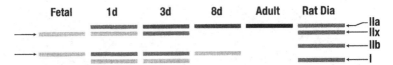

Figure 11.6. Schematic of SDS polycrylamide gel electrophoresis results showing the expression of transitional myosin heavy chain isoforms (arrows) that are expressed in fetal and early neonatal *Myotis lucifugus* but that are lost in progressively older neonates, particularly after age 8 days. Type IIa myosin heavy chain isoform is first expressed in 1- to 6-day-old neonates and is retained as the adult isoform. It migrates at the same rate or slightly slower than rat type IIa myosin heavy chain isoform. We concluded that functional myosin isoforms were present two to three weeks in advance of normal first time of volancy. Transitional myosin heavy chain isoforms are normal components of the ontogeny of mammalian muscle. (After Schutt *et al.* 1994.)

be a dangerous place. Neonatal *Myotis austroriparius* in Florida were altricial relative to more precocial *Tadarida brasiliensis* neonates that shared the same maternity roost. The *M. austroriparius* juveniles had significantly higher mortality, an observation based on examination of carcasses found on the roost floor (Hermanson & Wilkins 1986). These deaths were attributed to predation by insect larvae that were concentrated in the guano deposits on the roost floor: neonates were immobilized and consumed by large numbers of larvae shortly after falling to the ground. *Tadarida brasiliensis* neonates, in contrast, exhibited tenacious clustering ability that was also described by Pagels & Jones (1974). Hermanson & Wilkins hypothesized that hindlimb muscles of *T. brasiliensis* were relatively more precocial than those of the southeastern brown bats, facilitating clustering behavior and allowing the bats to remain secure within safe parts of the roost. Mortality was most significant in the youngest bats (age 1–5 days). Thus, enhanced strength and coordination of fore- and hindlimbs seemed to reduce predation on animals more than 5 days old. At the very least, Hermanson & Wilkins (1986) reported little mortality in juveniles greater than 5 days old and especially less mortality in the *M. austroriparius* once their hindlimb musculoskeletal systems achieved some threshold level of functionality.

To understand the ontogeny of flight, it is useful to have data about the correlated development of the musculoskeletal system. Powers *et al.* (1991) did an elegant job of correlating the growth of muscle as well as growth of the wing airfoil with the various behavioral stages exhibited by little brown bats. They showed that wing loading decreased throughout postnatal development, a factor that would reduce power requirements for flight. This was particularly significant in that adult wing loading values were achieved in many of the postnatal bats by age 15 days, a time when forward flapping flight was first practiced, although the bats were not capable of prolonged target-oriented flight (see Fig. 9.9, this volume). Powers *et al.* also indicated that the pectoralis and acromiodeltoideus muscles, both flight muscles with differing roles in the coordination of flight, demonstrated a significant increase in fiber cross-sectional area. There was a 2.9-fold increase in fiber cross-sectional area during the first 15 days of postnatal life. This contrasted with the condition of the quadriceps femoris, a knee extensor considered important in roosting behavior and in extending the uropatagium. The quadriceps muscles tended to have larger muscle fibers, relative to the flight muscles, at birth, and to show a steady but less impressive growth of fiber size during the postnatal period. In fact, the quadriceps muscle fibers were barely at the lower range of the cross-sectional area for adult quadriceps muscle fibers at the time of full flight onset, at approximately 25 days of age. The fibers of pectoralis and acromiodeltoideus muscles tended to achieve adult fiber size values before or at the time of full volancy. Measurements of fiber diameter showed that juvenile muscles overlapped the range of adult values at about 15 days of age (Schutt *et al.* 1994).

The pectoralis and acromiodeltoideus showed little variation in myosin ATPase characters during the postnatal ontogeny (Powers *et al.* 1991; Schutt *et al.* 1994) however, the α-GPD pathway tended to show differentiation among individual fibers during the period from day 10 to day 25, and to lose this differentiation in the adult condition (Powers *et al.* 1991). It has been argued that small insectivores such as *M. lucifugus* do not need, or cannot maintain, anaerobic fibers because such fibers require heavier glycogen stores to fuel the flight muscles (Yacoe *et al.* 1982; Powers *et al.* 1991; and see parallel argument for bird muscles by Goldspink 1977). It is not evident what function the relatively high glycolytic fibers would serve in postnatal animals 'practicing' wing movements in the roost, or during the first few trial flights adjacent to the roost. The expression of high glycolytic fibers during this period may represent some level of ontogenetic conservatism, similar to the expression of multiple transitional myosin isoforms in the same neonatal and older postnatal muscle fibers. Or, the expression of high-GPD fibers may represent some necessary developmental pathway, such as might occur during the formation of mitochondria, during the postnatal development of chiropteran flight muscles. Another alternative is that the high GPD levels observed in stage III bats (Powers *et al.* 1991) may power the short bursts of wing flapping seen in juveniles of this age. This latter point may underlie why juvenile bats do not, or can not, fly for long periods of time.

In a similar analysis on the pectoralis muscle of house sparrows, *Passer domesticus*, Jones (1982) observed rapid accumulation of intracellular lipid droplets in the myofibers of pre-fledgling birds from the ages of 9 to 15 days, but she did not find glycogen being stored in the myofibers. Newly fledged *P. domesticus* were unable to fly as well as adults, either in terms of speed or sustained length of flight, possibly because the myofibers were only about 60% of their adult diameters at the time of fledging and the muscles were incapable of generating adult force levels (Jones 1982).

One of the characters prominent in the primary flight muscles of bats (pectoralis, subscapularis, serratus anterior) is the homogeneous size and shape of the muscles. Fibers tend to be relatively smaller than seen in equivalently sized mammals, apparently because of the focus on oxidative metabolism in these fibers. The smaller size of the fibers has been thought to facilitate better diffusion of oxygen into the fibers, and of metabolic byproducts out of the fibers and into the blood (Mathieu-Costello *et al.* 1992; Lexell 1997). An additional feature of bat flight muscle is the presence of numerous capillaries surrounding each fiber. Obviously, a relatively high number of capillaries per fiber facilitates the exchange of metabolites from the blood to the fiber and back. Mathieu-Costello and colleagues concluded that the greater capacity of flight muscle to exchange metabolites was primarily facilitated by an increase in the size of the capillary to fiber interface. Greater efficiency of exchange, then, is due to an increase of absolute capillaries rather than simply a decrease in fiber size. Additional adaptations for oxidative metabolism include reported high density of mitochondria throughout the muscle fibers, as well as a significantly high deposition of lipid droplets between myofibrils of individual myofibers (Ohtsu *et al.* 1978). The high fat content is assumed to be necessary to sustain the high aerobic demands of flight placed upon the flight muscles. Correct determination of mitochondrial density and lipid distribution, as provided by Ohtsu *et al.*, requires the resolution provided by electron microscopy.

PRENATAL AND POSTNATAL EFFECTS ON BODY MASS AND
MUSCLE ONTOGENY

The data of Powers *et al.* (1991; Fig. 11.4) suggest that neonatal bats had body masses ranging from about 1.25 to 2.25 g. Further, there is a relative lack of synchronicity of births of *M. lucifugus* (Tuttle & Stevenson 1982). Births in colonies of *M. lucifugus* in Ithaca, New York, ranged over about 14 days (Wimsatt 1945; Taylor 1997). The peak birthing period in the two Ithaca studies deviated by 14 days, with Wimsatt's study reporting births during the last two weeks of June and early July and the Taylor study reporting births in the second and third weeks of July. The late dates of the birthing period observed by Taylor were attributed to cool temperatures during the preceding months, because temperature has a strong effect on the length of gestation (Tuttle & Stevenson 1982). These two points of variance lead to a question of how the timing of birth, length of gestation, and prenatal factors such as maternal nutrition, might effect fetal outcome and later postnatal growth and development. One study of *Pipistrellus subflavus* showed a lengthened period of gestation as a result of cold weather conditions, or a low abundance of insect prey (Hoying & Kunz 1998). Offspring of these pipistrelles were larger (indicated by body mass or by forearm length) at birth than animals born a year earlier when conditions were more favorable.

Low birth weight in domestic mammal species is well known to correlate with smaller adult weights. For example, in lambs, low-birth-weight animals result from pregnancies with twins or triplets. The lambs resulting from such pregnancies will grow normally but will have increased fat deposits (Villette & Theriez 1981). In pigs, low-birth-weight pigs grow more slowly, and have increased fat and reduced muscle mass relative to normal birthweight animals (Powell & Aberle 1980). Experimental analysis in lambs of the effects of placental insufficiency, resulting from larger than normal litter sizes, suggest that low-birth-weight lambs will have reduced numbers of myonuclei, a factor that causes a reduction in the rate and quality of muscle growth (Greenwood *et al.* 2000). These low-birth-weight lambs grew to smaller adult stature and had higher fat content in their muscles despite being fed a normal ration of food. In addition, the consequences of placental insufficiency were most pronounced on neonatal muscle growth when the challenge was presented during the last one-third of the gestation period.

Thus, based on observations of growth in lambs and piglets, one might expect that bats should exhibit some intrinsic control over the size of their offspring at birth, or have some mechanism that might mitigate environmental factors. Although the length of the gestational period of most mammals is fixed, bats exhibit a wide range of gestational length which may enable them to accommodate fluctuations in the environment (O'Farrell & Studier 1973; Racey 1973; Tuttle & Stevenson 1982). A spring gestational period characterized by prolonged cold weather might diminish the supply of insects available to pregnant *M. lucifugus* causing the bats to lengthen the gestational period, or delay fertilization, by entering into frequent periods of torpor and maintaining low body temperature. Despite these adaptive shifts of gestational period, if a wide range of neonatal body masses is evident, it is possible that the adult population will include a number of bats that exhibited low body mass at birth, may include animals with a higher fat content, or may not include these individuals because of selection against the survival of low-birth-weight animals. These are areas of research where it may be productive to investigate the effects of prenatal and postnatal factors on normal growth and development, and to determine the degree of variance in adult body mass, or muscle composition that results from low birth weight or other distinguishing characters. It is easy to assume that all *M. lucifugus* in a population are equally adept at securing insects: all bats should fly equally fast and would be equally maneuverable. However, a look at the range of athletic talents in the human population, or in athletic animals such as racehorses should cause caution about an assumption that all bats are 'created equal.' Similarly, juvenile bats may exhibit different foraging patterns based on interindividual differences in maneuverability. Such differences could be related to wing morphology such as aspect ratio or wing loading. During the first few weeks of true flight, young *M. lucifugus* show a correlation between age and increasing tendency to forage in cluttered but insect-rich environments (Adams 1996). Thus, bats that have low birth weight, or that were born later in the season, may be expected to reach the adult foraging niche relatively later in the season that normal-birth-weight animals, or animals that were born earlier in the season. A nice parallel to this was provided by Tuttle (1976) who noted that juvenile gray bats (*Myotis grisescens*) that had to fly longer distances from their cave roosts to foraging sites not only took longer to attain adult body mass, but actually suffered a significant body mass loss compared to bats with short transport distances during the 20-day period following attainment of volancy. I suggest that a combination of long transport distance (roost to foraging site) and low birth weight may be detrimental to survivorship of some bat species.

Another vespertilionid bat, *Pipistrellus pipistrellus*, has similar developmental patterns to *M. lucifugus* in terms of postnatal growth and time to flight (Hughes *et al.* 1995). The pipistrelles achieved volancy at about 28 days postnatal age. Analysis of flight ontogeny demonstrated that early flights were characterized by rapid wingbeats (about 12.5 wingbeats per second), while more mature animals settled down to a slower wingbeat (about 10.5 wingbeats per second) that presumably correlated with more efficient muscle use of power. Despite the decrease in wingbeat frequency, more mature bats flew faster even with comparable wingbeat amplitudes. Multiple factors could be invoked, including achievement of final adult wing morphometry and neuromuscular coordination.

Other bat species may have a more lengthy period of neonatal development. For example, *Tadarida brasiliensis cynocephala* has precocial features as a neonate (discussed above and see Hermanson & Wilkins 1986), but does not achieve volancy until about 38 days postnatal age (Pagels & Jones 1974). Kunz & Robson (1995) studied *T. b. mexicana* and noted first flights at 'the beginning of the sixth week,' after adult forearm dimensions were achieved and body mass approached 80% of adult female postpartum weight. These authors reiterated that most bats, including *T. brasiliensis*, are neither altricial nor precocial in the true sense of the term (perhaps as compared to definitions coined for birds by Nice 1962), because bats are large at birth (up to 25% of the maternal post-parturient weight: Kurta & Kunz 1987), which is a precocial character. In contrast, *T. brasiliensis* will grow slowly and not achieve functional loco-motion until three to six weeks after birth, a consequence of the rela-tively long period of growth observed in the pectoral limbs. This latter characteristic is considered highly altricial. Hermanson & Wilkins (1986) described *T. brasiliensis cyanocephala* as precocial in contrast to *Myotis aus-troriparius* because of their surefootedness within the roost immediately after birth. Perhaps we need to clarify this definition better. We are left with several important questions. Firstly, is there something about the nature of parental care of the offspring (see discussion in Chapter 12) or the nature and stability of food resources that facilitates a lengthened period of lactation? Secondly, do the flight muscles have an intrinsic delay in their maturation that might impose the late onset of flight on this species? I suspect that answers may derive from the first question. For instance, Williams *et al.* (1973) demonstrated that *T. brasiliensis* individuals routinely fly 50 km (one-way) from roosts in the American southwest to their nightly feeding locations, but that pregnant and lactating females will maintain loyalty to a single roost for most of the summer season (see also Geluso *et al.* 1976). Flights of 50 to 100 km length would seem difficult to sustain unless the musculoskeletal system was nearly complete in its development of adult-like qualities (flight speed, fatigue resistance, effi-ciency of control such as might result from practiced flight).

Young *Phyllostomus hastatus* (Phyllostomidae) require approxi-mately 56 days to achieve volancy (Stern *et al.* 1997). These omnivorous bats did not fly until wing area, wingspan, and aspect ratio reached nearly 90%, or more, of their adult dimensions. Stern *et al.* hypothesized that the lengthy period of maternal care expressed in these bats, relative to insectivorous bats, might be caused by the less stringent demands of flight control than might be observed in bats forced to fly in complex environments and forced to pursue elusive insect prey. They listed multi-ple arguments for this scenario, including the need for bats, in general, to permit nearly complete mineralization of their long bones to protect against shear stresses (see Swartz *et al.* 1992; Papadimitriou *et al.* 1996).

Although Stern *et al.* (1997) suggest that muscle development may preclude early flight, no evidence specific to the ontogeny of muscle was available although they derived several models suggesting that *P. hastatus* should exhibit low flight speeds and low cost of transport (P_{mp}) at the time of weaning. *P. hastatus* as an adult exhibits two muscle fiber types in the pectoralis (Hermanson *et al.* 1998). Thus, ontogeny of the fiber type phenotypes of these bats needs to be examined to compare and contrast with data for unitypic pectoralis muscles of vespertilionid bats. Do these bats exhibit an adult myofiber count or fiber type composition in the pectoralis at birth? Why is the period of maternal care (or the time to first flight) twice as long in *P. hastatus* compared to *Myotis* spp.? It would be of interest to learn if the lengthy preflight period was constrained by muscle development, neural development, or perhaps by nutritional characters.

Other areas of study include analysis of ontogeny of the musculoskeletal system in emballonurids which, as adults, tend to roost in exposed areas where the adaptive pressure for immediate onset of flight is critical. It would seem likely that their offspring should have precocial locomotory development to facilitate rapid escape from predators. How long is it before juvenile emballonurids achieve volancy? These questions would be interesting from two perspectives. Firstly, they would broaden our knowledge about the ontogeny of bats representing several divergent clades. Secondly, the distinction of unitypic pectoralis muscles (as seen in molossids and vespertilionids) from bitypic pectoralis msucles (seen in phyllostomids and apparently in emballonurids as well) allows comparison of ontogenetic trajectories towards distinct adult functional muscle designs. Studies of the development of the musculoskeletal system of these, and other bats, are lacking at present.

ACKNOWLEDGMENTS

I wish to thank Bill and Janet Schutt for their part in many of the field studies on neonatal little brown bats. Many of the data on muscle growth and development in little brown bats stemmed from Bill's doctoral research. Bill Schutt, Marketa Lillard, Jackie Petrie Bentley, and Matt Cobb were skilled assistants who kept me laughing even when the data made no sense. Diane Kelly provided a critical review of an earlier draft of this chapter. All work was performed in accordance with protocols approved by the Institutional Animal Care and Use Committee of Cornell University, and with the cooperation of the New York State Department of Environmental Conservation. Artwork by M. A. Simmons.

REFERENCES

Adams, R. A. (1996) Size-specific resource use in juvenile little brown bats, *Myotis lucifugus* (Chiroptera: Vespertilionidae): is there an ontogenetic shift? *Canadian Journal of Zoology*, **74**, 1204–10.

Altenbach, J. S. & Hermanson, J. W. (1987) Bat flight muscle function and the scapulo-humeral lock. In *Recent Advances in the Study of Bats*, ed. M. B. Fenton, P. A. Racey, & J. M. V. Rayner, pp. 100–18. Cambridge: Cambridge University Press.

Anthony, E. L. P. & Kunz, T. H. (1977) Feeding strategies of the little brown bat, *Myotis lucifugus*, in southern New Hampshire. *Ecology*, **58**, 775–86.

Ariano, M. A., Armstrong, R. B. & Edgerton, V. R. (1973) Hindlimb muscle fiber populations of five mammals. *Journal of Histochemistry and Cytochemistry*, **21**, 51–5.

Armstrong, R. B. & Phelps, R. O. (1984) Muscle fiber type composition of the rat hindlimb. *American Journal of Anatomy*, **171**, 259–72.

Armstrong, R. B., Iannuzo, C. D. & Kunz, T. H. (1977) Histochemical and biochemical properties of flight muscle fibers in the little brown bat, *Myotis lucifugus*. *Journal of Comparative Physiology*, **119**, 141–54.

Armstrong, R. B., Saubert, C. W. IV, Seeherman, H. J. & Taylor, C. R. (1982) Distribution of fiber types in locomotory muscles of dogs. *American Journal of Anatomy*, **163**, 87–98.

Ashmore, C. R., Robinson, D. W., Rattray, P. & Doerr, L. (1972) Biphasic development of muscle fibers in the fetal lamb. *Experimental Neurology*, **37**, 241–55.

Bottinelli, R., Betto, R., Schiaffino, S. & Reggiani, V. (1994) Maximum shortening velocity of myosin heavy chain isoforms in single skinned fast fibres of rat skeletal muscle. *Journal of Muscle Research and Cell Motility*, **15**, 413–19.

Brigham, R. M., Ianuzzo, C. D. & Fenton, M. B. (1990) Histochemical and biochemical plasticity of muscle fibers in the little brown bat *(Myotis lucifugus)*. *Journal of Comparative Physiology B*, **160**, 183–6.

Brooke, M. H. & Kaiser, KK. (1970) Three 'myosin adenosine triphosphate' systems: The nature of the pH lability and sulfhydryl dependence. *Journal of Histochemistry and Cytochemistry*, **18**, 670–2.

Buchler, E. R. (1980) The development of flight, foraging, and echolocation in the little brown bat *(Myotis lucifugus)*. *Behavior, Ecology, Sociobiology*, **6**, 211–18.

Buller, A. J. & Lewis, D. M. (1965) Further observations on the differentiation of skeletal muscles in the kitten hindlimb. *Journal of Physiology*, **176**, 355–70.

Close, R. (1964) Dynamic properties of fast and slow skeletal muscle of the rat during development. *Journal of Physiology*, **173**, 74–95.

Condon, K., Silberstein, L., Blau, H. M. & Thompson, W. J. (1990) Development of muscle fiber types in the prenatal rat hindlimb. *Developmental Biology*, **138**, 256–74.

Crow, M. T. & Stockdale, F. E. (1986) Myosin expression and specialization among the earliest muscle fibers of the developing avian limb. *Developmental Biology*, **113**, 238–54.

Denny-Brown, D. (1929) On the nature of postural reflexes. *Proceedings of the Royal Society of London B*, **104**, 253–301.

Dial, K. P. (1992) Avian forelimb muscles and nonsteady flight: can birds fly without using the muscles in their wings? *Auk*, **109**, 874–85.

Dill, D. B., Edwards, H. T. & Talbott, J. H. (1932) Studies in muscular activity. VII. Factors limiting the capacity for work. *Journal of Physiology*, **77**, 49–62.

Engel, W. K. & Karpati, G. (1968) Impaired skeletal muscle maturation following neonatal neurectomy. *Developmental Biology*, **17**, 713–23.

English, A. W., & Letbetter, W. D. (1992) A histochemical analysis of identified compartments of cat lateral gastrocnemius muscle. *Anatomical Record*, **204**, 123–30.

Estes, R. D. (1991) *The Behavior Guide to African Mammals*. Los Angeles: University of California Press.

Fenton, M. B. & Barclay, R. M. R. (1980) *Myotis lucifugus. Mammalian Species*, **142**, 1–8.

Fitts, R. H., Brimmer, C. J., Heywood-Cooksey, A. & Timmerman, R. J. (1989) Single muscle fiber enzyme shifts with hindlimb suspension and immobilization. *American Journal of Physiology*, **256**, C1082–91.

Fitts, R. H., Metzger, J. M., Riley, D. A. & Unsworth, B. R. (1985) Models of disuse: a comparison of hindlimb suspension and immobilization. *Journal of Applied Physiology*, **60**, 1946–53.

Foehring, R. C. & Hermanson, J. W. (1984) Morphology and histochemistry of flight muscles in free-tailed bats, *Tadarida brasiliensis. Journal of Mammalogy*, **65**, 388–94.

Geluso, K. N., Altenbach, J. S. & Wilson, D. E. (1976) Bat mortality: pesticide poisoning and migratory stress. *Science*, **194**, 184–6.

Goldspink, G. (1977) Mechanics and energetics of muscle in animals of different sizes, with particular reference to the muscle fiber composition of vertebrate muscle. In *Scale Effects in Animal Locomotion*, ed. T. J. Pedley, pp. 37–55. London: Academic Press.

Greenwood, P. L., Hunt, A. S., Hermanson, J. W. & Bell, A. W. (2000) Effects of birth weight and postnatal nutrition on neonatal sheep: 2. Skeletal muscle and growth. *Journal of Animal Science*, **78**, 50–61.

Guth, L. & Samaha, F. J. (1969) Qualitative differences between actomyosin ATPase of slow and fast mammalian muscle. *Experimental Neurology*, **25**, 138–52.

Guth, L. & Samaha, F. J.. (1970) Procedure for the histochemical demonstration of actomyosin ATPase. *Experimental Neurology*, **28**, 365–7.

Hauschka, S. D. (1994) The embryonic origin of muscle. In *Myology. Basic and Clinical*, 2nd edn, ed. A. G. Engel & C. Franzini-Armstrong, pp. 3–73. New York: McGraw-Hill.

Hermanson, J. W. (1998) Chiropteran muscle biology: a perspective from molecules to function. In *Bat Biology and Conservation*, ed. T. H. Kunz & P. A. Racey, pp. 127–39. Washington: Smithsonian Institution Press.

Hermanson, J. W. & Altenbach, J. S. (1981) Functional anatomy of the primary downstroke muscles in a bat, *Antrozous pallidus. Journal of Mammalogy*, **62**, 795–800.

Hermanson, J. W. & Altenbach, J. S. (1985) Functional anatomy of the shoulder and arm of the fruit-eating bat, *Artibeus jamaicensis. Journal of Zoology (London)*, **205**, 157–77.

Hermanson, J. W. & Foehring, R. C. (1988) Morphology and histochemistry of the flight muscles of *Artibeus jamaicensis*: implications for motor control. *Journal of Morphology*, **196**, 353–62.

Hermanson, J. W. & Wilkins, K. T. (1986) Pre-weaning mortality in a Florida nursery roost of *Myotis austroriparius* and *Tadarida brasiliensis. Journal of Mammalogy*, **67**, 751–4.

Hermanson, J. W., LaFramboise, W. A. & Daood, M. J. (1991) Uniform myosin isoforms in the flight muscles of little brown bats, *Myotis lucifugus. Journal of Experimental Zoology*, **259**, 174–80.

Hermanson, J. W., Ryan, J. M., Cobb, M. A., Bentley, J. & Schutt, W. A., Jr. (1998) Histochemical and electrophoretic analysis of the primary flight muscle of several phyllostomid bats. *Canadian Journal of Zoology*, **76**, 1983–99.

Hoying, K. M. & Kunz, T. H. (1998) Variation in size at birth and postnatal growth in the insectivorous bat *Pipistrellus subflavus* (Chiroptera: Vespertilionidae) *Journal of Zoology (London)*, **245**, 15–27.

Hughes, P. M., Rayner, J. M. V. & Jones, G. (1995) Ontogeny of 'true' flight and other aspects of growth in the bat *Pipistrellus pipistrellus. Journal of Zoology (London)*, **235**, 291–318.

Jones, C. (1967) Growth, development, and wing loading in the evening bat, *Nycticeius humeralis* (Rafinesque). *Journal of Mammalogy*, **48**, 1–19.

Jones, M. M. (1982) Growth of the pectoralis muscle of the house sparrow *(Passer domesticus). Journal of Anatomy*, **135**, 719–31.

Kelly, A. M. & Zacks, S. I. (1969) The histogenesis of rat intercostal muscle. *Journal of Cell Biology*, **42**, 135–53.

Kugelberg, E. (1973) Histochemical composition, contraction speed, and fatiguability of rat soleus motor units. *Journal of the Neurological Sciences*, **20**, 177–98.

Kunz, T. H. & Anthony, E. L. P. (1982) Age estimation and post-natal growth in the bat, *Myotis lucifugus. Journal of Mammalogy*, **63**, 23–32.

Kunz, T. H. & Robson, S. K. (1995) Postnatal growth and development in the Mexican free-tailed bat *(Tadarida brasiliensis mexicana)*: birth size, growth rates, and age estimation. *Journal of Mammalogy*, **76**, 769–83.

Kurta, A. & Kunz, T. H. (1987) Size of bats at birth and maternal investment during pregnancy. *Symposia of the Zoological Society of London*, **57**, 79–106.

LaFramboise, W. A., Daood, M. J., Guthrie, R. D., Moretti, P., Schiaffino, S. & Ontell, M. (1990) Electrophoretic separation and immunological identification of type 2X myosin heavy chain in rat skeletal muscle. *Biochimica Biophysica Acta*, **1035**, 109–12.

LaFramboise, W. A., Daood, M. J., Guthrie, R. D., Schiaffino, S., Moretti, P., Brozanski, B., Ontell, M. P., Butler-Browne, G. S., Whalen, R. G. & Ontell, M. (1991) Emergence of mature myosin phenotype in rat diaphragm muscle. *Developmental Biology*, **144**, 1–15.

Lexell, J. (1997) Muscle capillarization: morphological and morphometrical analyses of biopsy samples. *Muscle & Nerve*, suppl. **5**, S110–12.

Maier, A., McEwan, J. C., Dodds, K. G., Fischman, D. A., Fitzsimmons, R. B. & Harris, A. J. (1992) Myosin heavy chain composition of single fibers and their origins and distribution in developing fascicles of sheep tibialis cranialis muscles. *Journal of Muscle Research and Cell Motility*, **13**, 551–72.

Martins, T. L. F. (1997) Fledging in the common swift, *Apus apus*: weight-watching with a difference. *Animal Behavior*, **54**, 99–108.

Mathieu-Costello, O., Szewczak, J. M., Logemann, R. B. & Agey, P. J. (1992) Geometry of blood-tissue exchange in bat flight muscle compared with bat hindlimb and rat soleus muscle. *American Journal of Physiology*, **262**, R955–65.

McClearn, D., Medville, R. & Noden, D. M. (1995) Muscle cell death during development of head and neck muscles in the chick embryo. *Developmental Dynamics*, **202**, 365–77.

Mech, L. D. (1970) *The Wolf*. New York: Natural History Press.

Narusawa, M., Fitzsimmons, R. B., Izumo, S., Nadal-Girard, B., Rubinstein, N. A. & Kelly, A. M. (1987) Slow myosin in developing rat skeletal muscle. *Journal of Cell Biology*, **104**, 447–59.

Nemeth, P. & Pette, D. (1981) Succinate dehydrogenase activity in fibres classified by myosin ATPase in three hind limb muscles of rat. *Journal of Physiology*, **320**, 73–80.

Nice, M. M. (1962) Development of behavior in precocial birds. *Transactions of the Linnean Society of New York*, no. 8.

Norberg, U. M. (1976) Some advanced flight manoeuvres of bats. *Journal of Experimental Biology*, **64**, 489–95.

Nyström, B. (1968) Histochemistry of developing cat muscles. *Acta Neurologica Scandinavica*, **44**, 405–39.

O'Farrell, M. J. & Studier, E. H. (1973) Reproduction, growth, and development in *Myotis thysanodes* and *Myotis lucifugus* (Chiroptera: Vespertilionidae). *Ecology*, **54**, 18–30.

Ohtsu, R., Möri, T. & Uchida, T. A. (1978) Electron microscopical and biochemical studies of the major pectoral muscles of bats. *Comparative Biochemistry and Physiology*, **61A**, 101–207.

Ontell, M. & Dunn, R. F. (1978) Neonatal muscle growth: a quantitative study. *American Journal of Anatomy*, **152**, 539–56.

Ontell, M. & Kozeka, K. (1984) The organogenesis of murine striated muscle: a cytoarchitectural study. *American Journal of Anatomy*, **171**, 133–48.

Ontell, M., Bourke, D. & Hughes, D. (1988) Cytoarchitecture of the fetal murine soleus muscle. *American Journal of Anatomy*, **181**, 267–78.

Pagels, J. F. & Jones, C. (1974) Growth and development of the free-tailed bat, *Tadarida brasiliensis cynocephala* (LeConte). *Southwestern Naturalist*, **19**, 267–76.

Papadimitriou, H. M., Swartz, S. M. & Kunz, T. H. (1996) Ontogenetic and anatomic variation in mineralization of the wing skeleton of the Mexican free-tailed bat, *Tadarida brasiliensis*. *Journal of Zoology (London)*, **240**, 411–26.

Peter, J. B., Barnard, R. J., Edgerton, V. R., Gillespie, C. A. & Stempel, K. E. (1972) Metabolic properties of three fiber types of skeletal muscle in guinea pigs and rabbits. *Biochemistry*, **11**, 2627–33.

Peters, S. (1983) Postnatal development of gait behaviour and functional allometry in the domestic cat (*Felis catus*). *Journal of Zoology (London)*, **199**, 461–86.

Pette, D. & Staron, R. S. (1990) Cellular and molecular diversities of mammalian skeletal muscle fibers. *Reviews in Physiology, Biochemistry and Pharmacology*, **116**, 1–76.

Powell, S. E. & Aberle, E. D. (1980) Effects of birth weight on growth and carcass composition of swine. *Journal of Animal Science*, **50**, 860–8.

Powers, L. V., Kandarian, S. C. & Kunz, T. H. (1991) Ontogeny of flight in the little brown bat, *Myotis lucifugus*: behavior, morphology, and muscle histochemistry. *Journal of Comparative Physiology A*, **168**, 675–85.

Racey, P. A. (1973) Environmental factors affecting the length of gestation in heterothermic bats. *Journal of Reproduction and Fertility*, Suppl. **19**, 175–89.

Reiser, P. J., Moss, R. L., Giulian, G. G. & Greaser, M. L. (1985) Shortening velocity in single fibers from adult rabbit soleus muscles is correlated with myosin heavy chain composition. *Journal of Biological Chemistry*, **260**, 9077–80.

Schaller, G. (1972) *The Serengeti Lion*. Chicago: University of Chicago Press.

Schutt, W. A., Jr., Cobb, M. A., Petrie, J. L. & Hermanson, J. W. (1994) Ontogeny of the pectoralis muscle in the little brown bat, *Myotis lucifugus*. *Journal of Morphology*, **220**, 295–305.

Snow, D. H., Billeter, R., Mascarello, F., Carpenè, E., Rowlerson, A. & Jenny, E. (1982) No classical type IIb fibers in dog skeletal muscle. *Histochemistry*, **75**, 53–65.

Stern, A. A., Kunz, T. H. & Bhatt, S. S. (1997) Seasonal wing loading and the ontogeny of flight in *Phyllostomus hastatus* (Chiroptera: Phyllostomidae). *Journal of Mammalogy*, **78**, 1199–209.

Swartz, S. M., Bennett, M. B. & Carrier, D. R. (1992) Wing bone stresses in free flying bats and the evolution of skeletal design for flight. *Nature*, **359**, 726–9.

Sypert, G. W. & Munson, J. B. (1981) Basis of segmental motor control: motor neuron size or motor unit type? *Neurosurgery*, **8**, 608–21.

Talmadge, R. J., Grossman, E. J. & Roy, R. R. (1996) Myosin heavy chain composition of adult feline (*Felis catus*) limb and diaphragm muscles. *Journal of Experimental Zoology*, **275**, 413–20.

Taylor, A. R. (1997) Characteristics of two central New York maternity colonies of *Myotis lucifugus*. Unpublished Honors thesis, Cornell University, Ithaca, New York.

Taylor, C. R. (1978) Why change gaits? Recruitment of muscles and muscle fibers as a function of speed and gait. *American Zoologist*, **18**, 153–61.

Taylor, C. R. & Rowntree, V. J. (1973) Temperature regulation and heat balance in running cheetahs: a strategy for sprinters. *American Journal of Physiology*, **224**, 848–51.

Thompson, W. J., Condon, K. & Astrow, S. H. (1989) The origin and selective innervation of early muscle fiber types in the rat. *Journal of Neurobiology*, **21**, 212–22.

Tuttle, M. D. (1976) Population ecology of the gray bat (*Myotis grisecens*): factors influencing growth and survival of newly volant young. *Ecology*, **57**, 587–95.

Tuttle, M. D. & Stevenson, D. (1982) Growth and survival of bats. In *Ecology of Bats*, ed. T. H. Kunz, pp. 105–50. New York: Plenum Press.

Vaughan, T. A. (1959) Functional morphology of three bats: *Eumops, Myotis, Macrotus*. *University of Kansas Publications, Museum of Natural History*, **12**, 1–153.

Villette, Y. & Theriez, M. (1981) Influence of birth weight on lamb performances. II. Carcass and chemical composition of lambs slaughtered at the same weight. *Annals Zootechnology*, **30**, 169–82.

Walmsley, B., Hodgson, J. A. & Burke, R. E. (1978) The forces produced by medial gastrocnemius and soleus muscles during locomotion in freely moving cats. *Journal of Neurophysiology*, **41**, 1203–16.

Warrick, D. G. & Dial, K. P. (1998) Kinematic, aerodynamic and anatomical mechanisms in the slow, maneuvering flight of pigeons. *Journal of Experimental Biology*, **201**, 655–72.

Welty, J. C. (1975) *The Life of Birds*. Philadelphia: W. B. Saunders.

Whalen, R. G., Johnstone, D., Bryers, P. S., Butler-Browne, G. S., Ecob, M. S. & Jaros, E. (1984) A developmentally regulated disappearance of slow myosin in fast-type muscles of the mouse. *FEBS Letters*, **177**, 51–6.

Whalen, R. G., Sell, S. M., Butler-Browne, G. S., Schwartz, K., Bouveret, P. & Pinset-Harstrom, I. (1981) Three myosin heavy-chain isozymes appear sequentially in rat muscle development. *Nature*, **295**, 805–9.

Whalen, R. G., Toutant, M., Butler-Browne, G. S. & Watkins, S. C. (1985) Hereditary pituitary dwarfism in mice affects skeletal and cardiac myosin isozyme transitions differently. *Journal of Cell Biology*, **101**, 603–9.

Wigston, D. J. & English, A. W. (1991) Fiber-type proportions in mammalian soleus muscle during postnatal development. *Journal of Neurobiology*, **23**, 61–70.

Williams, T. C., Ireland, L. C. & Williams, J. M. (1973) High altitude flights of the free-tailed bat, *Tadarida brasiliensis*, observed with radar. *Journal of Mammalogy*, **54**, 807–21.

Williams, T. M., Dobson, G. P., Mathieu-Costello, O., Morsbach, D., Worley, M. B. & Phillips, J. A. (1997) Skeletal muscle histology and biochemistry of an elite sprinter, the African cheetah. *Journal of Comparative Physiology B*, **167**, 527–35.

Wilson, S. J., McEwan, J. C., Sheard, P. W. & Harris, A. J. (1992) Early stages of myogenesis in a large mammal: formation of successive generations of myotubes in sheep tibialis cranialis muscle. *Journal of Muscle Research and Cell Motility*, **13**, 534–50.

Wimsatt, W. A. (1945) Notes on breeding behavior, pregnancy, and parturition in some vespertilionid bats of the eastern US. *Journal of Mammalogy*, **26**, 23–33.

Yacoe, M. E., Cummings, J. W., Myers, P. & Creighton, G. K. (1982) Muscle enzyme profile, diet and flight in South American bats. *American Journal of Physiology*, **242**, R189–94.

Yokoyama, K. & Uchida, T. A. (1979) Ultrastructure of postembryonic development of the pectoral muscles in the Japanese lesser horseshoe bat, *Rhinolophus cornutus cornutus* from the standpoint of adaptation for flight. *Journal of the Faculty of Agriculture, Kyushu University*, **24**, 49–63.

Yokoyama, K., Ohtsu, R. & Uchida, T. A. (1979) Growth and LDH isozyme patterns in the pectoral and cardiac muscles of the Japanese lesser horseshoe bat, *Rhinolophus cornutus cornutus* from the standpoint of adaptation for flight. *Journal of Zoology (London)*, **187**, 85–96.

Zhang, M. & McLennan, I. S. (1998) Primary myotubes preferentially mature into either the fastest or slowest muscle fibers. *Developmental Dynamics*, **213**, 147–57.

Zomlefer, M. R., Zajac, F. E. & Levine, W. S. (1997) Kinematics and muscular activity of cats during maximum height jumps. *Brain Research*, **126**, 563–6.

12

The ontogeny of behavior in bats: a functional perspective

INTRODUCTION

Social organization in bats varies enormously. At one extreme, animals may remain solitary for much of the time. Several megachiropteran species are found solitarily or as mother–young pairs for most of the year (Bradbury 1977). On the other hand, roosts of the Brazilian free-tailed bat *Tadarida brasiliensis* encompass some of the largest aggregations of vertebrates known. Maternity colonies of this species often contain millions of individuals, with up to 30–50 million bats estimated at one site earlier this century (Wilson 1997). Most species have varied social organizations according to season, and according to the age and sex of the individuals concerned (Bradbury 1977). Most bat species clearly have ample opportunities for social interactions, both at roosts and away from them, for example at foraging sites.

Several authors (e.g., Bradbury 1977; Fenton 1985; McCracken 1987; Wilkinson 1995) have reviewed aspects of social organization and communication in bats. Although much attention has been paid to describing different patterns of social organization, and to communication among adult bats and mother–young pairs, there are few data on the ontogeny of social behavior. Most studies at maternity colonies have concentrated on behavior of females and on mother–young interactions, rather than on the development of social behavior in infants. In this chapter I will attempt to fill this gap by reviewing several recent studies that have addressed the ontogeny of behavior, especially behaviors that have a social context.

I begin by examining the scant literature on social interactions among pups in maternity colonies. I will then address social interactions between mothers and pups. The main questions that I ask are: how do infants shift from becoming largely dependent on maternal behaviors to becoming independent? I consider all ontogenetic stages of development to be specialized and adapted to their ecological niches. The neonate is adapted to the needs of feeding and predator avoidance imposed by its environment, and metamorphoses into an adult which can potentially reproduce, and which may possess quite different mechanisms for feeding and hazard avoidance from the neonate (Coppinger & Smith 1990). I arbitrarily use the term 'infant' to describe youngsters that have not left maternity sites to feed by themselves. After their first flights from maternity sites, youngsters are termed 'juveniles.' In this context, how do schedules of grooming and suckling change during development? Is allomaternal care widespread, and, if not, how do mothers avoid misdirecting maternal care? What evidence exists for social learning during development? I will describe the emergence behavior of juveniles, and try to explain its significance in a functional context. Do young learn about the location of foraging areas and how to forage from their mothers, or are they left to fend for themselves? Finally, I will ask whether significant parental care extends beyond weaning. Nomenclature follows Corbet & Hill (1991).

SOCIAL INTERACTIONS AMONGST PUPS

Females of most species are colonial, and parturition is often fairly synchronous. For example, in the fringed myotis *Myotis thysanodes*, all births may occur within two weeks (O'Farrell & Studier 1973). Infants of similar age therefore occur in groups, but little is known about interactions among infants at maternity colonies. Infants may huddle together in clusters for thermoregulatory advantages (Tuttle & Stevenson 1982). In the evening bat *Nycticeius humeralis*, for example, the extent of clustering in the maternity colony increases with decreasing temperature, and neonates are found at the centers of clusters (Watkins & Shump 1981). Tuttle & Stevenson (1982) argued that species inhabiting the coldest roosts show the greatest synchrony of parturition, and perhaps it would be in these species that the greatest potential for social interactions among pups occurs. Kunz (1982) points out that bats in the centers of clusters would be buffered more from ambient temperatures than bats in peripheral positions. In most cases the cluster will generate heat, and bats in central positions will be warmer, which may increase growth rates in infants (Tuttle & Stevenson 1982). It would be interesting to determine how cluster position varies with developmental state in bats, and how important competition and co-operation are in determining the most favorable cluster position in infants and adults. Overall, it is clear that research on social interactions among infant bats has to date focused on thermoregulatory advantages of clustering behavior. Future research should also consider behavioral factors such as the effects of position in the cluster on predation risk and access to suckling females.

PATERNAL CARE

In the majority of bats, as in many mammals (Clutton-Brock 1991), parental care seems to be given exclusively by females. Milk production has been noted in males of the Dyak fruit bat *Dyacopterus spadiceus* (Megachiroptera) (Francis *et al.* 1994), but it is not clear if this was aberrant or of functional significance. Harem-holding male Seba's short-tailed bats *Carollia perspicillata* may guard infants at roosts from intrusions by other males, so may show some paternal care (Porter 1979). Indirect paternal care may be important in monogamous species such as the yellow-winged bat *Lavia frons*, where both sexes defend foraging territories in which the young learns to feed (Vaughan & Vaughan 1987). Males and unrelated female bats in maternity roosts may also bring indirect benefits to infants by, for example, increasing roost temperature. A male *Vampyrum spectrum* has been observed bringing prey to its young at the roost (G. S. Wilkinson, personal communication), and a male in captivity brought food to a roost, where the prey were eaten by a female who was present with her pup (J. W. Bradbury, in Vehrencamp *et al.* 1977). Paternal behavior in bats may be rare, but it is certainly understudied. Fruitful areas for research may include guarding of infants by males, and the provision of food by carnivorous species. There is a surprisingly large number of seemingly monogamous bat species (Bradbury 1977; Heller *et al.* 1993), and paternal care may be most prevalent in these. I will next concentrate on direct maternal care, where considerable research has been performed.

MATERNAL CARE

Direct maternal care typically begins at parturition, when infants may be born headfirst or in breech position (Kleiman 1969). Infant mortality is typically low during early postnatal growth, highlighting the efficacy of maternal care. Juvenile mortality may increase during the first weeks of leaving the roost in some species, when bats begin foraging on their own (e.g., Tuttle & Stevenson 1982; Ransome 1990; Stern & Kunz 1998).

Suckling

Infant bats are born in an altricial state, and grow rapidly (Kunz 1987; Kunz & Stern 1995). For some time the infants depend on milk for their growth. Dry matter and energy concentration in bat milk increase between early and mid-lactation (Kunz et al. 1995a). Weaning times show considerable inter- and intraspecific variation, and are reviewed by Tuttle & Stevenson (1982). Lactation may last for 45 days in greater horseshoe bats *Rhinolophus ferrumequinum*, although the young emerge from the roost and begin foraging between 25–30 days (Jones et al. 1995). Thus this species may mix self-foraging and suckling for 15–20 days after first leaving the roost (the so-called 'grace period' – Bradbury 1977). Suckling may be protracted in some species (up to 9–10 months in common vampire bats *Desmodus rotundus*; Schmidt & Manske 1973). Changes in suckling behavior with infant age have been described in captive groups of common pipistrelles *Pipistrellus pipistrellus* (Hughes 1990) and brown long-eared bats *Plecotus auritus* (McLean & Speakman 1996). In both species all infants remain attached to their mothers for 100% of observations in the roost for the first 3–5 days, with attachment probabilities declining with infant age (Fig. 12.1). No *Pipistrellus pipistrellus* were observed attached to mothers in the roost by 35 days (by which time infants were eating mealworms), and only 5% of *Plecotus auritus* infants were attached at 36–40 days. These data suggest that any potential for allosuckling may increase with infant age, since infants are more likely to wander and hence encounter other females as they develop.

There is no evidence for nipple preference in *Pipistrellus pipistrellus*, or for structured sequential use of either nipple in *P. pipistrellus* (Hughes 1990) or *Plecotus auritus* (De Fanis & Jones 1995). Infant *Tadarida brasiliensis* may alternate between teats up to five times during a single meal (McCracken & Gustin 1991). Milk secretion in *Pipistrellus pipistrellus* is regulated by the suckling schedule, and frequent suckling may promote the rate of milk production (Wilde et al. 1995). In *T. brasiliensis*, females roost apart from their infants and suckle them according to a regular daily schedule. Conversely, female *Myotis lucifugus* and *M. velifer* roost with their pups and seem to suckle them on demand (Kunz et al. 1995a).

Carrying of young by mothers

Mothers carry non-volant young in several species (e.g., Fenton 1969; Davis 1970; Ansell 1986; Baumgarten & Vieira 1994). The Cape serotine *Pipistrellus* (formerly *Eptesicus*) *capensis* has been recorded carrying twins

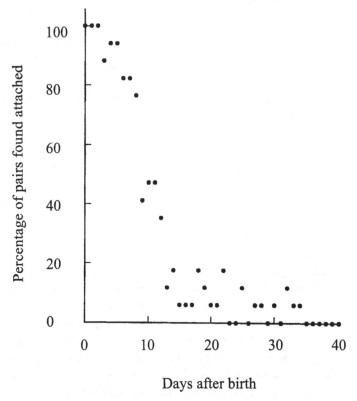

Days after birth

Figure 12.1. The percentage of mother–infant pairs that were found attached on examination each morning in a captive colony of *Pipistrellus pipistrellus*. The number of intact pairs falls sharply after the day of birth (day 0), so that fewer than 50% of pairs are attached by day 10. Pipistrelles were of the 55 kHz phonic type (Jones & Parijs 1993). (From Hughes 1990.)

that equalled 90% of the mother's mass (Ansell 1986). Carrying of young may be in response to disturbance at the roost (Fenton 1969), or can occur if the mother changes roost naturally. The high cost of flight associated with carrying loads that would correspond to the masses of young (Hughes & Rayner 1991, 1993) make it unlikely that mothers would carry young on foraging flights, unless the risks of leaving them at the maternity roost are exceptionally high. However, Tuttle (1986) claims that epauletted fruit bats *Epomorphorus* spp. may carry twins to feeding sites, even when the young are capable of flying by themselves. At feeding sites the mother and young may fly together, and the young are then carried home.

Some authors have argued that frugivorous species may be more likely to carry young to foraging sites than are insectivorous species, because frugivorous bats often forage when not volant (Fenton 1969). This hypothesis has not been tested rigorously, and indeed Schneider's leaf-nosed bat *Hipposideros speoris* may carry volant young to foraging areas, even though it is insectivorous (Marimuthu 1988; Radhamani *et al.* 1990). Greater false vampire bat, *Megaderma lyra*, mothers may carry young to night roosts while they forage (Marimuthu 1988), perhaps to minimize risks associated with the detection of their day roosts by predators. Carrying young to foraging areas could also assist the young by allowing for the learning of foraging area topography, or in learning foraging skills (Radhamani *et al.* 1990). Direct evidence for these benefits is lacking, however, and may not explain why neonates are carried to foraging areas. In some species young may be carried only until they attain a critical pro-portion of maternal mass – e.g., 50–60% in Geoffroy's tail-less bat *Anoura geoffroyii* (Baumgarten & Vieira 1994). It would be informative to relate the percentage of mass carried by the mother to maternal wing loading in cross-species comparisons, to determine whether species with relatively lower wing loadings carry young more often or for longer periods. The necessity of load-carrying during pregnancy, and sometimes during lac-tation, has presumably been a major selective force explaining why female bats are often bigger than males (thus having larger wings) (Myers 1978; Williams & Findley 1979). Pubic nipples are found in bats in the families Rhinopomatidae, Craseonycteridae, Megadermatidae, Rhino-lophidae, and Hipposideridae, and almost certainly facilitate the carry-ing of infants by mothers (Simmons 1993).

Grooming

Experimental studies of grooming in ungulates show that ectopara-site loads increase if grooming is restricted (Mooring *et al.* 1997). Ectoparasites can reduce growth rates in birds (e.g., Brown & Brown 1986) and in cattle (Norval *et al.* 1989). It is likely, therefore, that grooming func-tions in limiting ectoparasite damage in bats. Grooming may be con-ducted by the mother (a form of allogrooming), or by the infant itself (autogrooming). McLean & Speakman (1997) have studied the ontogeny of grooming behavior in captive *Plecotus auritus*. In *P. auritus*, allogrooming by the mother declines rapidly with infant age (as it does in *Nyctalus noctula*: Kleiman 1969). Self-grooming increases in a linear manner (Fig. 12.2). Thus, maternal investment in the grooming of infants declines as infants grow. Overall, grooming frequency appears to increase with infant age, though it remains possible that infant grooming may not be as efficient as maternal grooming. The relationship between parasite

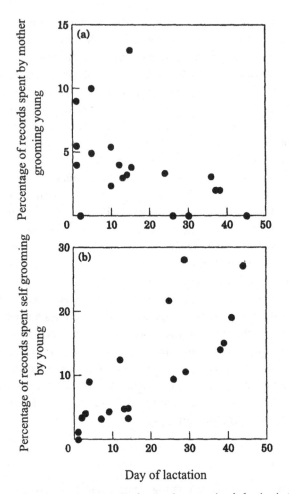

Figure 12.2. Ontogenetic changes in grooming behavior in *Plecotus auritus*. The percentage of records spent (a) by mothers grooming infants and (b) autogrooming by the infant in relation to infant age (day of lactation). (From McLean & Speakman 1997.)

load and grooming behavior remains unknown in *P. auritus*. In many bats, grooming may be conducted orally, or by scratching. To my knowledge there are no studies that document the relative frequencies of these events, or their functional significance.

Although there is strong evidence that grooming reduces ectoparasite loads in some mammals (see above), there is no evidence for this in bats. Experimental studies that manipulate parasite loads in bats and examine grooming responses are needed. It would also be instructive to know the energy cost of grooming (see McLean & Speakman 1997). Correlative studies on *Rhinolophus ferrumequinum* show that infestation with blood-feeding spinturnicid wing mites, *Eyndhovenia euryalis*, decreases as infants develop, and that infant autogrooming increases with age (G. Jones, unpublished). These results suggest that grooming reduces parasite load, as opposed to it occurring as a response to high levels of ectoparasitism (though see Wilkinson 1986). In several bat species, juveniles carry higher ectoparasite loads than adults (Gannon & Willig 1995), and lactating females often carry high loads (Dietz & Walter 1995; G. Jones, unpublished), at a time when grooming is reduced (McLean & Speakman 1997). Alternative explanations, such that ectoparasites prefer to feed on young infants, and move off them onto other young as the host ages, need to be ruled out in studies investigating the function of grooming, however. Infant *Nyctalus noctula* call when groomed by their mothers, and calling appears to elicit further grooming (Kleiman 1969).

Thermoregulation

Neonates may be enclosed in the wings of mothers, perhaps for thermoregulatory reasons (Marimuthu 1988). Thermoregulation of neonates may be especially important, since young little brown bats, *Myotis lucifugus*, and fringed myotis, *M. thysanodes*, are unable to thermoregulate by themselves before 9½ and 4½ days respectively (Studier & O'Farrell 1972). The grey-headed flying fox, *Pteropus poliocephalus*, may possess some thermoregulatory abilities at 2 days, but cannot thermoregulate as efficiently as adults at ambient temperatures of 10°C until 1 month old (Bartholomew *et al.* 1964). In *Plecotus auritus*, contact between mother and infant declines as infants age also (McLean & Speakman 1997). Extended contact with very young infants may therefore be important for thermoregulatory reasons, since young infants are inefficient at maintaining a constant body temperature, and thus maternal contact may promote infant growth.

ALLOMATERNAL CARE

Allomaternal care, whereby a female cares for young that are not her offspring, is fairly widespread in mammals (Packer *et al.* 1992). In this section I will briefly review the categories of allomaternal care that occur in bats, and comment on their extent, and on why they may evolve.

Allosuckling, or suckling of offspring born to other females, seems to be relatively rare in bats (Fenton 1985). The majority of species show selective and exclusive suckling of offspring. Lactation, especially during its middle to later phases, is expensive in energetic terms (e.g., Kurta *et al.* 1989; Kunz *et al.* 1995b), and the costs of misdirected maternal care may therefore be high. In a few species where allosuckling does occur, it is relatively infrequent. Low levels of allosuckling have been recorded in captive groups of *Pipistrellus pipistrellus* (Kleiman 1969; Eales *et al.* 1988; De Fanis & Jones 1996), though exclusive suckling of offspring has been recorded in other captive groups of this species (Hughes *et al.* 1989), and has been deduced from genetic analyses (Bishop *et al.* 1992). It is often unclear whether allosuckling is the consequence of unusual conditions experienced in captivity (e.g., high mortality of mothers), or how much of it is simply due to brief bouts of 'milk stealing' by infants against the interests of the suckling female.

The evolutionary importance of allosuckling can only be appreciated fully from field studies with minimal disturbance to the bats. Allosuckling in *Tadarida brasiliensis* may seem possible because of the potential for misdirected maternal care in the huge colonies of these bats: densities of pups in McCracken & Gustin's (1991) study averaged 4000 pups/m². Nevertheless, allosuckling is rare, short-term, and against the interest of mothers in this species. Genotype studies estimated allosuckling to occur in 17% of mother–young pairs in the gregarious *T. brasiliensis* (McCracken 1984), but subsequent studies suggested that mispairings of mothers and pups may be the consequence of short-term milk stealing (<0.5% of all suckling observed), or could result from the investigators' disturbance at the colony (McCracken & Gustin 1991).

Allosuckling has been found in a natural colony of *Nycticeius humeralis* (Wilkinson 1992a). Suckling is directed almost always towards offspring until two weeks before weaning (Fig. 12.3). Over 18% of nursing bouts then involve infants other than the mother's own, and 31% of marked females and 24% of marked infants were involved. Allosuckling bouts were as long in duration as maternal bouts. Moreover, allosuckling was not directed at pups from the same matriline, and it mainly involved female infants.

There are two obvious functional explanations for the evolution of allosuckling in bats. If females are philopatric to their natal sites, and show high levels of relatedness, allosuckling may evolve through kin-selection if the benefits of allosuckling outweigh the costs (Hamilton 1964). Allosuckling may also evolve through reciprocity (Trivers 1971; Axelrod & Hamilton 1981). Neither of these explanations applies to *T. brasiliensis* or *N. humeralis*. Genetic analyses of both species show (*N. humeralis*: Wilkinson 1992a) or suggest (*T. brasiliensis*: McCracken *et al.* 1994;

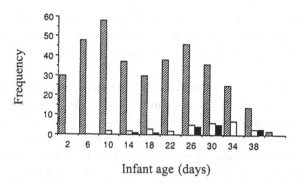

Infant age (days)

Figure 12.3. The number of nursing bouts involving descendant infants (hatched bars), nondescendant infants (open bars) and rejected nursing attempts (solid bars) in relation to infant age in *Nycticeius humeralis*. (From Wilkinson 1992a.)

McCracken & Gassel 1997) low levels of within-colony relatedness or little population structure at maternity colonies. No evidence of reciprocity was obtained in either study.

So why has allosuckling evolved in *N. humeralis*? Wilkinson (1992a) gave the following explanation. Females may benefit directly by dumping excess milk to nondescendant pups. The females will benefit because the future production of milk may be maintained or improved, and because their flight costs for foraging that night would be reduced by flying with a lower body mass. Night-to-night variation in the abundance of aerial insect prey eaten by *N. humeralis* is large, and Wilkinson speculated that it was the females who return after a successful night's foraging who would nurse nondescendant infants. The bias towards the suckling of female nondescendant pups could perhaps be explained by considering delayed benefits. Males disperse, and females show natal philopatry. Females may benefit if colony size in the future is larger (so they suckle female nondescendants that are likely to return to their natal colony), since the potential for information transfer about feeding and roosting sites (Wilkinson 1992b) may then be greater. Allosuckling in *N. humeralis* therefore seems to be the results of immediate and delayed benefits to the donor female. Whether *N. humeralis* is unusual in showing these behaviors deserves further study. Indeed, the annual cycle of *N. humeralis* seems typical of many temperate insectivorous bats, so maybe further examples of allomaternal care will be found. Allosuckling and regurgitation of blood to nondescendant young may be widespread in common vampire bats, *Desmodus rotundus* (Schmidt et al. 1980). *D. rotundus* has an exceptionally long duration of juvenile development and exhibits a range of behaviors that probably evolved through kin-selection and reciprocal altruism (Wilkinson 1984,1985a, b). Feeding of nondescendant young may even ensure survival of orphaned infants, at least in captivity (Schmidt et al. 1980).

Other forms of allomaternal care in bats may include 'baby-sitting' for infants other than offspring. 'Guarding' and retrieval of nondescendant young has been postulated for Myotis thysanodes (O'Farrell & Studier 1973), but whether females that remain in the roost really engaged in 'altruistic' behaviors towards nondescendant infants did not receive rigorous study because individuals were not marked. In Carollia perspicillata, new harem-holding males may guard infants fathered by previous harem males, so may show allopaternal care. It is probably in the interests of the new harem males to adopt youngsters sired by other males, however, as it may result in the mothers of these infants remaining in the harems of the new males in the long term (Porter 1979). Females may assist nondescendant young by providing thermoregulatory and metabolic benefits (Trune & Slobodchikoff 1976), and by helper-assisted birth (Kunz et al. 1994). Many adult bats give 'distress calls' when held captive close to their roosts (e.g., Fenton et al. 1976; August 1979; Russ et al. 1998) or when their young are held (O'Farrell & Studier 1973), and these calls may attract conspecifics which may potentially mob potential predators (Russ et al. 1998). Distress calls may therefore function in indirect alloparental care if they promote the survival of nondescendants. The extent and true nature of these observations in wild populations deserves further study.

MECHANISMS FOR AVOIDING THE MISDIRECTION OF MATERNAL CARE

Current evidence therefore suggests that mothers in most species direct care to their own infants. When allomaternal care occurs, it may be explained by mutualism or by milk stealing. How then do mothers locate and identify their offspring in maternity colonies, some of which contain extremely high densities of infants?

The evolution of coloniality in many bat species means that mothers may face formidable tasks in locating their infants on return to maternity colonies. Mother bats use several methods to locate their offspring at maternity colonies. First, they land at sites close to where they left their offspring, and probably use locational cues to achieve this (e.g., Tadarida brasiliensis: McCracken 1993). Removal of pups from the crèche (and hence removal of vocal cues) still results in the mother returning to the site where she last fed her infant. Infant T. brasiliensis are relatively

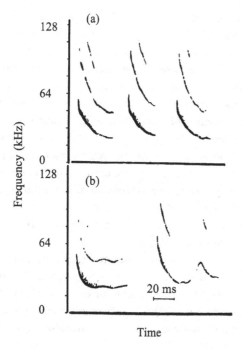

Figure 12.4. Sonograms of isolation calls produced by 15-day-old infant *Pipistrellus pipistrellus* showing within-individual consistency and between-individual variation at a fixed age: (a) three consecutive calls from one infant, (b) single calls from two other infants. Pipistrelles were of the 55 kHz phonic type (Jones & Parijs 1993). (From Jones *et al.* 1991.)

immobile at maternity sites, and locational memory narrows down the task of finding offspring (McCracken 1993). Second, infants have individually distinctive vocalizations that attract mothers when they are separated from them (isolation calls, e.g., pale spear-nosed bats *Phyllostomus discolor*: Rother & Schmidt 1985; *T. brasiliensis*: Gelfand & McCracken 1986; *Pipistrellus pipistrellus*: Jones *et al.* 1991 [Fig. 12.4]; *Nycticeius humeralis*: Scherrer & Wilkinson 1993; *Plecotus auritus*: De Fanis & Jones 1995), and mothers respond to playbacks of calls from their own infants in preference to those of other infants in choice experiments (*T. brasiliensis*: Balcombe 1990; *Pipistrellus pipistrellus*: De Fanis & Jones 1996). In *N. humeralis* isolation calls are variable among and repeatable within infants, and contain diagnostic patterns that persist throughout development (Scherrer & Wilkinson 1993). Vocal signatures may not occur in solitary species such as the hoary bat, *Lasiurus cinereus* (Koehler & Barclay 1988).

Females also prefer the scent of their offspring to that from other infants (*T. brasiliensis*: Gustin & McCracken 1987; probably *Plecotus auritus*:

De Fanis & Jones 1995). Scent may be of limited importance for offspring recognition in *Pipistrellus pipistrellus* (De Fanis & Jones 1996), and females only begin responding to offspring odors when the infants are older than seven days. This could be in response to convergence in the structure of the isolation calls emitted by infants as they develop (De Fanis & Jones 1996). If calls become less reliable indicators of identity, the bats may start using other cues such as scent, which may be more time-invariant, especially in species where stereotyped echolocation calls develop from variable isolation calls (Fenton 1994). However, the echolocation calls of the big brown bat, *Eptesicus fuscus*, seem to become more distinctive between 60 and 170 days, suggesting increasing self-consistency in echolocation calls after weaning in this species (Masters *et al.* 1995).

Experiments on *T. brasiliensis* suggest that mothers mark their pups with scent glands, and that females can distinguish odors from their own muzzles from those of other females. Females may thus distinguish offspring by matching against their own odor, or by recognition of an odor that is similar to their own. The relative importance of external markers (e.g., from scent marks), internal markers (e.g., from milk), and individual pup odors in influencing female choice of pups remains unresolved (Loughry & McCracken 1991).

In many bats, mother–offspring reunions also involve recognition of mothers by offspring. In *P. pipistrellus*, infants may develop preferences for maternal odors as they approach weaning (De Fanis & Jones 1996). All of five infant *Plecotus auritus* chose the scent and sequences of echolocation calls from their mothers in preference to those of other lactating females (De Fanis & Jones 1995). Conversely, infant *T. brasiliensis* are unable to distinguish the echolocation calls of their mothers from those of other mothers (Balcombe 1990), but are able to recognize maternal odors at least at some stages of their development (Loughry & McCracken 1991). In maternity colonies, infant *T. brasiliensis* move towards their mothers even when the distance between the pair is 40 cm, and recognition appears to be mutual (McCracken & Gustin 1991). Mothers emit 'directive calls' that appear to show individual differences in structure, and infants are attracted to directive calls. It is not known whether pups are able to identify the directive calls of their mother from those of other females (Balcombe & McCracken 1992).

SOCIAL LEARNING DURING DEVELOPMENT

All work on social learning in bats during their development has concerned the possible learning of vocalizations. For a general review of social learning in bats, the reader is referred to Wilkinson (1995). A review of vocal learning in mammals is provided by Janik & Slater (1997). Vater in Chapter 5 (this volume) considers the development of the bat cochlea. Here I will consider the evidence for the social learning of vocalizations by infant bats.

Echolocation calls

Echolocation calls of juveniles may fluctuate in frequency for some months (or even years) before stabilizing at typical adult values. In five species studied to date (the lesser horseshoe bat *Rhinolophus hipposideros*: Jones *et al.* 1992; *R. ferrumequinum*: Jones & Ransome 1993; the trident bat *Asellia tridens*: Jones *et al.* 1993; Daubenton's bat *Myotis daubentonii*: Jones & Kokurewicz 1994; and *M. lucifugus*: Pearl & Fenton 1996), first-year bats emit calls of lower frequency than older individuals. That this relationship exists among bats having different call structures, and different phylogenetic histories, suggests that a common process may link maturation with the development of call frequency in bats (Jones 1995).

Deafened (Woolf 1974), or hand-reared (Gould 1975), infant bats develop adult-like sonar sounds, as do infants reared by mothers whose voices were modified experimentally (Gould 1975). Echolocation calls clearly have a strong genetic component. However, fine-tuning of calls could be developed by social learning, and requires subtle analysis. Indeed, deafened infant horseshoe bats, *Rhinolophus rouxi*, develop adult-like echolocation calls, though the dominant constant frequency (CF) component of the call differs by up to 14 kHz from that of undeafened infants, and the harmonic structure of the call is affected by deafening (Rübsamen & Schäfer 1990).

Masters *et al.* (1995) quantified the heritabilities of frequency-modulated (FM) echolocation calls emitted by *Eptesicus fuscus* in a standardized laboratory set-up. All mothers in the study gave birth to twins. High values of heritabilities were obtained for several features of pulse design. Start frequency of the fundamental, period sweep rate, and several measures of the relative amplitude of harmonics showed similarity due to relatedness in mother–offspring regressions. Start frequency was also heritable in inter-sibling correlations. Reported measures of heritabilities will not necessarily reflect additive genetic variance however, since maternal effects (e.g., social learning) cannot be eliminated in observational studies. Genetic influences on call frequency are confounded also in species that give birth to twins, because the twins may have different fathers (e.g., Mayer 1995). Multiple paternity may partly explain the

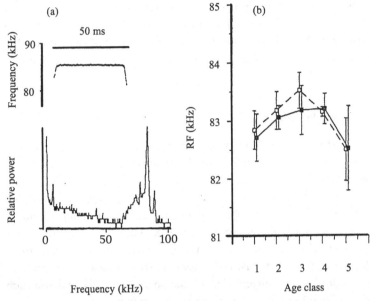

Figure 12.5. (a) Sonagram (upper) and power spectrum (lower) of an echolocation call produced by *Rhinolophus ferrumequinum*. (b) Resting frequencies (RFs – see text for definition) of different age classes at a hibernation site. Age classes are: 1) bats in their 1st year; 2) 2nd year; 3) 3rd year; 4) bats in their 4th to 10th years; 5) bats in more than their 10th year. Open circles are females, filled squares males. Means ± standard deviations illustrated. (From Jones & Ransome 1993.)

greater similarity between parameters of calls from mother and offspring than between litter-mates in *E. fuscus*.

Masters *et al.* (1995) argued that signal learning plays some role in the development of echolocation calls in *E. fuscus*, especially given the inflated values of heritabilities reported for call parameters in mother–offspring regressions. Convincing evidence for maternal effects influencing the fine-tuning of call design come from studies on *R. ferrumequinum* (Jones & Ransome 1993), a species that emits calls with a long CF component. The echolocation calls of *R. ferrumequinum* are relatively simple in structure (Fig. 12.5). Because most of the energy in the calls is concentrated in the CF portion, they are ideally suited for accurate frequency resolution from sonagraphic analysis, unlike species that emit FM calls, where frequency parameters are more difficult to quantify. In CF bats, the cochlea and auditory neurons are closely tuned to the frequency emitted when the bats are stationary (the resting frequency, or RF). The frequency of adult greater horseshoe bats varies according to their age in

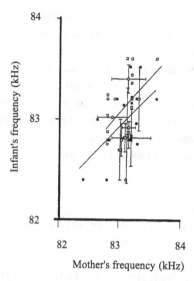

Figure 12.6. Echolocation call frequency (RF – see text for definition) of juveniles is correlated with that of their mothers in *Rhinolophus ferrumequinum*, both for female (open squares) and male (filled squares) offspring. Error bars are standard deviations for > one offspring/mother. (From Jones & Ransome 1993.)

a predictable way: RF increases rapidly between years 1 and 2, slightly between years 2 and 3, and remains relatively stable during mid-life. RF decreases when the bats become relatively old, however, with the decline being especially rapid between 10 and 23 years (Fig. 12.5).

The RF of juveniles remains relatively stable between 5 and 7 weeks, and is then correlated with the RF of their mothers, both in male and female juveniles (Fig. 12.6). The correlations could result from the effects of learning or heredity, but the high heritabilities associated with call frequency imply that learning is important. The most compelling evidence for the importance of learning comes from the relationship of juvenile RF with maternal age. Old females raise young of both sexes with lower RFs than do young females. If RF were determined in a purely genetic manner, then there would be no reason to expect the association between juvenile RF and maternal age. Age-related changes in maternal RF, and associated learning of maternal RF by the young, are the most parsimonious explanations for this association (Jones & Ransome 1993).

Jones & Ransome (1993) argued that although the basic structure of echolocation calls in *R. ferrumequinum* is likely to be determined genetically, the fine-tuning of the call seems to have a learnt component.

Auditory feedback seems to be important in the proper development of echolocation calls in *R. rouxi* (Rübsamen & Schäfer 1990), and feedback from maternal calls could be important in that species. The importance of social learning in the development of echolocation call design in bats needs experimental study. Cross-fostering studies would be valuable, though given the scarcity of allosuckling in bats (see above), they would be difficult. Progress may also be possible through studies of isolated young that are subjected to experimental playbacks of calls, as has been performed in studies of the development of social vocalizations (see below).

Isolation calls

In many bat species, infant isolation calls differ in structure from vocalizations produced by mothers, so learning of call parameters from mothers is not likely. Rasmuson & Barclay (1992) found no greater similarity in the structure of isolation calls of *Eptesicus fuscus* between twins than between calls of less related bats, and argued that the structure of isolation calls lacked any obvious genetic component (but see criticisms by Wilkinson 1995). Infants had individual vocal signatures, but these could presumably not signal kinship affiliations.

In contrast, Scherrer & Wilkinson (1993) found that isolation calls contained information about family identity in *Nycticeius humeralis*. *N. humeralis* typically produces twins or triplets under natural conditions, and so is ideal for studies on familial similarities in vocalizations. Calls from twins could be assigned to the correct family on 68% of occasions using a discriminant function model, and three components of calls were heritable. Similarities between isolation calls of bats from the same family can be appreciated from the sonagrams in Fig. 12.7. *N. humeralis* cannot learn isolation calls from mothers because the mothers emit no equivalent sounds, and because the calls are emitted at birth. Learning from other pups is unlikely as there is no convergence in call structure among infants over time. Scherrer & Wilkinson (1993) argued that the isolation calls of *N. humeralis* had a strong genetic component, though the extent of the additive component of this variation was not clear. The functional importance of heritable variation in isolation call structure could be investigated experimentally. It would be interesting to know, for example, whether mothers can distinguish between kin from differences in isolation calls, and whether acoustic discrimination could be a possible mechanism for resource allocation during milk feeding between sibs. The entire field of sibling competition for maternal milk has indeed received little attention to date.

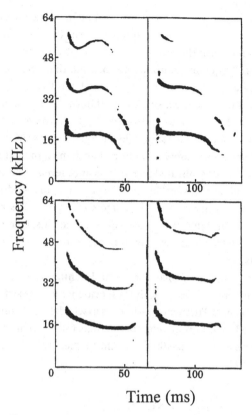

Time (ms)

Figure 12.7. Similarity between isolation calls of sibling infants in two families (upper and lower panels) of *Nycticeius humeralis*. (From Scherrer & Wilkinson 1993.)

Kin recognition between mother and offspring would be possible if mothers could compare their own calls with those of offspring, if calls contained heritable cues, and if phenotypic matching occurred (Scherrer & Wilkinson 1993). These criteria have yet to be fully met in studies on bats.

Acoustic learning may be important in the development of isolation calls in *Phyllostomus discolor*. In this species mothers produce directive calls which are individually distinctive, especially in their patterns of frequency modulation. Mothers and infants exchange directive and isolation calls over time, and the isolation calls of infants gradually come to resemble the directive calls of their mothers (Esser & Schmidt 1989). Esser (1994) used an experimental technique to determine whether infants learnt characteristics of directive calls in *P. discolor*. Pups were isolated from conspecifics at birth, and hand-reared. One group was played a maternal directive call with 10 frequency minima and maxima, the other received no acoustic stimulation. At 50 and 100 days, the isolation calls of the two groups differed, with the group receiving playbacks of directive calls producing more frequency minima and maxima than the control group. Additional evidence for learning of temporal and frequency patterns of directive calls was obtained, suggesting that the structure of isolation calls from pups that received acoustic playbacks converged on the structure of the playback reference signal (Esser 1994). Esser's statistical analyses were performed on numbers of calls rather than on individuals, however, so data are pseudoreplicated and significance levels are unreliable. Experimental evidence for vocal learning of calls used to co-ordinate foraging movements of other bats in the same social group has been found recently in adult and juvenile greater spear-nosed bats, *P. hastatus* (Boughman 1998). Vocal learning may therefore occur in several contexts in phyllostomid bats.

Overall, unequivocal evidence for vocal learning is scarce in mammals (Janik & Slater 1997). Some of the best evidence for its existence comes from studies on bats. Further experiments in which larger samples of infants are tutored by different reference signals (rather than when one group receives no auditory feedback) would further strengthen the case for vocal learning.

THE ONTOGENY OF EMERGENCE TIMING

In several colonial species (e.g., cave myotis, *Myotis velifer*: Kunz 1974; large mouse-eared bat, *M. myotis*: Audet 1990; *M. lucifugus*: Kunz & Anthony 1996; *R. ferrumequinum*, northern bat, *Eptesicus nilssonii*: Duvergé et al., in press), newly volant juveniles time their evening emergence from maternity roosts until after their mothers leave, perhaps because their slow flight (Hughes et al. 1995) renders them vulnerable to predation by diurnal and crepuscular raptorial birds (Fenton et al. 1994). Moreover, because newly volant bats are often still receiving some milk from their mothers (Jones et al. 1995), any benefits of early emergence that may give earlier access to insect food are probably small (Duvergé et al., in press). As juvenile greater horseshoe bats develop, their evening emergence becomes earlier, until by about 60–80 days, when it is similar to that of adults (Duvergé et al., in press). Juvenile emergence in *M. lucifugus* becomes indistinguishable from that of adults within two weeks of volancy (Kunz & Anthony 1996). In *E. nilssonii*, newly volant juveniles again emerge later than their mothers, but the mothers return to the roost when it is dark, and call, seemingly to communicate with juveniles. Such behavior probably reduces predation risks to the juveniles (J. Rydell, personal communication). Juvenile *E. nilssonii* may be unable to practice flight within the roost in the way that, for example, *R. ferrumequinum* can, and juveniles may then be especially vulnerable to predation during their first flights. Juvenile bats probably do not emerge after adults to perfect flight and echolocation with minimal acoustic clutter from adults (Kunz 1974; Buchler 1980), because in both *R. ferrumequinum* and *E. nilssonii* the adults feed at sites far away from those used initially by young. Minimizing predation risk is by far the most consistent explanation for the late emergence of juvenile bats given that the visual acuity of some raptorial birds (falcons) declines sharply with decreases in luminance (Fox *et al.* 1976).

THE DEVELOPMENT OF FORAGING BEHAVIOR

Many bat species produce a single young each year, and mothers should invest extensively in offspring care to maximize their own fitness. The development of flight morphology (Adams, Chapter 9, this volume) and flight muscles (Hermanson, Chapter 11, this volume) are reviewed elsewhere in this volume. Here I will concentrate on whether juvenile bats learn about where and how to forage from their mothers or from other bats, or if juveniles are left to develop foraging skills by themselves.

Species that show evidence of tuition from mothers

Lavia frons is a monogamous, territorial bat in the family Megadermatidae. A territorial male and female typically roost within 2 m of each other, and the young cling to their mothers for a week before they make their own foraging flights. The young practice flapping while clinging to their mothers, and associate with either parent during their first foraging flights. This species is very unusual among bats in showing paternal investment. *L. frons* hunts by making sallies from perches, and the young often fly to one of their parents and take over its foraging perch. Parents and young share feeding territories (Vaughan & Vaughan 1987). Newly volant young of the related heart-nosed bat, *Cardioderma cor,* also hunt from perches, and associate with mothers at foraging sites. Young may perch within a few centimeters of their mothers, or up to 20 m away, and contact between the pair may be maintained by female song (Vaughan 1976). There are anecdotal reports of female lesser *Megaderma spasma* and Australian *Macroderma gigas* false vampires bringing prey items back to the roost for their offspring to eat (cited in Bradbury 1977).

G. S. Wilkinson (personal communication) observed a male *Vampyrum spectrum* bringing back a dove to the roost for offspring to eat. Such behavior may provide a search image of particular prey items for juveniles, as well as giving nutritional investment. In general then, parental assistance in the development of foraging skills in megadermatid bats may be extensive, and deserves further research.

The lesser bulldog bat, *Noctilio albiventris*, shows interesting associations between mother and offspring (Brown *et al.* 1983). In a captive colony mothers fed their young with masticated fish or mealworms prior to weaning. Mother and young may leave the roost to travel to feeding grounds together, and may stay in contact away from the roost by duetting vocally (Brown *et al.* 1983). Mother and young pallid bats, *Antrozous pallidus*, may maintain contact close to the roost by vocal duetting in flight (Vaughan & O'Shea 1976).

Brigham & Brigham (1989) found a mother and juvenile (aged 36–45 days) *Eptesicus fuscus* foraging in close proximity of one another and sharing night roosts, but because only one mother–young pair was followed, it was not clear if the juvenile was associating specifically with its mother, or perhaps with other bats from the same colony. After 2 months, juvenile *Desmodus rotundus* are fed regurgitated blood by their mothers, before beginning to drink blood by themselves at 4 months (Schmidt *et al.* 1980). Female *D. rotundus* may share wound sites simultaneously with their daughters during their first year (Wilkinson 1985*a*), and Wilkinson (1995) suggested that social learning may be important in the development of foraging skills in species with difficult feeding techniques. Recently volant *Nycticeius humeralis* seem to follow adult females at evening emergence, but juveniles may follow unrelated females more than their own mothers (Wilkinson 1995). Newly volant bats of this species follow adults to new roosts after roost exclusion experiments (Wilkinson 1992*b*).

Species that may lack maternal tuition about foraging

In many species, juveniles probably receive no maternal tuition about foraging during early volancy (*Myotis lucifugus*: Buchler 1980; *Pipistrellus pipistrellus*: Racey & Swift 1985; *M. myotis*: Audet 1990; probably *Phyllostomus hastatus* [though there is no difference in emergence times of young and adults]: Stern *et al.* 1997). *Rhinolophus ferrumequinum* serves as an example: first foraging flights are close to the maternity roost, and juveniles expand their foraging range with increasing age, until by about 50 days they travel as far as adults (Fig. 12.8; Jones *et al.* 1995). Weaning is

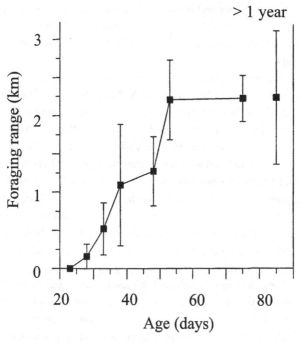

Figure 12.8. The expansion of foraging range (straight-line distance from maternity roost to furthest point reached) in juvenile *Rhinolophus ferrumequinum*. Data points are means for bats in 5-day age classes ± standard deviations, with the point on the right-hand side representing the mean foraging range of bats older than 1 year. (From Jones *et al.* 1995.)

estimated to occur at 45 days, before which juveniles return to the roost during the night to suckle. Having a small foraging range reduces travel costs to the roost, but may also be a consequence of poor flight performance (see Hermanson, Chapter 11, this volume) and/or navigational skills. Food intake increases rapidly between days 29–55 (Jones *et al.* 1995), and foraging time and efficiency (intake/foraging time) increase with age (Duvergé 1996). Foraging time increases with juvenile age in *P. hastatus* (Stern *et al.* 1997) and *M. myotis*, in which juvenile returns to the maternity colony are slower but as direct as those of adults (Audet 1990). In *Pipistrellus pipistrellus* foraging range and attempted feeding rates (terminal buzzes/minute) increase with age, until, within about 18 days after first leaving the maternity roost, their attempted feeding rate was comparable with the maximum rate recorded for adults (Racey & Swift 1985). One fruitful line of research would be to investigate how juvenile bats find their way back to the roost if they are foraging by themselves. Do they return via the route they left by, or do they take 'shortcuts' along previously unknown routes that may indicate the development and use of path integration, or recognition of familiar landmarks from novel landmarks (Bennett 1996)?

There was no difference in habitat selection between juvenile R. ferrumequinum and their mothers radio-tracked over the same time period, although mothers foraged further from the roost (Duvergé 1996). Juveniles mainly eat *Aphodius* beetles (Scarabaeidae) or crane flies (Tipulidae), which are probably relatively easy to capture. Their mothers eat moths, at least in good weather (Ransome 1996), and these are presumably more profitable prey (Jones 1990), though they may be more difficult to capture. Juveniles do not eat moths until about 42 days, and this may coincide with the full development of Doppler-shift compensation in echolocation (Konstantinov 1989). Thus dietary segregation between mothers and offspring occurs, except in poor feeding conditions when mothers may be forced to eat lower-quality prey when moths are scarce (Ransome 1996). Dietary segregation between mothers and juveniles has also been noted in *Lasiurus cinereus* (Rolseth et al. 1994), where juveniles during their first week post- fledging eat many more chironomid flies than do adults. Adults eat more dragonflies (Odonata), perhaps because they are more adept at pursuing these large, rapidly flying prey, or because adults emerge earlier, flying at a time when Odonata are more abundant. Juvenile *E. fuscus* eat softer prey compared with adults, which eat mainly beetles (Hamilton & Barclay 1998). Adult *M. lucifugus* forage in more cluttered habitats than newly volant juveniles (Adams 1997), and juveniles with the smallest wings forage in the least cluttered habitats (Adams 1996). How resource partitioning varies with wing size during development is discussed further by Adams in Chapter 9 (this volume).

MATERNAL CARE AFTER WEANING

Although newly volant bats in several species may receive little or no maternal tuition about foraging skills and sites from their mothers, associations and social learning may be important later in life. In *Desmodus rotundus*, for example, adult females have been found together with yearling offspring and adult yearling female grandoffspring at foraging sites (Wilkinson 1985a). The status of mothers in determining the status of their offspring in social groups has also received little attention since Bradbury's (1977) preliminary observations on *Phyllostomus discolor*, where offspring of nomadic females are more likely to become nomadic than are the progeny of harem residents. In *Rhinolophus ferrumequinum*, survival of first-year offspring between breeding seasons depends critically on the survival of their mothers (Ransome 1995). Juvenile mortality peaks 10 days after weaning, and the reproductive success of breeding females is strongly correlated with how long they spend with their offspring in the maternity roost after birth. Thus, although juveniles are left to forage for themselves (Jones *et al.* 1995), some sort of maternal care appears to be important after weaning (Ransome 1995). One possibility is that acceptance into the maternity colony is facilitated by the presence of the mother, or perhaps sharing of information about foraging sites becomes important once the foraging ranges of juveniles extend as far as those of adults. Associations between mothers and yearling daughters may also occur in *R. ferrumequinum* at foraging sites and at night roosts (S. J. Rossiter, G. Jones & R. D. Ransome, unpublished). In some species it is therefore worth looking for kin-biased associations after the period of first volancy, perhaps when juveniles are able to attain the foraging distances reached by their mothers. In captive *D. rotundus*, young males stay in the main group until 20–27 months, and then leave to form temporary bachelor groups (Park 1991). Male dispersal may occur at 12–18 months in natural populations (Wilkinson 1985b). Although female philopatry makes it more likely for extended maternal care to benefit female offspring, it could benefit male offspring in species where male dispersal is delayed.

CONCLUSIONS

In conclusion, there has been considerable progress since Bradbury's (1977) statement that the ontogeny of social behavior has received only minor attention. Many rigorous and detailed studies have now been conducted on the early life of bats, and some of these have involved careful observations with large sample sizes, and even experimental manipulations. Nevertheless, there is still much to learn, and a considerable amount of our knowledge about the functional significance of the ontogeny of behavior remains anecdotal. In particular, many behaviors are assumed to be forms of parental care, yet whether they actually increase offspring fitness remains unknown (Clutton-Brock 1991). I hope that this review will inspire researchers to to fill in some gaps, and to conduct well-designed observational, comparative, and experimental studies on the significance of the developmental of social behavior in bats. The longevity, philopatry, and coloniality of some bats allows great potential for the evolution of co-operative behaviors, and

studies of the development of such behaviours will have consequences
that extend beyond bat biology.

ACKNOWLEDGMENTS

I thank Gary McCracken and Jerry Wilkinson for helpful comments
on an earlier draft.

REFERENCES

Adams, R. A. (1996). Size-specific resource use in juvenile little brown bats, *Myotis lucifugus* (Chiroptera: Vespertilionidae): is there an ontogenetic shift? *Canadian Journal of Zoology*, **74**, 1204–10.
Adams, R. A. (1997). Onset of volancy and foraging patterns of juvenile little brown bats, *Myotis lucifugus*. *Journal of Mammalogy*, **78**, 239–46.
Ansell, W. F. H. (1986). Records of bats in Zambia carrying non-volant young in flight. *Arnoldia Zimbabwe*, **9**, 315–18.
Audet, D. (1990). Foraging behavior and habitat use by a gleaning bat, *Myotis myotis* (Chiroptera: Vespertilionidae). *Journal of Mammalogy*, **71**, 420–7.
August, P. X. V. (1979). Distress calls in *Artibeus jamaicensis*: ecology and evolutionary implications. In *Vertebrate Ecology in the Northern Neotropics*, ed. J. F. Eisenberg, pp. 151–60. Washington: Smithsonian Institution Press.
Axelrod, R. & Hamilton, W. D. (1981). The evolution of cooperation. *Science*, **211**, 1390–6.
Balcombe, J. P. (1990). Vocal recognition of pups by mother Mexican free-tailed bats, *Tadarida brasiliensis mexicana*. *Animal Behaviour*, **39**, 960–6.
Balcombe, J. P. & McCracken, G. F. (1992). Vocal recognition in Mexican free-tailed bats: do pups recognize mothers? *Animal Behaviour*, **43**, 79–87.
Bartholomew, G. A., Leitner, P. & Nelson, J. (1964). Body temperature, oxygen consumption, and heart rate in three species of Australian flying foxes. *Physiological Zoology*, **37**, 179–98.
Baumgarten, J. E. & Vieira, E. M. (1994). Reproductive seasonality and development of *Anoura geoffroyi* (Chiroptera: Phyllostomidae) in central Brazil. *Mammalia*, **58**, 415–22.
Bennett, A. T. D. (1996). Do animals have cognitive maps? *Journal of Experimental Biology*, **199**, 219–24.
Bishop, C. M., Jones, G., Lazarus, C. M. & Racey, P. A. (1992). Discriminate suckling in pipistrelle bats is supported by DNA fingerprinting. *Molecular Ecology*, **1**, 255–8.
Boughman, J. W. (1998). Vocal learning by greater spear-nosed bats. *Proceedings of the Royal Society London*, **265B**, 227–33.
Bradbury, J. W.(1977). Social organization and communication. In *Biology of Bats*, vol. 3, ed. W. A. Wimsatt, pp. 1–72. New York: Academic Press.
Brigham, R. M. & Brigham, A. C. (1989). Evidence for association between a mother bat and its young during and after foraging. *American Midland Naturalist*, **121**, 205–7.
Brown, C. R. & Brown, M. B. (1986). Ectoparasitism as a cost of coloniality in cliff swallows (*Hirundo pyrrhonota*). *Ecology*, **67**, 1206–18.
Brown, P. E., Brown, T. W. & Grinnell, A. D. (1983). Echolocation, development, and vocal communication in the lesser bulldog bat, *Noctilio albiventris*. *Behavioral Ecology and Sociobiology*, **13**, 287–98.

Buchler, E. R. (1980). The development of flight, foraging, and echolocation in the little brown bat (*Myotis lucifugus*). *Behavioral Ecology and Sociobiology*, **6**, 211–18.

Clutton-Brock, T. H. (1991). *The Evolution of Parental Care*. Princeton: Princeton University Press.

Coppinger, R. P. & Smith, C. K. (1990). A model for understanding the evolution of mammalian behavior. In *Current Mammalogy*, ed. H. Genoways, pp. 335–74. New York: Plenum Press.

Corbet, G. B. & Hill, J. E. (1991). *A World List of Mammalian Species*, 3rd edn. Oxford: Oxford University Press.

Davis, R. (1970). Carrying of young by flying female North American bats. *American Midland Naturalist*, **83**, 186–96.

De Fanis, E. & Jones, G. (1995). Post-natal growth, mother–infant interactions and development of vocalizations in the vespertilionid bat *Plecotus auritus*. *Journal of Zoology (London)*, **235**, 85–97.

De Fanis, E. & Jones, G. (1996). Allomaternal care and recognition between mothers and young in pipistrelle bats (*Pipistrellus pipistrellus*). *Journal of Zoology (London)*, **240**, 781–7.

Dietz, M. von & Walter, G. (1995). Zur Ektoparasitenfauna der Wasserfledermaus (*Myotis daubentoni* Kuhl, 1819) in Deutschland unter besonderer Berücksichtigung der saisonalen Belastung mit der Flughautmilbe *Spinturnix andegavinus* Deunff, 1977. *Nyctalus (N. F.)*, **5**, 451–68.

Duvergé, P. L. (1996). Foraging activity, habitat use, development of juveniles, and diet of the greater horseshoe bat (*Rhinolophus ferrumequinum* – Schreber 1774) in south-west England. PhD dissertation, University of Bristol.

Duvergé, P. L., Jones, G., Rydell, J. & Ransome, R. D. (in press) The functional significance of emergence timing in bats. *Ecography*.

Eales, L. A., Bullock, D. J. & Slater, P. J. B. (1988). Shared nursing in captive pipistrelle bats (*Pipistrellus pipistrellus*)? *Journal of Zoology (London)*, **216**, 584–7.

Esser, K.-H. (1994). Audio-vocal learning in a non-human mammal: the lesser spear-nosed bat *Phyllostomus discolor*. *NeuroReport*, **5**, 1718–20.

Esser, K.-H. & Schmidt, U. (1989). Mother–infant communication in the lesser spear-nosed bat *Phyllostomus discolor* (Chiroptera, Phyllostomidae) – evidence for acoustic learning. *Ethology*, **82**, 156–68.

Fenton, M. B. (1969). The carrying of young by females of three species of bats. *Canadian Journal of Zoology*, **47**, 158–9.

Fenton, M. B. (1985). *Communication in the Chiroptera*. Bloomington: Indiana University Press.

Fenton, M. B. (1994). Assessing signal variability and reliability: 'to thine ownself be true.' *Animal Behaviour*, **47**, 757–64.

Fenton, M. B., Belwood, J. J., Fullard, J. H. & Kunz, T. H. (1976). Responses of *Myotis lucifugus* (Chiroptera: Vespertilionidae) to calls of conspecifics and to other sounds. *Canadian Journal of Zoology*, **54**, 1443–8.

Fenton, M. B., Rautenbach, I. L., Smith, S. E., Swanepoel, C. M., Grosell, J. & Jaarsveld, J. van (1994). Raptors and bats: threats and opportunities. *Animal Behaviour*, **48**, 9–18.

Fox, R., Lehmkuhle, S. W. & Westendorf, D. H. (1976). Falcon visual acuity. *Science*, **192**, 263–5.

Francis, C. M., Anthony, E. L. P., Brunton, J. A. & Kunz, T. H. (1994). Lactation in male fruit bats. *Nature*, **367**, 691–2.

Gannon, M. R. & Willig, M. R. (1995) Ecology of ectoparasites from tropical bats. *Environmental Entomology*, **24**, 1495–1503.

Gelfand, D. L. & McCracken, G. F. (1986). Individual variation in the isolation calls of Mexican free-tailed bat pups (*Tadarida brasiliensis mexicana*). *Animal Behaviour*, **34**, 1078–86.

Gould, E. (1975). Experimental studies of the ontogeny of ultrasonic vocalizations in bats. *Developmental Psychobiology*, **8**, 333–46.

Gustin, M. K. & McCracken, G. F. (1987). Scent recognition between females and pups in the bat *Tadarida brasiliensis mexicana*. *Animal Behaviour*, **35**, 13–19.

Hamilton, W. D. (1964). The genetical theory of social behaviour. *Journal of Theoretical Biology*, **7**, 1–51.

Hamilton, I. M. & Barclay, R. M. R. (1998). Diets of juvenile, yearling, and adult big brown bats (*Eptesicus fuscus*) in southeastern Alberta. *Journal of Mammalogy*, **79**, 764–71.

Heller, K.-G., Achmann, R. & Witt, K. (1993). Monogamy in the bat *Rhinolophus sedulus*? *Zeitschrift für Säugetierkunde*, **58**, 376–7.

Hughes, P. M. (1990). Aspects of microchiropteran reproduction in relation to flight performance. PhD dissertation, University of Bristol.

Hughes, P. M. & Rayner, J. M. V. (1991). Addition of artificial loads to long-eared bats *Plecotus auritus*: handicapping flight performance. *Journal of Experimental Biology*, **161**, 285–98.

Hughes, P. M. & Rayner, J. M. V. (1993). The flight of pipistrelle bats *Pipistrellus pipistrellus* during pregnancy and lactation. *Journal of Zoology (London)*, **230**, 541–55.

Hughes, P. M., Rayner, J. M. V. & Jones, G. (1995). Ontogeny of 'true' flight and other aspects of growth in the bat *Pipistrellus pipistrellus*. *Journal of Zoology (London)*, **235**, 291–318.

Hughes, P. M., Speakman, J. R., Jones, G. & Racey, P. A. (1989). Suckling behaviour in the pipistrelle bat (*Pipistrellus pipistrellus*). *Journal of Zoology (London)*, **219**, 665–70.

Janik, V. M. & Slater, P. J. B. (1997). Vocal learning in mammals. *Advances in the Study of Behavior*, **26**, 59–99.

Jones, G. (1990). Prey selection by the greater horseshoe bat (*Rhinolophus ferrumequinum*): optimal foraging by echolocation? *Journal of Animal Ecology*, **59**, 587–602.

Jones, G. (1995). Variation in bat echolocation: implications for resource partitioning and communication. *Le Rhinolophe*, **11**, 53–9.

Jones, G. & Kokurewicz, T. (1994). Sex and age variation in echolocation calls and flight morphology of Daubenton's bats *Myotis daubentonii*. *Mammalia*, **58**, 41–50.

Jones, G. & Parijs, S. M. van (1993). Bimodal echolocation in pipistrelle bats: are cryptic species present? *Proceedings of the Royal Society London*, **251B**, 119–25.

Jones, G. & Ransome, R. D. (1993). Echolocation calls of bats are influenced by maternal effects and change over a lifetime. *Proceedings of the Royal Society London*, **252B**, 125–8.

Jones, G., Duvergé, P. L. & Ransome, R. D. (1995). Conservation biology of an endangered species: field studies of greater horseshoe bats. *Symposia of the Zoological Society London*, **67**, 309–24.

Jones, G., Gordon, T. & Nightingale, T. (1992). Sex and age differences in the echolocation calls of the lesser horseshoe bat, *Rhinolophus hipposideros*. *Mammalia*, **56**, 189–93.

Jones, G., Hughes, P. M. & Rayner, J. M. V. (1991). The development of vocalizations in *Pipistrellus pipistrellus* (Chiroptera: Vespertilionidae) during post-natal growth and the maintenance of individual vocal signatures. *Journal of Zoology (London)*, **225**, 71–84.

Jones, G., Morton, M., Hughes, P. M. & Budden, R. M. (1993). Echolocation, flight morphology and foraging strategies of some West African hipposiderid bats. *Journal of Zoology (London)*, **230**, 385–400.

Kleiman, D. G. (1969). Maternal care, growth rate, and development in the noctule (*Nyctalus noctula*), pipistrelle (*Pipistrellus pipistrellus*) and serotine (*Eptesicus serotinus*) bats. *Journal of Zoology (London)*, **157**, 187–211.

Konstantinov, A. I. (1989). The ontogeny of echolocation functions in horseshoe bats. In *European Bat Research 1987*, ed. V. Hanák, I. Horácek, & J. Gaisler, pp. 271–80. Prague: Charles University Press.

Koehler, C. E. & Barclay, R. M. R. (1988). The potential for vocal signatures in the calls of young hoary bats (*Lasiurus cinereus*). *Canadian Journal of Zoology*, **66**, 1982–5.

Kunz, T. H. (1974). Feeding ecology of a temperate insectivorous bat (*Myotis velifer*). *Ecology*, **55**, 693–711.

Kunz, T. H. (1982). Roosting ecology of bats. In *Ecology of Bats*, ed. T. H. Kunz, pp. 1–55. New York: Plenum Press.

Kunz, T. H. (1987). Post-natal growth and energetics of suckling bats. In *Recent Advances in the Study of Bats*, ed. M. B. Fenton, P. A. Racey & J. M. V. Rayner, pp. 395–420. Cambridge: Cambridge University Press.

Kunz, T. H. & Anthony, E. L. P. (1996). Variation in the timing of nightly emergence behavior in the little brown bat, *Myotis lucifugus* (Chiroptera: Vespertilionidae). In *Contributions in Mammalogy: A Memorial Volume Honoring Dr. J. Knox Jones, Jr.*, ed. H. H. Genoways & R. J. Baker, pp. 225–35. Lubbock: Texas Tech University Press.

Kunz, T. H. & Stern, A. A. (1995). Maternal investment and post-natal growth in bats. *Symposia of the Zoological Society, London*, **67**, 123–38.

Kunz, T. H., Allgaier, A. L., Seyjagat, J. & Caligiui, R. (1994). Allomaternal care: helper- assisted birth in the Rodrigues fruit bat, *Pteropus rodricensis* (Chiroptera: Pteropodidae). *Journal of Zoology (London)*, **232**, 691–700.

Kunz, T. H., Oftedal, O. T., Robson, S. K., Kretzmann, M. B. & Kirk, C. (1995a). Changes in milk composition during lactation in three species of insectivorous bats. *Journal of Comparative Physiology B*, **164**, 543–51.

Kunz, T. H., Whitaker, J. O., Jr. & Wadanoli, M. D. (1995b). Dietary energetics of the insectivorous Mexican free-tailed bat (*Tadarida brasiliensis*) during pregnancy and lactation. *Oecologia*, **101**, 407–15.

Kurta, A., Bell, G. P., Nagy, K. A. & Kunz, T. H. (1989). Energetics of pregnancy and lactation in free-ranging little brown bats (*Myotis lucifugus*). *Physiological Zoology*, **62**, 804–18.

Loughry, W. J. & McCracken, G. F. (1991). Factors influencing female–pup scent recognition in Mexican free-tailed bats. *Journal of Mammalogy*, **72**, 624–6.

Marimuthu, G. (1988). Mother–young relations in an insectivorous bat, *Hipposideros speoris*. *Current Science*, **57**, 983–7.

Masters, W. M., Raver, K. A. S. & Kazial, K. A. (1995). Sonar signals of big brown bats, *Eptesicus fuscus*, contain information about individual identity, age and family affiliation. *Animal Behaviour*, **50**, 1243–60.

Mayer, F. (1995). Genetic population structure of the noctule bat *Nyctalus noctula*: a molecular approach and first results. *Symposia of the Zoological Society, London*, **67**, 387–96.

McCracken, G. F. (1984). Communal nursing in Mexican free-tailed bat maternity colonies. *Science*, **223**, 1090–1.

McCracken, G. F. (1987). Genetic structure of bat social groups. In *Recent Advances in the Study of Bats*, ed. M. B. Fenton, P. A. Racey, & J. M. V. Rayner, pp. 281–98. Cambridge: Cambridge University Press.

McCracken, G. F. (1993). Locational memory and female–pup reunions in Mexican free- tailed bat maternity colonies. *Animal Behaviour*, **45**, 811–13.

McCracken, G. F. & Gassel, M. F. (1997). Genetic structure in migratory and nonmigratory populations of Brazilian free-tailed bats. *Journal of Mammalogy*, **78**, 348–57.

McCracken, G. F. & Gustin, M. K. (1991). Nursing behavior in Mexican free-tailed bat maternity colonies. *Ethology*, **89**, 305–21.

McCracken, G. F., McCracken, M. K. & Vawter, A. T. (1994). Genetic structure in migratory populations of the bat *Tadarida brasiliensis mexicana*. *Journal of Mammalogy*, **75**, 500–14.

McLean, J. A. & Speakman, J. R. (1996). Suckling behaviour in the brown long-eared bat (*Plecotus auritus*). *Journal of Zoology (London)*, **239**, 411–16.

McLean, J. A. & Speakman, J. R. (1997). Non-nutritional maternal support in the brown long-eared bat. *Animal Behaviour*, **54**, 1193–204.

Mooring, M. S., McKenzie, A. A. & Hart, B. L. (1996). Grooming in impala: role of oral grooming in removal of ticks and effects of ticks in increasing grooming rate. *Physiology and Behavior*, **59**, 965–71.

Myers, P. (1978). Sexual dimorphism in size of vespertilionid bats. *American Naturalist*, **112**, 701–12.

Norval, R. A. I., Sutherst, R. W., Jorgensen, O, G., Gibson, J. D. & Kerr, J. D. (1989). The effect of bont tick (*Amblyomma hebraeum*) on the weight gain of Africander steers. *Veterinary Parasitology*, **33**, 329–34.

O'Farrell, M. J. & Studier, E. H. (1973). Reproduction, growth, and development in *Myotis thysanodes* and *M. lucifugus* (Chiroptera: Vespertilionidae). *Ecology*, **54**, 18–30.

Packer, C., Lewis, S. & Pusey, A. (1992). A comparative analysis of non-offspring nursing. *Animal Behaviour*, **43**, 265–81.

Park, S.-R. (1991). Development of social structure in a captive colony of the common vampire bat, *Desmodus rotundus*. *Ethology*, **89**, 335–41.

Pearl, D. L. & Fenton, M. B. (1996). Can echolocation calls provide information about group identity in the little brown bat (*Myotis lucifugus*)? *Canadian Journal of Zoology*, **74**, 2184–92.

Porter, F. L. (1979). Social behavior in the leaf-nosed bat, *Carollia perspicillata*. I. Social organization. *Zeitschrift für Tierpsychologie*, **49**, 406–17.

Racey, P. A. & Swift, S. M. (1985). Feeding ecology of *Pipistrellus pipistrellus* (Chiroptera: Vespertilionidae) during pregnancy and lactation. I. Foraging behaviour. *Journal of Animal Ecology*, **54**, 205–15.

Radhamani, T. R., Marimuthu, G. & Chandrashekaran, M. K. (1990). Relationship between infant size and carrying of infants by hipposiderid mother bats. *Current Science*, **59**, 602–3.

Ransome, R. D. (1990). *The Natural History of Hibernating Bats*. London: Christopher Helm.

Ransome, R. D. (1995). Does significant maternal care continue beyond weaning in greater horseshoe bats? *Bat Research News*, **36**, 102–3.

Ransome, R. D. (1996). *The Management of Feeding Areas for Greater Horseshoe Bats*, English Nature Research Report no. 174. Peterborough: English Nature.

Rasmuson, T. M. & Barclay, R. M. R. (1992). Individual variation in the isolation calls of newborn big brown bats (*Eptesicus fuscus*): is variation genetic? *Canadian Journal of Zoology*, **70**, 698–702.

Rolseth, S. L., Koehler, C. E. & Barclay, R. M. R. (1994). Differences in the diets of juvenile and adult hoary bats, *Lasiurus cinereus*. *Journal of Mammalogy*, **75**, 394–8.

Rother, G. & Schmidt, U. (1985). Die ontogenetische Entwicklung der Vokalisation bei *Phyllostomus discolor* (Chiroptera). *Zeitschrift für Säugetierkunde*, **50**, 17–16.

Rübsamen, R. & Schäfer, M. (1990). Audiovocal interactions during development? Vocalisation in deafened young horseshoe bats vs. audition in vocalisation impaired bats. *Journal of Comparative Physiology A*, **167**, 771–84.

Russ, J. M., Racey, P. A. & Jones, G. (1998). Intraspecific responses to distress calls of the pipistrelle bat, *Pipistrellus pipistrellus*. *Animal Behaviour*, **55**, 705–13.

Scherrer, J. A. & Wilkinson, G. S. (1993). Evening bat isolation calls provide evidence for heritable signatures. *Animal Behaviour*, **46**, 847–60.

Schmidt, U. & Manske, U. (1973). Die Jugendentwicklunge der Vampirfledermäuse (*Desmodus rotundus*). *Zeitschrift für Säugetierkunde*, **38**, 14–33.

Schmidt, C., Schmidt, U. & Manske, U. (1980). Observations of the behavior of orphaned juveniles in the common vampire bat (*Desmodus rotundus*). In *Proceedings of the Fifth International Bat Research Conference*, ed. D. E. Wilson & A. L. Gardner, pp. 105–11. Lubbock: Texas Tech Press.

Simmons, N. B. (1993). Morphology, function, and phylogenetic significance of pubic nipples in bats (Mammalia: Chiroptera), *American Museum Novitates*, **3077**, 1–35.

Stern, A. A. & Kunz, T. H. (1998). Intraspecific variation in postnatal growth in the greater spear-nosed bat. *Journal of Mammalogy*, **79**, 755–63.

Stern, A. A., Kunz, T. H. & Bhatt, S. S. (1997). Seasonal wing loading and the ontogeny of flight in *Phyllostomus hastatus* (Chiroptera: Phyllostomidae). *Journal of Mammalogy*, **78**, 1199–1209.

Studier, E. H. & O'Farrell, M. J. (1972). Biology of *M. thysanodes* and *M. lucifugus* (Chiroptera: Vespertilionidae) I. Thermoregulation. *Comparative Biochemistry and Physiology*, **41A**, 567–95.

Trivers, R. L. (1971). The evolution of reciprocal altruism. *Quarterly Review of Biology*, **46**, 35–57.

Trune, D. R. & Slobodchikoff, C. N. (1976). Social effects of roosting on the metabolism of the pallid bat *Antrozous pallidus*. *Journal of Mammalogy*, **57**, 656–63.

Tuttle, M. D. (1986). Gentle fliers of the African night. *National Geographic Magazine*, **169**, 540–58.

Tuttle, M. D. & Stevenson, D. (1982). Growth and survival of bats. In *Ecology of Bats*, ed. T. H. Kunz, pp. 105–50. New York: Plenum Press.

Vaughan, T. A. (1976). Nocturnal behavior of the African false vampire bat (*Cardioderma cor*). *Journal of Mammalogy*, **57**, 227–48.

Vaughan, T. A. & O'Shea, T. J. (1976). Roosting ecology of the pallid bat, *Antrozous pallidus*. *Journal of Mammalogy*, **57**, 19–42.

Vaughan, T. A. & Vaughan, R. P. (1987). Parental behavior in the African yellow-winged bat (*Lavia frons*). *Journal of Mammalogy*, **68**, 217–23.

Vehrencamp, S. L., Stiles, F. G. & Bradbury, J. W. (1977). Observations on the foraging behavior and avian prey of the Neotropical carnivorous bat, *Vampyrum spectrum*. *Journal of Mammalogy*, **58**, 469–78.

Watkins, L. C. & Shump, K. A., Jr. (1981). Behavior of the evening bat *Nycticeius humeralis* at a nursery roost. *American Midland Naturalist*, **105**, 258–68.

Wilde, C. J., Kerr, M. A., Knight, C. H. & Racey, P. A. (1995). Lactation in vespertilionid bats. *Symposia of the Zoological Society, London*, **67**, 139–49.

Wilkinson, G. S. (1984). Reciprocal food sharing in vampire bats. *Nature*, **309**, 181–4.

Wilkinson, G. S. (1985a). The social organization of the common vampire bat. I. Pattern and cause of association. *Behavioral Ecology and Sociobiology*, **17**, 111–21.

Wilkinson, G. S. (1985b). The social organization of the common vampire bat. II. Mating system, genetic structure, and relatedness. *Behavioral Ecology and Sociobiology*, **17**, 123–34.

Wilkinson, G. S. (1986). Social grooming in the common vampire bat, *Desmodus rotundus*. *Animal Behaviour*, **34**, 1880–9.

Wilkinson, G. S. (1992a). Communal nursing in the evening bat, *Nycticeius humeralis*. *Behavioral Ecology and Sociobiology*, **31**, 225–35.

Wilkinson, G. S. (1992b). Information transfer at evening bat colonies. *Animal Behaviour*, **44**, 501–18.

Wilkinson, G. S. (1995). Information transfer in bats. *Symposia of the Zoological Society, London*, **67**, 345–60.

Williams, D. F. & Findley, J. S. (1979). Sexual size dimorphism in vespertilionid bats. *American Midland Naturalist*, **102**, 113–26.

Wilson, D. E. (1997). *Bats in Question*. Washington: Smithsonian Institution Press.

Woolf, N. (1974). The ontogeny of bat sonar sounds: with special emphasis on sensory deprivation. PhD dissertation, Johns Hopkins University, Baltimore.

Index